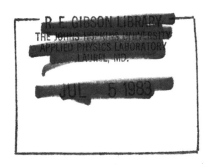

REMOTE SENSING
OF ATMOSPHERES
AND OCEANS

Academic Press Rapid Manuscript Reproduction

PROCEEDINGS OF THE INTERACTIVE WORKSHOP
ON INTERPRETATION OF REMOTELY SENSED DATA
HELD IN WILLIAMSBURG, VIRGINIA
MAY 23-25, 1979

Co-Sponsored by the Office of Naval Research, under the Department of
Navy Research Grant N00014-79-G-0039, and the Institute for
Atmospheric Optics and Remote Sensing (IFAORS),
in cooperation with NASA-Langley Research Center.
The United States Government has a royalty-free license throughout the
world in all copyrightable material contained herein.

REMOTE SENSING OF ATMOSPHERES AND OCEANS

edited by

ADARSH DEEPAK

*Institute for Atmospheric Optics
and Remote Sensing
Hampton, Virginia*

ACADEMIC PRESS
A Subsidiary of Harcourt Brace Jovanovich, Publishers

New York   London   Toronto   Sydney   San Francisco   **1980**

ACADEMIC PRESS, INC.
111 Fifth Avenue, New York, New York 10003

United Kingdom Edition published by
ACADEMIC PRESS, INC. (LONDON) LTD.
24/28 Oval Road, London NW1 7DX

Library of Congress Cataloging in Publication Data

Main entry under title:

Remote sensing of atmospheres and oceans.

Includes index.
1. Atmosphere—Remote sensing—Congresses.
2. Oceanography—Remote sensing—Congresses.
I. Deepak, Adarsh.
QC871.R45   551.5'028   80-18881
ISBN 0-12-208460-8

PRINTED IN THE UNITED STATES OF AMERICA

80 81 82 83   9 8 7 6 5 4 3 2 1

CONTENTS

Interpretation of Aerosol Sounding—J. Lenoble, *Chairman*

Gaseous Constituent Retrievals—A. E. S. Green, *Chairman*

Remote Sounding by Microwaves—D. H. Staelin, *Chairman*

REMOTE SOUNDING OF WINDS

Wind Sounding—E. R. Westwater, Chairman

REMOTE SOUNDING OF OCEAN PARAMETERS

Ocean Parameter Sounding—H. E. Fleming and M. T. Chahine, Chairmen

INTERPRETATION OF RECENT RESULTS FROM SPACE

Recent Results from Space—M. P. McCormick and B. J. Conrath,
Chairmen

PARTICIPANTS

Mian M. Abbas, *Department of Physics and Atmospheric Science, Disque Hall, Room 911, Drexel University, Philadelphia, Pennsylvania 19104*

James Baily, *Office of Naval Research, 800 North Quincy Street, Arlington, Virginia 22217*

Bruce Barkstrom, *George Washington University and NASA-Langley Research Center, MS 423, Hampton, Virginia 23665*

Jeffrey Baron, *Lockheed Missiles and Space Company, 0/62-41 B/562, P.O. Box 504, Sunnyvale, California 94086*

Pawan K. Bhartia, *Systems and Applied Sciences, 5809 Annapolis Road, Hyattsville, Maryland 20784*

Gail P. Box, *Institute of Atmospheric Physics, University of Arizona, Tucson, Arizona 85721*

Michael A. Box, *Institute of Atmospheric Physics, University of Arizona, Tucson, Arizona 85721*

James J. Buglia, *NASA-Langley Research Center, MS 423, Hampton, Virginia 23665*

Dale M. Byrne, *United Technologies Research Center, West Palm Beach, Florida 33402*

Joseph C. Casas, *Old Dominion University, 17 Research Drive, Hampton, Virginia 23666*

Moustafa T. Chahine, *JPL/California Institute of Technology, MS 183-301, 4800 Oak Grove Drive, Pasadena, California 91103*

Bob Chase, *JPL/California Institute of Technology, 4800 Oak Grove Drive, Pasadena, California 91103*

William P. Chu, *NASA-Langley Research Center, MS 234, Hampton, Virginia 23665*

Barney J. Conrath, *NASA-Goddard Space Flight Center, Code 622, Greenbelt, Maryland 20771*

Robert C. Costen, *NASA-Langley Research Center, MS 423, Hampton, Virginia 23665*

David S. Crosby, *NOAA/NESS, OA/S321/DSC, Washington, DC 20233*

William T. Davis, *NASA-Langley Research Center, MS 364, Hampton, Virginia 23665*

Adarsh Deepak, *Institute for Atmospheric Optics and Remote Sensing, 17 Research Drive, Hampton, Virginia 23666*

Douglas DePriest, *Office of Naval Research, Code 436, 800 North Quincy Street, Arlington, Virginia 22217*

David B. Evans, *Lawrence Berkeley Laboratory, Building 90, Room 2024 N, Berkeley, California 94720*

Henry E. Fleming, *Department of Meteorology, Naval Postgraduate School, Monterey, California 93940*

Howard R. Gordon, *Department of Physics, University of Miami, Coral Gables, Florida 33124*

Alex E. S. Green, *Department of Physics and Astronomy, University of Florida, Gainesville, Florida 32601*

Richard N. Green, *NASA-Langley Research Center, MS 423, Hampton, Virginia 23665*

Norman C. Grody, *NOAA/NESS/S31, World Weather Building, Room 703, Washington, DC 20233*

Patrick J. Hamill, *Systems and Applied Sciences, 17 Research Drive, Hampton, Virginia 23666*

Lawrence H. Hoffman, *NASA-Langley Research Center, MS 423, Hampton, Virginia 23665*

William F. Johnson, *Air Weather Service, Scott Air Force Base, Illinois 62225*

Lewis D. Kaplan, *NASA-Goddard Space Flight Center, 22-G42, Code 911, Greenbelt, Maryland 20771*

Ashok Kaveeshwar, *Systems and Applied Sciences, 5809 Annapolis Road, Hyattsville, Maryland 20784*

Lloyd S. Keafer, *NASA-Langley Research Center, MS 364, Hampton, Virginia 23665*

James F. Kibler, *NASA-Langley Research Center, MS 423, Hampton, Virignia 23665*

William G. Knorr, *ITT Aerospace Optical Division, 3700 East Pontiac Street, Fort Wayne, Indiana 46815*

Ed Koenig, *ITT Aerospace Optical Division, 3700 East Pontiac Street, Fort Wayne, Indiana 46815*

Jacqueline Lenoble, *Laboratoire d'Optique Atmos., Université de Lille I, B.P. 36, 59650 Villeneuve d'Ascq., France*

William H. Mach, *Department of Meteorology, Florida State University, Tallahassee, Florida 32303*

M. Patrick McCormick, *NASA-Langley Research Center, MS 234, Hampton, Virginia 23665*

Larry M. McMillin, *NOAA/NESS, OA/S321/KEC, Washington, DC 20233*

Harry D. Orr, III, *NASA-Langley Research Center, MS 401A, Hampton, Virginia 23665*

Theodore J. Pepin, *Department of Physics and Astronomy, University of Wyoming, P.O. Box 3095, University Station, Laramie, Wyoming 82071*

Walter Planet, *NOAA/NESS, Code S321/MS B, Suitland, Maryland 20233*

John P. Rahlf, *TRW, One Space Park, Redondo Beach, California 90278*

Pamela Livingstone-Rarig, *Systems and Applied Sciences, 17 Research Drive, Hampton, Virginia 23666*

John A. Reagan, *Department of Electrical Engineering, Engineering Building #20, University of Arizona, Tucson, Arizona 85721*

Ellis E. Remsberg, *NASA-Langley Research Center, MS 401B, Hampton, Virginia 23665*

Rolando Rizzi, *Instituto DiFisica, Via Irnerio 46, Bologna, 40126, ITALY*

Clive Rodgers, *Clarendon Laboratory, Oxford University, OXI 3PU ENGLAND*

Arieh Rosenberg, *NASA-Goddard Space Flight Center, Code 911, Greenbelt, Maryland 20771*

P. W. Rosenkranz, *Research Laboratory of Electronics/MIT, Cambridge, Massachusetts 02139*

James M. Russell, III, *NASA-Langley Research Center, MS 401A, Hampton, Virginia 23665*

Glen W. Sachse, *NASA-Langley Research Center, MS 235A, Hampton, Virginia 23665*

H. J. P. Smith, *VISIDYNE, 19 Third Avenue, Northwest Industrial Park, Burlington, Massachusetts 01803*

David H. Staelin, *Research Laboratory of Electronics/MIT, Cambridge, Massachusetts 02139*

Richard G. Strauch, *NOAA/ERL/WPL, Meteorological Studies, Boulder, Colorado 80303*

Joel Susskind, *NASA-Goddard Space Flight Center, Code 911, Greenbelt, Maryland 20771*

Thomas J. Swissler, *Systems and Applied Sciences, 17 Research Drive, Hampton, Virginia 23666*

Steven L. Taylor, *Systems and Applied Sciences, 5809 Annapolis Road, Hyattsville, Maryland 20784*

P. M. Toldalagi, *Research Laboratory of Electronics/MIT, Cambridge, Massachusetts 02139*

D. W. Toomey, *Raytheon Company, 430 Boston Post Road, Wayland, Massachusetts 01778*

J. T. Twitty, *Systems and Applied Sciences, 17 Research Drive, Hampton, Virginia 23666*

H. Andrew Wallio, *NASA-Langley Research Center, MS 401A, Hampton, Virginia 23665*

Charles Walton, *NOAA/NESS, OA/S14/CW, Washington, DC 20233*

Edward J. Wegman, *Office of Naval Research, 800 North Quincy Street, Arlington, Virginia 22217*

Ed R. Westwater, *NOAA/WPL, R45x4, Boulder, Colorado 80303*

Patricia A. Winters, *NASA-Langley Research Center, MS 125, Hampton, Virginia 23665*

Andrew Zardecki, *Department of Physics, Laval University, Quebec, Canada G1K 7P4*

Workshop Speakers and Chairmen (left to right): J. Lenoble, U. de Lille, France; M. M. Abbas, Drexel U.; H. R. Gordon, U. of Miami; A. Deepak, IFAORS; C. Rodgers, U. of Oxford, UK; D. H. Staelin, MIT; A. E. S. Green, U. of Florida. (Second Row): L. D. Kaplan, NASA-GSFC; M. T. Chahine, JPL; P. W. Rosenkranz, MIT; H. J. P. Smith, Visidyne, Inc.; D. DePriest, Office of Naval Res.; E. Koenig, ITT; D. S. Crosby, NOAA/NESS. (Third Row): J. Susskind, NASA-GSFC4 B. Barkstrom, NASA-LaRC; P. M. Toldalagi, MIT; H. E. Fleming, NOAA/ NESS; B. J. Conrath, NASA-GSFC; E. R. Westwater, NOAA/ERL; L. M. McMillin, NOAA/NESS; R. G. Strauch, NOAA/ERL/WPL.

Not included in the photograph: P. Hamill, Sys. & Appl. Sci.; S. L. Taylor, Sys. & Appl. Sci.; D. W. Toomey, Raytheon Co.; N. C. Grody, NOAA/NESS; C. Walton, NOAA/NESS; M. P. McCormick, NASA-LaRC; T. J. Pepin, U. of Wyoming; J. M. Russell III, NASA-LaRC.

PREFACE

This volume contains the technical proceedings of the Interactive Workshop on Interpretation of Remotely Sensed Data, held in Williamsburg, Virginia, May 23–25, 1979.

The workshop was organized to provide an interdisciplinary forum to assess the state-of-the-art in the interpretation of measurements obtained in remote sounding of various atmospheric and ocean parameters, and to identify those important problems in which further research efforts are needed. Seventy scientists from the industry, universities, government agencies, and research laboratories attended the workshop, in which thirty papers were presented. Complete texts of twenty-five of these papers, and their discussions, are included in this volume.

The workshop program was divided into ten sessions, each covering a specific topic and chaired by the following scientists: A. Deepak, Recent Advances in Inversion Methods; L. D. Kaplan, Atmospheric Temperature Sounding; J. Lenoble, Interpretation of Aerosol Sounding; A. E. S. Green, Gaseous Constituent Retrievals; D. H. Staelin, Remote Sounding by Microwaves; E. R. Westwater, Wind Sounding; H. E. Fleming and M. T. Chahine, Ocean Parameter Sounding; M. P. McCormick and B. J. Conrath, Recent Results from Space. The papers included the following topics: remote sounding of atmospheric temperature, trace gases, precipitation and aerosols, sea surface temperature, ocean color, and winds. The papers discussed the current state of knowledge, as well as the results of the latest investigations in their specific areas of research. Ample time was allowed for discussions following each paper. Discussions were recorded and the transcripts postedited.

The scope of the workshop included areas of research that were not discussed in the First Interactive Workshop on Inversion Methods in Atmospheric Remote Sounding, held in Williamsburg, Virginia, in December 1976, the proceedings for which were published by Academic Press in December 1977. Dr. Douglas DePriest, Office of Naval Research (ONR), in his introductory remarks, drew attention to the importance of mathematical and statistical methodologies in the remote sensing of oceanographic, terrestrial, and atmospheric quantities.

To ensure proper representation of major disciplines involved, a workshop program committee composed of the following scientists was set up: A. Deepak (Chairman), Institute for Atmospheric Optics and Remote Sensing (IFAORS); M. T. Chahine, Jet Propulsion Laboratory; D. J. DePriest, Office of Naval Research; H. E. Fleming, Naval Postgraduate School, Monterey, B. M. Herman, University of Arizona; M. P. McCormick, NASA – Langley Research Center; W. L. Smith, NOAA/University of Wisconsin; and D. Staelin, Massachusetts Institute of Technology.

The editor wishes to acknowledge the enthusiastic support and cooperation of the members of the Technical Program Committee, session chairmen, speakers, and participants for making this a stimulating and valuable workshop for everyone. Special thanks are due the authors for their cooperation in enabling a prompt publication of the workshop proceedings. It is a pleasure to acknowledge the valuable assistance of Mrs. M. D. Crotts and S. A. Allen, IFAORS, in organizing the workshop, and H. Malcahy and M. Goodwin, IFAORS, in preparing and typing the final manuscripts.

The Workshop was cosponsored by the Office of Naval Research and the Institute for Atmospheric Optics and Remote Sensing, in cooperation with NASA – Langley Research Center.

A. Deepak

A DIFFERENTIAL INVERSION METHOD FOR HIGH
RESOLUTION ATMOSPHERIC REMOTE SENSING

Mian M. Abbas

Department of Physics and Atmospheric Science
Drexel University
Philadelphia, Pennsylvania

*The spectral lines of atmospheric gases may be fully resolved
with high resolution observations by techniques, such as infrared
heterodyne spectrometers. An inversion method suitable for such
observations is discussed and is found to have several advantages
over conventional methods. The method is based on matching the
derivatives of the observed radiances or transmittances with the
calculated values for the modeled atmosphere. The proposed
method provides a significant narrowing of the weighting functions
and improvement in the overall accuracy of the retrieved profiles. The method is applied to inversion of ozone absorption
lines in the earth's atmosphere and the results are compared
with those obtained with a conventional method.*

I. INTRODUCTION

The spectral lines of atmospheric gases may be detected and

fully resolved by using high resolution techniques such as

infrared heterodyne spectrometers. An inversion method is

presented, which is suitable for such measurements and is found

to have several advantages over conventional methods.

The proposed method is based on matching the derivatives

of observed radiances or transmittances with the calculated

values of a modeled atmosphere and leads to a significant

1

narrowing of the weighting functions and an improvement in the
overall accuracy of the retrieved profiles. The method is
applied to inversion of ozone absorption lines in the earth's
atmosphere and the results are compared with those obtained by
using a conventional method.

Atmospheric remote sensing techniques are based on measuring
the outcoming radiation at a selected set of frequencies and
finding an inverse solution to the radiative transfer equation
in terms of either the concentration or the temperature profile
of the atmosphere. With low resolution instruments ($\Delta \nu > \alpha$),
only average spectral intensities are measured since individual
lines cannot be resolved. An evaluation of atmospheric profiles
is made through an analytic inversion over an absorption band,
where the observed radiances or transmittances generally
represent an average over several lines. With recent advances
in techniques of infrared heterodyning and tunable diode lasers,
however, where resolving powers as high as 10^6 to 10^7 may be
achieved, it is now possible to fully resolve individual
spectral lines with detection sensitivities approaching the quan-
tum detection limit (see, for example, Refs. 1 to 6).

An inversion of individual spectral lines provides more
accurate information about vertical distribution of stratospheric
constituents than is possible with lower resolution measurements
because of two factors. First, if it is assumed that the
spectral parameters and the lineshape function of the observed
line are accurately known (or have been determined by
laboratory measurements), the accuracy of the retrieved profiles
is higher because no averaging over a number of lines is
required. Second, the interference from other gases may be
virtually eliminated by a proper choice of the observed line.

An inversion method is discussed here which appears to
have several advantages over the usual methods, and is
applicable to ultra-high resolution measurements where the

lineshape is fully resolved. The inversion process is based on finding an inverse solution to the derivative of the radiative transfer equation with respect to frequency. In the iterative technique employed, the slope of the observed line is matched with the slope of the synthetic line computed for the retrieved atmospheric parameters.

II. DUSCUSSION OF METHOD

The outcoming spectral intensity from a nonscattering atmosphere is given by the radiative transfer equation

$$I_\nu(P,T) = B_\nu(T_s) \, \tau_\nu^s + \int_{y_s}^{y_t} B_\nu(T) \, K(P,T) \, dy \tag{1}$$

where $B_\nu(T_s)$ is the Planck function, τ_ν^s is the transmittance from the surface to the top of the atmosphere and $K(P,T) = \partial\tau_\nu/\partial y$ (with $y = -\ell n \, P$) is the weighting function. The atmospheric transmittance τ_ν is

$$\tau_\nu = \exp\left(- \int \sum_i k_{\nu i} \, du_i\right) \tag{2}$$

where $k_{\nu i}$ is the absorption coefficient and du_i is the element of column density of the absorbing gas.

The derivative of the spectral intensity with respect to frequency is given by

$$\dot{I}_\nu = B_\nu(T_s) \, \dot{\tau}_\nu^s + \int_{y_s} B_\nu(T) \, \dot{K}_\nu(P,T) \, dy \tag{3}$$

where the dot over a symbol refers to a derivative with respect to frequency, and it is assumed that the frequency interval is sufficiently small so that $\dot{B}_\nu = 0$.

For atmospheric observations in the solar occultation mode, the second term in Eqs. (1) and (3) is generally negligble so that the observed radiance and its derivative are

$$I_\nu = B_\nu(T_s)\ \tau_\nu^s \tag{4}$$

and

$$\dot{I}_\nu = B_\nu(T_s)\ \dot{\tau}_\nu^s \tag{5}$$

The differential quantities $\dot{I}_\nu$ and $\dot{\tau}_\nu$ may be measured directly in systems based on tunable diode lasers or they may be computed from a smoothed spectral line profile obtained from high resolution measurements.

The discussion in this paper is limited to inversion of ground-based solar occultation measurements of the earth's atmosphere for evaluation of concentration profiles which may be obtained through an inverse solution of Eq. (5). From Eq. (2)

$$\frac{\dot{\tau}_\nu}{\tau_\nu} = \frac{1}{g} \int_0^P \sum_i \dot{k}_{\nu i}\ (P,T)\ q_i(P)\ dP \tag{6}$$

Equation (6) may be solved through an iterative procedure with $\dot{k}_\nu$ as a weighting function. The plots of $\dot{k}_\nu$ as a function of height are expected to be narrower and at higher levels in the atmosphere than the corresponding functions k_ν. This effect may be seen from the expressions for k_ν and $\dot{k}_\nu$ in a pressure broadening regime where

$$k_\nu(P,T) = \frac{1}{\pi}\ \frac{S(T)\ \alpha(P,T)}{(\nu - \nu_o)^2 + \alpha^2(P,T)} \tag{7}$$

and

$$\dot{k}_\nu = \frac{-2\ S(T)\ \alpha(P,T)\ (\nu - \nu_o)}{\pi[(\nu - \nu_o)^2 + \alpha^2(P,T)]^2} \tag{8}$$

where S(T) is the line strength and $\alpha(P,T)$ is the half-width
given by

$$\alpha(P,T) = \alpha_o (P/P_o) \left(\frac{T_o}{T}\right)^{1/2}$$

with the subscript o referring to the reference values.   The
form of Eq. (8) indicates that the plots of $\dot{k}_\nu$ as a function of
pressure P are narrower than those for k_ν.

     The normalized weighting functions k_ν and $\dot{k}_\nu$ for inversion
of atmospheric ozone line with center ν_o = 1011.6670 cm^{-1} (S =
0.414 cm^{-1} (cm-atm)$^{-1}$, α = 0.07 cm^{-1}, E = 422.96 cm^{-1}) are
shown in Figs. 1 and 2.   A comparison of the two figures shows
that that $\dot{k}_\nu$ plots are narrower and peak at higher levels in
the atmosphere.   The weighting functions $\dot{k}_\nu$ have widths of $\approx$ 13 km
compared with widths of $\approx$ 18 km for k_ν.

     An example of the application of this method is given here
for an inversion of a synthetic ozone line (Fig. 3) with the
line center ν_o = 1011.6670 cm^{-1} calculated for a midlatitude
winter model of the earth's atmosphere.   The retrieved ozone
profile by using this method, along with the initial guess, and
the model profiles corresponding to the synthetic absorption
line are shown in Fig. 4a).   The retrieved profile has been
extrapolated below $\approx$ 7 km and above $\approx$ 35 km.   A comparison of
the retrieved and model ozone profiles indicates excellent
agreement in the range 7 km to 35 km.   An inversion of this
ozone line was also made by using the conventional method;
the retrieved profile being shown in Fig. 4b for comparison
(3,7).   The weighting functions for this case extend from 0 to
26 km, and the profile has been extrapolated below $\approx$ 7 km and
above $\approx$ 30 km.   A comparison of the results in Figs. 3 and 4
shows that the overall accuracy of the retrieved profile is
much better in the differential method.   The number of
iterations required for convergence in the differential method

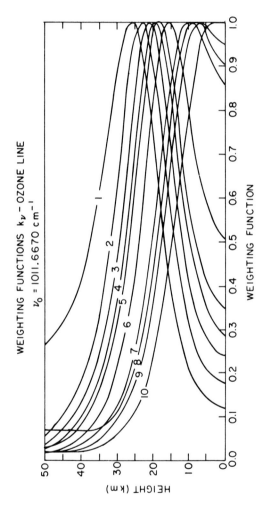

FIGURE 1. Normalized weighting functions k_ν for the ozone line with line center $\nu_o = 1011.6670$ cm^{-1} for frequencies with $\Delta\nu(MHz) = (\nu - \nu_o)$ given by (1) 30; (2) 60; (3) 90; (4) 120; (5) 150; (6) 240; (7) 480; (8) 540; (9) 600; and (10) 810.

6

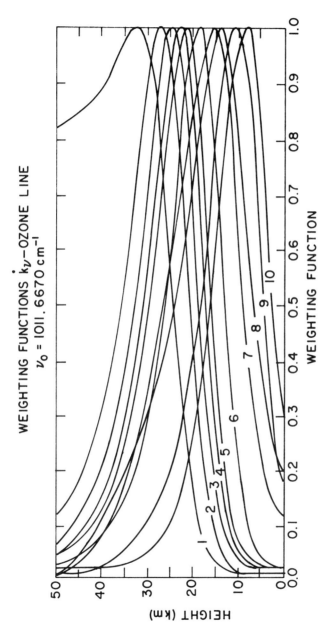

WEIGHTING FUNCTIONS $\dot{k}_\nu$-OZONE LINE
$\nu_o = 1011.6670\ cm^{-1}$

HEIGHT (km)

WEIGHTING FUNCTION

FIGURE 2. Normalized weighting functions k_ν for the ozone line with line center $\nu_o = 1011.6670$ cm^{-1} for frequencies with $\Delta\nu(MHz) = (\nu - \nu_o)$ given by (1) 30; (2) 60; (3) 90; (4) 120; (5) 150; (6) 240; (7) 480; (8) 540; (9) 960; and (10) 1260.

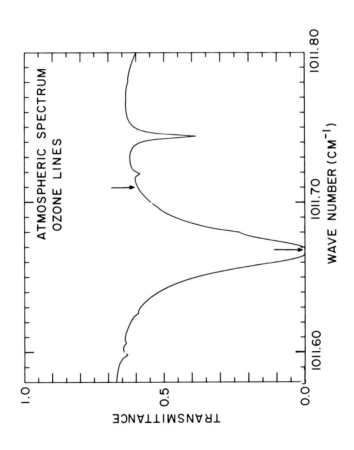

FIGURE 3. Synthetic atmospheric spectrum showing a moderately strong absorption line of ozone (20,3,26; 30,3,27). Line center $\nu_o = 1011.6670 \ cm^{-1}$; $\alpha_o = 0.07 \ cm^{-1}$; lower state energy $E'' = 422.96 \ cm^{-1}$.

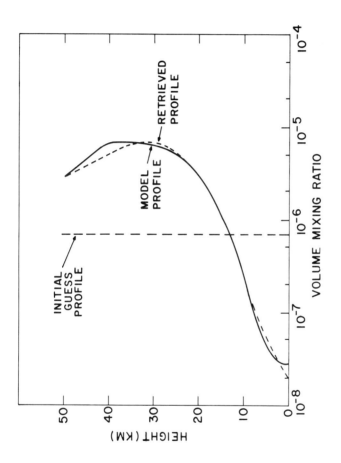

FIGURE 4. The retrieved volume mixing ratio profile of ozone obtained from an inversion of the synthetic line of Fig. 3. (a) Differential method. (b) Conventional method.

9

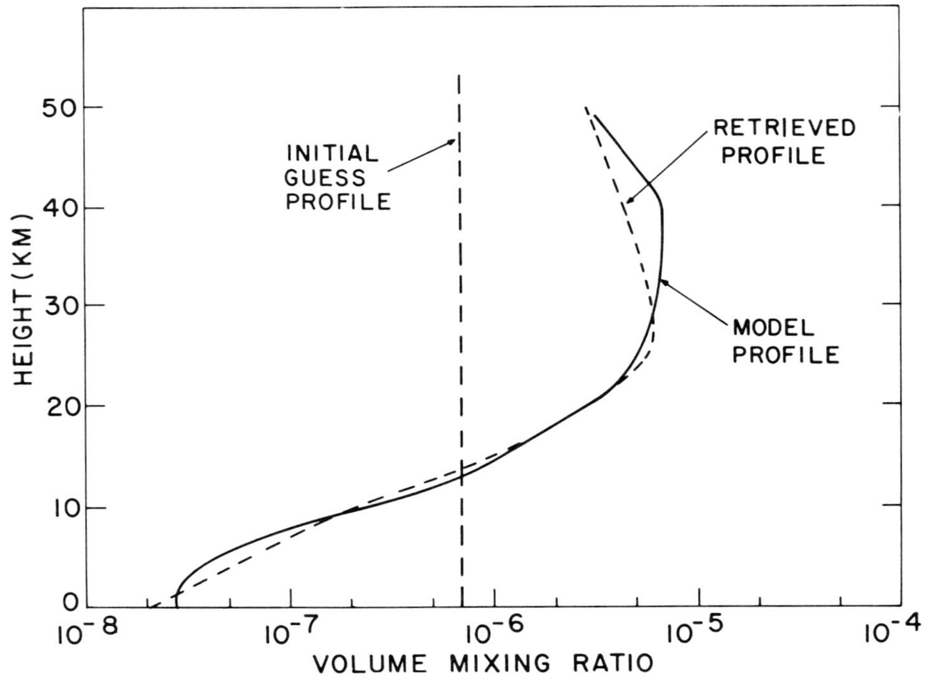

FIGURE 4b.

is 6 to 8, compared with 15 to 20 iterations in the conventional method. The additional time required to calculate the
derivatives, however, largely offsets any reduction in the
computation time.

The retrieved profiles shown in Figs. 3 and 4 are obtained
from a noise-free synthetic spectral line. The effect of noise
in the differential method depends upon whether the derivative
$\dot{I}_\nu$ is measured directly or is computed from the observed line
profiles. When $\dot{I}_\nu$ is measured directly, the observed data (with
noise in $\dot{I}_\nu$) may be analyzed directly or it may be smoothed for
a reduction in noise. However, when I_ν is measured, it is
necessary to compute the derivatives from a smoothed uniform
spectral line. The smoothing process improves the accuracy of
retrieval through a reduction in noise.

To study the effect of noise on retrievals in the differential inversion method, noise may be added to the synthetic line
profiles by using a random noise generator with a Gaussian
distribution. A uniform line profile is then obtained by
smoothing the noisy data by using least-square fitting methods.
Figure 5 shows the envelopes of retrieved profiles obtained
from several inversions of such a profile with random noise
corresponding to signal-to-noise ratio (SNR) = 10. For the
case shown in Figure 5, the effects of noise in the two
retrievals are seen to be comparable. For SNR's < 10, however,
the differential method was found to become more sensitive to
noise data and the usefulness of the method becomes dependent
on the sophistication of the smoothing technique employed in
obtaining a well structured line profile.

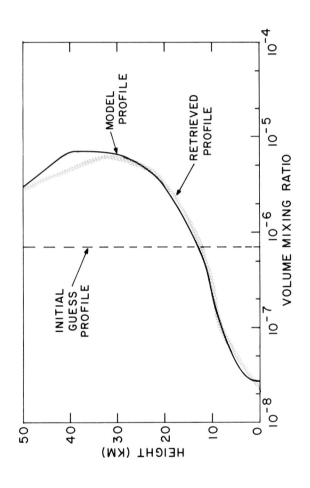

FIGURE 5. The envelope of the retrieved volume mixing ratio profile of ozone, from an inversion of the synthetic line of Fig. 3 with superimposed random noise of Gaussian distribution corresponding to a SNR ≈ 10. (a) Differential method. (b) Conventional method.

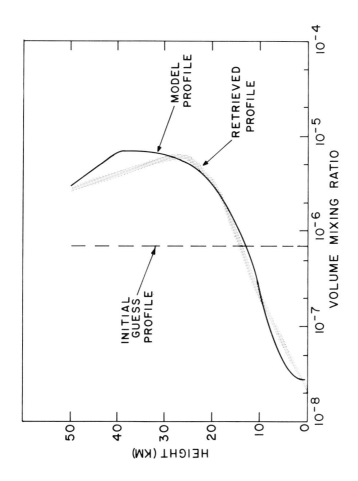

FIGURE 5b.

13

III.   CONCLUDING REMARKS

The method presented here thus appears to have several
advantages for inversion of high resolution measurements when
the spectral lines are fully resolved. The main features of
the suggested method are (1) narrower weighting functions
leading to an improved vertical resolution and a higher overall
accuracy; (2) higher vertical levels which may be probed;
and (3) more stable and faster convergence to the desired
solution.

REFERENCES

1.   Abbas, M. M., Mumma, M. J., Kostiuk, T., and Buhl, D.,
     Appl. Optics, 15, 427-436 (1976).
2.   Abbas, M. M., Kostiuk, T., Mumma, M. J., Buhl, D., Kunde,
     V. G., and Brown, L. W., *Geophys. Res. Letters, 5,* 317
     (1978).
3.   Abbas, M. M., Kunde, V. G., Mumma, M. J., Kostiuk, T.,
     Buhl, D., and Frerking, M. A., *J. Geophys. Res. 84,*
     2681-2690 (1979).
4.   Menzies, R. T., and Seals, R. K., Jr., *Science, 197,* 1275-
     1277 (1977).
5.   Peyton, B. J., Lange, R. A., Savage, M. G., Seals, R. K., Jr.,
     and Allario, F., AIAA 15th Aerospace Sciences Meeting,
     AIAA Paper No. 71-73 (January 1977).
6.   Teich, M. C., *in* "Infrared Detectors: Semiconductors and
     Semimetals," Vol. 5, p. 361. Academic Press, New York
     (1971).
7.   Chahine, M. T., *J. Atm. Sci. 29,* 741-747 (1972).

DISCUSSION

Planet: How, if at all, sensitive is your method to the
uncertainties in the spectroscopic parameters that are derived
in the laboratory; for example, the temperature dependence of the
half width, and the strength?

Abbas: The ultimate accuracy of measurements which may be
achieved by this method is limited by uncertainties in the
spectral parameters. The two parameters which are crucial are
the line strength and the half width. The measured total gas
content and the vertical gas mixing ratio profiles are thus
determined as a function of these two quantities. For absolute
values, the spectral parameters of the line to be observed have
to be determined accurately in a parallel program of laboratory
measurements with a heterodyne or diode laser system.

Planet: Hopefully you will look at the temperature dependence of
the half-width of the line all the way down to 200 K temperature.

Abbas: Yes.

Susskind: The form of the absorption coefficient that you showed
was based on a Lorentz line shape, and when you took the derivative the weighting function is getting sharper and its peak is
higher. Of course as you go higher in the atmospere and closer
to the line centers, you do not have a Lorentz line shape. You
are then going into a Voigt and Doppler regime. What is the
nature of the derivative for that type of line shape? Is it also
sharpened? The line shape that you showed is probably not the
appropriate one for those conditions.

Abbas: The line shapes which were actually used in the programs
were, of course, mixed line shapes. The two formulas which I
showed were just for demonstration purposes--just to show the
change in the form of the weighting functions in going over to
the derivatives, instead of using the original radiances. The
form of the weighting functions for mixed line shapes are more
complicated.

Susskind: But nevertheless, the same principle holds. The
derivative is still sharper.

Abbas: Yes, the weighting functions are sharper, but as we go
into the Doppler broadening regime at higher levels, we tend to
lose information. The remote sensing technique from the ground
is useful only up to regions where pressure broadening dominates.
Thus we can sense the atmosphere from the ground up to heights
of the order of 35 km.

Susskind: Well, you were showing some very nice results up to 50 km. It seemed to me that one area of big improvement of the derivative versus the nonderivative was above 35 km, and that is because your weighting functions, in fact, were moved up.

Abbas: The peaks of the weighting functions which I showed reached up to about 35 km. The retrieved profiles were extrapolated for heights above 35 km and below 6 km or so. The bottom weighting function that was used peaks about 6 or 7 km. So what you saw on those plots were curves which were extrapolated from 35 to 50 km and also from 6 km to the ground.

Susskind: I have some other comments. One, did the weighting functions have a considerable overlap?

Abbas: Yes.

Susskind: Were you using all of those observations? You know, you might have had 15 channels considerably overlapped. You are using all of those observations in your simulation?

Abbas: Yes.

Susskind: So you are really becoming very sensitive, then, to different sources of noise in the instrument, as you indicated, or other uncertainties, perhaps, as Dr. Planet mentioned in terms of the line shape itself. Another factor is the actual temperature profile. In such an experiment, you assumed the temperature profile is known. To what accuracy do you need the temperature profile to be able to reduce this source of noise, the uncertainty in the line strength due to temperature uncertainty to agree to the type of noise levels that you used? Or is it possible, to measure the temperature profile as part of your experiment by looking at temperature sensitive lines simultaneously or something like that?

Abbas: Yes, a well-planned experiment will involve a simultaneous measurement of the temperature profile. Perhaps by looking at a different line which is particularly suitable for this purpose, or through a different system or instrument, the temperature profile may be determined accurately. But to answer the first part of your question, how sensitive is the retrieval to inaccuracies in the temperature profile? It depends on the particular line being observed. Lines with very low values of the lower state energies will be less sensitive than lines with higher values of lower state energy. So a criterion in choosing a line which is most suitable for measurement of mixing ratio profiles will be to choose one with a very low value of lower state energy.

Wallio: How do you account for possible nonsymmetry in observed line shapes due to interference species?

Abbas: In general, the interference of other molecules has to be modeled, assuming that the mixing ratio profiles of the interfering species is known. The interference effects, however, maybe reduced by choosing a suitable line. In fact, one criterion which is used in choosing a suitable line is that the interference effects of other gases are negligible. With the spectral resolution available with the heterodyne technique, it is generally possible to choose a line so that the interference of other species is negligible.

Chahine: Let me ask you a very basic question. From information point of view by going from the function space to a derivative space, one cannot increase the information content of a problem. What has happened in your case in going from the function space to the derivative space is that you added one piece of information which is the knowledge of the line shape.

Abbas: That is correct.

Chahine: By the same token, then, we should be able to smooth the data using the line shape and obtain the same accuracies with the derivative as well as with the function of the kernel. Wouldn't this sound logical to you?

Abbas: I agree with your comment that by going over to the derivative space we are using additional information, that is, the knowledge of the line shape function. But as far as the smoothing is concerned, I skipped over some of the details. Although smoothing can be done in both cases, it is not necessary in inversions when radiances are used. The derivatives can be measured directly in systems based on the heterodyne technique. $\dot{i}_\nu$ is thus a directly measureable quantity and we do not have to do any smoothing. Smoothing in this case, however, will further improve the accuracy of the retrieved profiles, provided it is done over a sufficiently small spectral region.

Rodgers: I think my question has really just been answered. It seems to me intuitively that doing a linear transformation of the data should not make any difference, but of course, if you can measure the derivative directly rather than by taking differences, then obviously you are increasing the information. I am not sure if you are putting in any extra information about the line shape because that ought to be in there whether you are taking a derivative or not. You know what the line shape is whether you are measuring the intensity of the derivative.

Abbas: The derivative of intensity $\dot{I}_\nu$ is measured directly. This information, coupled with the information of the line shape leads to narrowing of the weighting functions. If the actual line shape deviates from the assumed form, a source of error will be introduced. This uncertainty, however, is common to both methods.

Westwater: In the presence of clouds, will this technique have more difficulties, less difficulties or the same difficulties as conventional IR techniques?

Abbas: I assume it would be the same.

SOME ADAPTIVE FILTERING TECHNIQUES APPLIED
TO THE PASSIVE REMOTE SENSING PROBLEM[1]

P. M. Toldalagi

Research Laboratory of Electronics
Massachusetts Institute of Technology
Cambridge, Massachusetts

*For several years now, statistical regression techniques have
been successfully applied to the point inversion problem of
temperature profiles by using microwave data. Significant
improvements in retrieval performance have been achieved by
introducing recursive statistical regression techniques which
incorporate past (Kalman filter) or past and future observations
(smoothing filter). A review of these new techniques is
presented with some results for the case of the TIROS-N/MSU and
NIMBUS-6/SCAMS experiments.*

*The lack of any satisfactory model for the medium being
sounded appears to be one major limitation to the performance
of current statistical retrieval techniques. A new and more
global approach is proposed which casts the retrieval problem
into the framework of adaptive filtering; to illustrate this
topic, one numerical implementation of such an adaptive system
is presented with a multilayer semi-spectral general circulation
model for the atmosphere used to fine-tune the sensor as well as
the dynamical equations of a Kalman filter. Furthermore, it is
shown that in this framework, the assimilation of radiometric
data appears as a relatively easy and natural subproblem.*

[1]*This work was supported in part by NASA Contract NAS5-21980
and NOAA Contract NO4-8-M01-1.*

I.   INTRODUCTION

Passive remote sensing of the atmosphere has grown, since the
launch of the first meteorological satellite in the early 1960s,
into a very active area of research. Several communities of
scientists such as physicists, meteorologists, earth and planetary
scientists, oceanographers, mathematicians, and electrical
engineers have been involved, each of them discovering new
problems and new areas of research. In the domain of passive
microwave remote sensing, which is used to illustrate this topic,
the emphasis has generally been on the design of better instruments and development of new methods for interpreting the radiometric measurements. Few scientists, however, have attempted to
understand and model the global meteorological mechanisms ruling
the evolution of the medium which they are sounding in order to
adapt their inversion techniques to actual weather activities.
This fact is probably one of the main reasons behind the reserva-
tions still expressed by many meteorologists and day-to-day
forecasters concerning satellite observations of the atmosphere.

II.   SIMPLIFIED DESCRIPTION OF A GENERAL PASSIVE REMOTE SENSING
      EXPERIMENT

In order to illustrate some of the problems arising in
an actual remote sensing experiment, consider a package of
satellite-borne radiometers at several different frequencies,
all viewing the same portion of the earth's atmosphere. Let
the gain and noise level of each radiometer be known and
furthermore assume that the object of such a mission is to
retrieve some atmospheric parameter p_0, which is known to be
one of several parameters influencing the radiation observed
by the different instruments. This standard situation is best
represented by the diagram of Fig. 1.

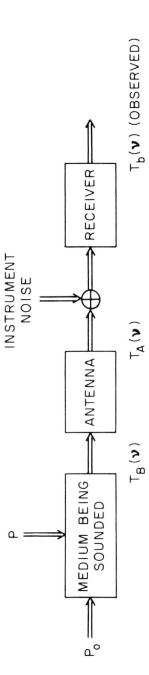

FIGURE 1. General remote sensing configuration.

In this figure, p_0 represents the parameter or set of parameters which are sought, p represents globally all other parameters possibly influencing the instrument reading in the same
region and, finally, ν is the channel frequency. $T_A(\nu)$ is the
antenna temperature, $T_B(\nu)$ the true brightness temperature, and
$T_b(\nu)$ the observed brightness temperature, which is an estimate
of $T_B(\nu)$,

Often, it is more practical to view T_b as the response of
the combined "system" receiver-instrument-medium to the parameter
p_0, for the certain atmospheric "state" or "regime" characterized
by p. This leads to the equivalent formulation:

$$T_b(\nu) = t(p_0, p, \nu) + \text{Noise} \qquad\qquad (1)$$

where the noise is now an equivalent noise corresponding to the
response of the instrument to the input noise.

For all scientists dealing with remote sensing, the problem
is to solve Eq. (1) or equivalently to answer the following
question: given a series of observations $T_b(\nu)$ above the same
geographical location, what is the value of p_0 which best
explains such an observation?

It is very important to realize that the previous question
leads either to a joint solution of Eq. (1) in terms of p and
p_0, or to a solution only in p_0, if one can assume *a priori* some
value for p, using another source of information about the
system, independent of the instruments.

In the first case, the complexity of the problem, if solvable
at all, is increased dramatically and a very large number of
instruments of all kinds is required to measure all components
of p and p_0. In the second case, however, the retrieval soft-
ware must contain some decision mechanism determining the right
regime of the system from a data base of precomputed values of p.

In operational systems, for instance, where large amounts of
satellite data are to be processed, the satellite is constantly
moving and sounding different portions of the atmosphere. There

is no reason to assume, in general, that p remains constant from one region of the globe to another. As a consequence, the response of the system instrument-atmosphere may be drastically modified and a retrieval system "adapted" to these modifications is required.

Unfortunately, although large numbers of technical papers on inversion methods have been published in the remote sensing literature with many different applications in mind, very few clearly state how they actually handle the problem of determining p and how the inversion method should be adapted to the variations of p.

In the following sections, a description of how this problem is handled in the case of the temperature profile retrieval problem using SCAMS and TIROS-N microwave measurements is presented. Several years of experience have now proven that microwave temperature retrievals using statistical methods perform surprisingly well in view of the relatively low number of channels which are used (3 or 4 channels only), and, therefore, deterministic methods will not be discussed.

III. NOTATION

Before discussing some of the assumptions, as well as the final implementation of such statistical methods, the notation which will be used throughout this paper is introduced. Vector quantities are represented with a single underbar, whereas matrix quantities use double underbars.

Figure 2 represents the geometry for a scanning microwave sounding experiment. The notation used for this experiment is as follows:

$\underline{T}_{b_i}$ vector of brightness temperatures observed for spot i

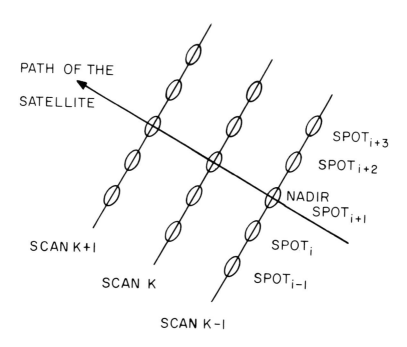

FIGURE 2. Footprints or "spots" of a scanning microwave instrument on the ground.

$\underline{\underline{W}}_i$ observation matrix, the rows of which are temperature weighting functions $\underline{w}^T(\nu)$ for different frequencies

$\underline{T}_i$ discretized temperature profile at spot i

$\underline{w}_i$ vector of additive observation noises

This notation leads in the case of microwave soundings to the following equation:

$$\underline{T}_{b_i} = \underline{\underline{W}}_i \cdot \underline{T}_i + \underline{w}_i \qquad (2)$$

Finally, the superscript T when used denotes a transposition.

In the case of the SCAMS and TIROS-N instruments, $\underline{\underline{W}}$ is a matrix almost completely independent of the temperature profile $\underline{T}$ but somewhat dependent on the surface emissivity and,

to a lesser extent, on the atmospheric water vapor content of
the region which is viewed. The regime of such a system is
thus determined by the set of all these parameters. However,
because Eq. (2) has a stochastic meaning, p is also characterizing the whole set of statistical assumptions needed to solve
Eq. (2).

IV. THE D-MATRIX APPROACH

Historically, the first statistical inversion method introduced in the remote sensing literature is a statistical regression technique called the D-matrix approach. (See Refs. 1-3.)

Assumptions: For each particular "regime" of the atmosphere:

$$\underline{T}_{b_i} = \underline{\underline{W}}_i \cdot \underline{T}_i + \underline{w}_i \tag{2}$$

where $\underline{\underline{W}}_i$ is independent of $\underline{T}_i$ (linearity assumption); $\underline{w}_i$ is a
zero-mean additive white noise, independent of the profiles $\underline{T}_i$;

$$\underline{\underline{R}}_{T_b, T_b} = E\left[(\underline{T}_b - \overline{\underline{T}}_b) \cdot (\underline{T}_b - \overline{\underline{T}}_b)^T \right]$$

$$\underline{\underline{R}}_{T, T_b} = E\left[(\underline{T} - \overline{\underline{T}}) \cdot (\underline{T}_b - \overline{\underline{T}}_b)^T \right]$$

$\overline{\underline{T}}_b$, $\overline{\underline{T}}$ are known and only regime dependent.

Problem: Find the minimum mean square estimate (m.m.s.e.)
$\hat{\underline{T}}_i$ of $\underline{T}_i$ given the observation $\underline{T}_{b_i}$ at the same spot i.

Solution:

$$\hat{\underline{T}}_i = \underline{\underline{D}}_i \cdot \begin{vmatrix} 1 \\ \underline{T}_{b_i} \end{vmatrix} \tag{3}$$

or equivalently:

$$\hat{\underline{T}}_i = \overline{\underline{T}} + \underline{\underline{R}}_{T,T_b} \cdot \left[\underline{\underline{R}}_{T_b,T_b}\right]^{-1} \cdot (\underline{T}_b - \overline{\underline{T}}_b) \tag{4}$$

where overbars denote expected values (i.e., means) and where matrices of the form $\underline{\underline{R}}_{X,Y}$ represent the cross covariance matrix of $\underline{X}$ and $\underline{Y}$.

The D-matrix approach is thus essentially a pointwise inversion technique in the sense that it does not make any assumption about the spatial structure of the atmosphere. For example, if a measurement is performed over a certain position i, only this information combined with other climatological information is used to infer the temperature profile prevailing in the same region. However, as it is noted from the assumptions as well as from Eqs. (3) and (4), the matrix $\underline{\underline{D}}$ is a function only of the local regime, and thus it can be precomputed in some data base. Figure 3 gives a description of an actual inversion procedure.

For the TIROS-N microwave spectrometer a data base of D-matrices was precomputed for different seasons, latitude bands, atmospheric water vapor contents, and emissivities. In parallel, an approximate map of the globe is stored for any value of the satellite position, to indicate whether the satellite instruments are viewing land, sea, ice, or some coastal region. This positional information together with that of the window channel is used to determine the corresponding surface emissivity of land or the approximate water vapor content of the atmosphere over ocean. Finally, the regime having been characterized, the corresponding D-matrix is selected and multiplied by the observed brightness temperatures to yield the corresponding temperature profile estimate $\hat{\underline{T}}$.

Preliminary results in Table I are listed for the case of the TIROS-N experiment. Two different methods were implemented

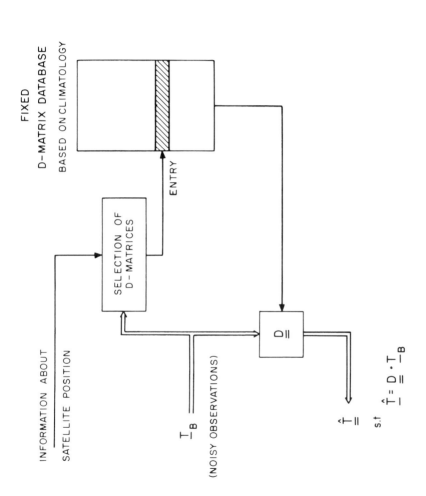

FIGURE 3. D-matrix technique implementation.

TABLE I. TIROS-N/MSU D-matrix Inversion (Nadir), February
25-26, 1973.

	4 Channels				3 Channels			
	Smooth		Discrete		Smooth		Discrete	
Pres-sure (mb)	Mean (K)	RMS (K)	Mean (K)	RMS (K)	Mean (K)	RMS (K)	Mean (K)	RMS (K)
1000	-8.5	8.3	-8.5	8.8	-0.6	4.1	-0.6	4.0
850	-0.2	2.9	-0.2	3.0	0.5	2.7	0.4	2.8
700	2.4	1.9	2.4	2.0	0.0	2.9	0.2	2.9
500	2.2	1.8	2.2	2.1	-0.2	2.3	-0.2	2.3
400	1.5	2.1	1.5	2.3	-0.8	2.1	-0.8	2.2
300	-0.2	2.4	-0.2	2.5	-1.4	2.1	-1.5	2.1
250	-2.2	2.8	-2.2	2.6	-1.9	3.3	-2.0	3.3
200	-5.4	3.7	-5.4	3.8	-3.0	4.4	-3.1	4.4
150	-4.6	3.0	-4.6	3.3	-2.5	2.7	-.23	2.9
100	-1.4	1.9	-1.4	2.2	-0.8	1.8	-0.4	2.2

both using the window channel to estimate surface emissivity.
One uses only the upper 3 channels, the second one uses all 4
channels including the window channel to estimate temperature
profiles. In each case, D-matrices were compared for "discrete"
geographical regions or for "smoothed" regions (spline interpolation of the statistics between regions). No correction had
been made at that time to account for biases of the instruments.
Finally, by comparison of the 3 and 4 channel cases, it seems
that the window channel has an important damaging effect on the
root mean square (rms) error of the first two levels, with
only a modest improvement at the other levels. This degradation
probably resulted because errors committed in estimating
surface emissivities are multiplied to correlated temperature

profile errors; thus the rms is increased in the lower levels
where the window channel has the most effect. In principle,
however, when separated from the temperature profile estimation,
the window channel should improve the performance, particularly
when the actual surface emissivity or humidity departs significantly from expected values.

V. THE KALMAN FILTERING APPROACH

Kalman filtering has been remarkably successful since its
appearance in the systems and control literature in the late
1950s, not only in the field of aeronautics and astronautics
but also in a wide range of other engineering fields. The first
attempt at using such a technique for temperature and water
vapor soundings, using the SCAMS instruments flown on NIMBUS-6,
was made by Ledsham and Staelin (4). Later, Gustavson and
Ledsham adapted it with some success to infrared soundings
too (5).

Consider the retrieval problem of N temperature profiles at
the same time, corresponding to N different spots within the
same scan. Such a system is commonly referred to as an "N-spot
filter." The following notation will also be useful:

$$\underline{x}^{k^T} = \left[(\underline{T}_1^k - \underline{\bar{T}})^T, \ (\underline{T}_2^k - \underline{\bar{T}})^T, \ \dots \ (\underline{T}_N^k - \underline{\bar{T}})^T \right]$$

$\underline{x}^k$ is called the state vector in scan k. It consists of the
concatenation of the deviations of the discrete temperature
profiles $\underline{T}_i$, i = 1 ... N, from their mean,

$$\underline{y}^{k^T} = \left[(\underline{T}_{b_1}^k - \underline{\bar{T}}_b)^T, \ (\underline{T}_{b_2}^k - \underline{\bar{T}}_b)^T, \ \dots \ (\underline{T}_{b_N}^k - \underline{\bar{T}}_b)^T \right]$$

$\underline{y}^k$ is called the "observation vector in scan k." It consists

of the concatenation of all measurements at all frequencies
referred to their local mean,

$$\underline{\underline{C}}_k = \begin{bmatrix} \underline{\underline{W}}_1 & & & 0 \\ & \underline{\underline{W}}_2 & & \\ & & \cdot & \\ 0 & & \cdot & \\ & & & W_N \end{bmatrix} \quad \text{is the "observation" matrix, and}$$

$$\underline{w}^{k^T} = \begin{bmatrix} \underline{w}_1^{k^T}, & \underline{w}_2^{k^T}, & \ldots & \underline{w}_N^{k^T} \end{bmatrix} \quad \text{represents the total noise}$$

in scan k.

From Eq. (2):

$$\underline{y}^k = \underline{\underline{C}}_k \cdot \underline{x}^k + \underline{w}^k \tag{5}$$

Assumptions on the spatial structure of the atmosphere are
introduced by considering that the process $\underline{x}^k$, indexed by the
scan number k, is a Markovian process. This new assumption is
usually expressed as:

$$\underline{x}^{k+1} = \underline{\underline{\Phi}}_k^{k+1} \underline{x}^k + \underline{u}^k \tag{6}$$

where $\underline{u}^k$ is an input noise generating the process $\underline{x}^k$ from an
initial value $\underline{x}^o$, and $\underline{\underline{\Phi}}_k^{k+1}$ is a matrix called the "state
transition" matrix or simply the "dynamics" of the process $\underline{x}^k$.
In fact it can be shown that the elements of $\underline{\underline{\Phi}}_k^{k+1}$ are related
to the correlation lengths of the temperature field being
measured.

The introduction of such an assumption leads to a surprisingly "clean" formulation of the temperature retrieval problem.

Assumptions: Let

$$\underline{x}^{k+1} = \underline{\underline{\Phi}}_k^{k+1} \cdot \underline{x}^k + \underline{u}^k$$

$$\underline{Y}^k = \underline{\underline{C}}_k \cdot \underline{x}^k + \underline{w}^k$$

where $\underline{u}^k$, $\underline{w}^k$, $\underline{x}^o$ are independent, zero-mean processes with only climatologically dependent covariances $\underline{\underline{Q}}_k$, $\underline{\underline{R}}_k$, and $\underline{\underline{\Sigma}}_0$, respectively. (See Eqs. (5) and (6).)

Problem: If $\Gamma^k = E(\underline{Y}^k, \underline{Y}^{k-1}, \ldots)$ represents the set of all observations collected along the orbit up to scan k, find the minimum-mean-square-estimate (m.m.s.e.) $\underline{\hat{x}}_{k|k}$ of $\underline{x}^k$ given $\underline{\Gamma}^k$.

Solution:

$$\underline{\hat{x}}_{k|k} = E(\underline{x}^k | \Gamma^k)$$

$$\underline{\hat{x}}_{k+1|k} = \underline{\underline{\Phi}}_k^{k+1} \cdot \underline{\hat{x}}_{k|k}$$

$$\underline{\hat{x}}_{k|k} = \underline{\hat{x}}_{k|k-1} + \underline{\underline{K}}_g^k \cdot (\underline{Y}^k - \underline{\underline{C}}_k \cdot \underline{\hat{x}}_{k|k-1}) \qquad (7)$$

$$\underline{\hat{x}}_{0|-1} = 0$$

Kalman Gain:

$$\underline{\underline{K}}_g^k = \underline{\underline{P}}_{k|k-1} \cdot \underline{\underline{C}}_k^T (\underline{\underline{C}}_k \cdot \underline{\underline{P}}_{k|k-1} \cdot \underline{\underline{C}}_k^T + \underline{\underline{R}}_k)^{-1} \qquad (8)$$

Covariance update:

$$\underline{\underline{P}}_{k+1|k} = \underline{\underline{\Phi}}_k^{k+1} \cdot \underline{\underline{P}}_{k|k} \cdot \underline{\underline{\Phi}}_k^{k+1^T} + \underline{\underline{Q}}_k$$

$$\underline{\underline{P}}_{k|k} = \underline{\underline{P}}_{k|k-1} - \underline{\underline{K}}_g^k \cdot \underline{\underline{C}}_k \cdot \underline{\underline{P}}_{k|k-1} \qquad (9)$$

$$\underline{\underline{P}}_{0|-1} = E(\underline{x}^o \cdot \underline{x}^{o^T}) = \underline{\underline{\Sigma}}_0$$

where $\underline{\underline{P}}_{k|k}$ is the error covariance of $\underline{\hat{x}}_{k|k}$ and $\underline{\underline{P}}_{k|k-1}$ is the error covariance of $\underline{\hat{x}}_{k|k-1}$.

A block diagram for implementing a Kalman filter is presented in Fig. 4. In the upper part of this figure, the structure of the assumed system is shown.

As in the D-matrix case, a data base must be precomputed with all relevant statistics stored. At any time, the position of the satellite is used to select the right climatological means, covariances of the noises and the initial covariance $\underline{\underline{\Sigma}}_o$. The lowest channels are used to estimate the surface emissivity and the atmospheric water vapor. This information combined with the current temperature estimate $\hat{\underline{X}}_{k|k-1}$ is used to fine-tune the weighting functions. By using the relevant statistics, the gain can be computed for the current scan.

Many algorithms are available in the literature for efficiently computing Eqs. (7), (8), and (9). In this case, where the dimension of the state can reach large numbers, depending on the number of spots processed at one time (75 for a 5-spot filter), a square root type of algorithm is necessary to ensure the positivity of the covariance matrices $\underline{\underline{P}}_{k|k}$, $\underline{\underline{P}}_{k|k-1}$ for all k.

Table II shows comparative results obtained by Ledsham in the case of the SCAMS instrument for 2 full days of data (6). A definite improvement is shown over the classical D-matrix approach.

To conclude this section, two major advantages of the Kalman filtering technique are summarized:

1. A spatial smoothing is introduced in a dynamic fashion along the satellite orbit; thus neighboring spots are statistically coupled and the retrieval rms error is improved.

2. Weighting functions can be fine-tuned according to changes in the state $\underline{X}^k$ itself. In the case where the new linearizing point is directly related to the current state estimate, the filter is called an "extended" Kalman filter.

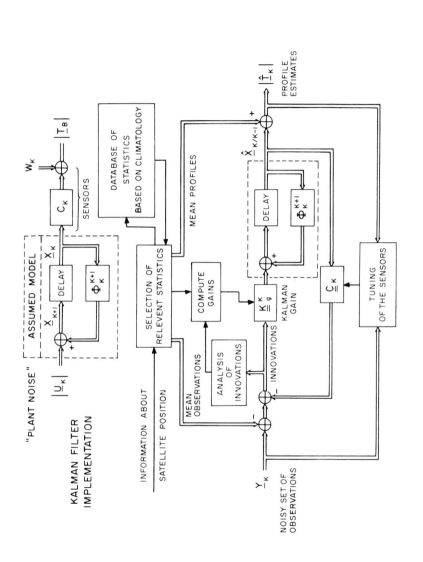

FIGURE 4. *Kalman filter implementation.*

TABLE II. Comparison of the Performances of the D-Matrix
and Kalman Filtering Techniques for SCAMS, January 24-25, 1976
(6).

	D-Matrix				Kalman			
	Nadir		Extreme		Nadir		Extreme	
Pressure	Mean	Var	Mean	Var	Mean	Var	Mean	Var
(mb)	(K)	(K)	(K)	(K)	(K)	(K)	(K)	(K)
1000	5.7	8.1	8.2	11.6	1.7	5.1	2.2	6.3
850	2.3	3.9	2.0	4.6	2.2	3.2	0.8	2.5
700	1.6	2.3	1.1	3.0	1.8	2.0	0.6	1.1
500	- 0.5	2.6	- 0.7	2.8	0.1	1.3	-0.1	1.6
400	- 1.3	3.3	- 1.2	3.0	0.5	2.0	0.9	1.4
300	- 2.8	4.0	- 1.7	3.5	-0.3	3.1	1.1	2.4
250	- 4.1	4.6	- 1.9	3.9	-1.9	4.0	-0.3	3.6
200	- 1.3	3.0	0.9	2.8	-2.9	2.9	-1.6	3.0
150	2.5	2.3	3.1	1.9	-1.4	1.4	-1.5	1.9
100	5.9	3.7	5.6	3.3	3.3	2.8	2.1	1.5
70	2.6	4.1	2.1	3.2	1.5	3.5	0.3	2.4
50	2.1	4.5	1.8	3.3	1.9	3.7	0.7	2.2
30	- 1.5	5.2	- 1.2	4.6	0.2	4.2	-0.5	3.5
10	-12.3	5.9	-10.1	6.6	-5.4	7.5	-5.5	7.2

3. Finally, mean errors are reduced too, because a tuning
of the weighting functions can be shown to be equivalent to
a dynamical tuning of the means.

VI. A MORE GENERAL ADAPTIVE SYSTEM

The type of adaptation in the D-matrix and the Kalman filter
is reviewed. In both cases, a data base is constructed by
using climatological arguments. The incoming observations are

first compared with some climatologically expected behavior, and
later are processed to be interpreted in meteorological terms.
Only in the case of the extended Kalman filter where the sensor
$\underset{=}{C}_k$ is modified with the observations, is the tuning process
dependent on *actual* observations. However, the local dynamics
$\underset{=}{\Phi}_k^{k+1}$ is still adapted in some "average" fashion.

One way of obtaining a better adaptation of the retrieving
process to the observed phenomenon is to replace the static
memory of the system, i.e., the climatological data base, with
a mathematical model which can be modified dynamically. Such
a system is represented in Fig. 5.

In passive remote sensing there is no other way of gathering
information about an unknown system than by collecting actual
observations and organizing them. In order to interpret these
observations, the entire geophysical system and sensor must be
represented by some mathematical model that is able to simulate
observations. The only possible verification of such a model
is by comparison of true and synthesized observations. If this
difference is not zero, the model must be fine-tuned with the
hope of improving the next prediction.

The previous considerations led to the design of an
adaptive system currently being tested; it will retrieve
temperature profiles from the TIROS-N microwave sensing unit.
A 5-level, hemispheric, spectral General Circulation Model
(GCM) similar to those of Bourke, and Daley, Girard and
Henderson was constructed (7,8). A rhomboidal spectral truncation scheme is used with wave numbers up to $J = \pm 15$. Spherical
harmonics are used to represent the horizontal extent of each
field, whereas finite differences are computed in the vertical.
Orszag's transform method was followed to compute all nonlinear
terms and because of the agreed limitation of forecasting no
more than 24 or 36 hours, most of the physics could be neglected
(diffusion processes, condensation, etc.) (9). Finally,

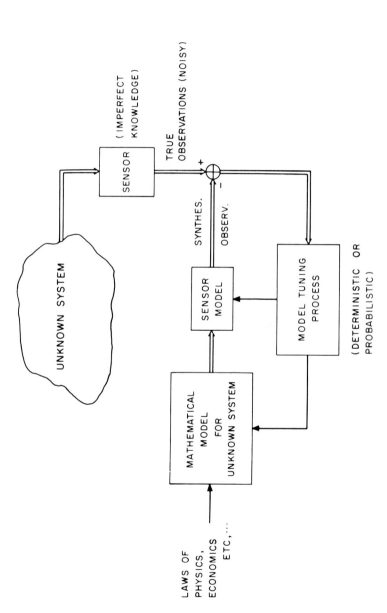

FIGURE 5. General adaptive filtering scheme.

the use of a semi-implicit time-stepping strategy allows consideration of time steps of almost 1 hour, roughly equivalent to one-half orbit.

Figure 6 contains a block diagram of this system with an initialization cycle of 24 hours. In the upper left corner, the GCM is loaded every 24 hours with actual NMC gridded analysis fields (temperatures, surface pressures, winds and so on). Every hour a forecast is computed by the GCM and transmitted to the lower part of the figure containing the temperature retrieving system. This forecast, together with the current satellite location, is combined to tune the Kalman filter performing the actual inversions.

At the end of half an orbit, the GCM is "tuned" by assimilating the result of the previous observations. Satellite observations, even in the case of microwave instruments, have a horizontal resolution much higher than the one acceptable to the GCM. As a consequence, a low pass filter is necessary to smooth the observations to the required resolution before assimilation.

Although much more complex than a Kalman filter, early results show that such a system is quite feasible for large-scale operations. Furthermore, its extreme modularity seems to make it an excellent experimental device for analyzing a wide variety of problems, ranging from the effect of remote sensing data on actual forecasts to the theoretical comparison of several instrument designs.

VII. SUMMARY

To summarize, the development of statistical retrieval techniques for deriving atmospheric parameters from remote sensing data began with single-point regression analyses and has evolved into multipoint regressions, Kalman filtering, and

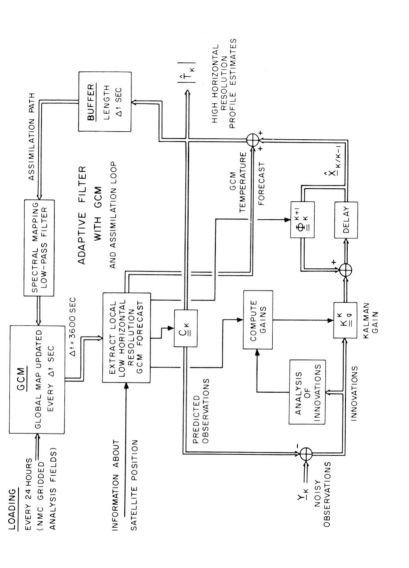

FIGURE 6. *Adaptive filter with GCM and assimilation loop.*

Kalman filtering augmented by relatively complete physical
models of the atmospheric circulation.

ACKNOWLEDGMENT

    The author wishes to acknowledge Prof. D. H. Staelin and
Dr. P. W. Rosenkranz from the Massachusetts Institute of
Technology for their comments.

SYMBOLS

p	all atmospheric parameters other than p_o influencing the measurements
p_o	some atmospheric parameters to be retrieved
$t(p_o,p,\nu)$	transfer function of the medium
$\underline{u}^k$	input noise
$\underline{w}_i$	vector of additive observation noises
$\underline{w}^k$	observation noise vector of a Kalman filter
$\underline{\underline{C}}^k$	observation matrix
E	expected value
$\underline{\underline{Q}}_k$	input noise covariance
$\underline{\underline{R}}_k$	observation noise covariance
$\underline{\underline{R}}_{X,Y}$	cross-covariance matrix of any vector process $\underline{X}$, with a vector process $\underline{Y}$
$T_b(\nu)$	observed brightness temperature
T_{b_i}	vector of brightness temperatures observed for spot i
$\underline{T}_i$	true discretized temperature profile at spot i
$T_A(\nu)$	antenna temperature for frequency ν
$T_B(\nu)$	true brightness temperature

$\underline{\underline{W}}_i$ observation matrix for spot i

$\overline{\underline{X}}$ expected value of any $\underline{X}$

$\underline{x}^k$ state vector for scan k in a Kalman filter

$\underline{X}^T$ transposition of any vector $\underline{X}$

$\underline{y}^k$ observation vector in scan k

Γ^k set of all observations $\underline{y}^j$ collected up to k, that is $j \leq k$

ν frequency of observation

$\underline{\underline{\phi}}_k^{k+1}$ "state transition matrix" of $\underline{x}^k$ from scan k to scan k + 1

$\underline{\underline{\Sigma}}_0$ initial state covariance

REFERENCES

1. Rodgers, C. D., "Satellite Infrared Radiometer: A Discussion of Inversion Methods." Memo. No. 66-13, University of Oxford, England (1966).
2. Westwater, E. R., and Strand, O. N., *J. Assoc. Comput. Mach.* 15, 100-114 (1968).
3. Rosenkranz, P. W., *et al.*, *J. Geophys. Res.* 77, 5833-5844 (1972).
4. Ledsham, W. H., and Staelin, D. H., *in* "Remote Sounding of the Atmosphere from Space" (H. J. Bolle, ed.), pp. 149-157. Pergamon Press, Oxford (1978).
5. Gustavson, D., and Ledsham, W. H., "Application of Estimation Theory to Inverse Problems in Meteorology." IEEE Conference on Decision and Control, San Diego, California (January 1979).

6. Ledsham, W. H., "Optimum Retrieval Techniques in Remote
    Sensing of Atmospheric Temperature, Liquid, Water and
    Water Vapor." Ph.D. Thesis, MIT EECS Department (June
    1978).
7. Bourke, W., *in* "Modeling for the First GARP Global
    Experiment," pp. 206-217. GARP Pub. Series No. 14,
    ICSU/WMO, Geneva (1974).
8. Daley, R., Girard, C., and Henderson, J., *Atmosph. 14(2)*,
    98-134 (1976).
9. Orszag, S., *J. Atmos. Sci. 27*, 830-835 (1970).

DISCUSSIONS

Susskind: I think I understood from talking to you before, that
you do not have to make *a priori* zenith angle corrections but
when you are modeling your observations here in terms of the
matrix, it is calculated directly.

Toldalagi: Exactly.

Susskind: So you have gotten away from that problem with the D
matrix.

Toldalagi: Exactly.

Susskind: Another thing. I am intrigued by what you are trying
to do. What are the things that are influencing your observations
besides the temperature? Humidity? Clouds? Rainfall? Are you
attempting to try and put this type of information into your
forecast model and predict?

Toldalagi: Certainly in the future we will add equations for
humidity and moisture. In the present implementation we have
only a temperature equation. This structure is fairly modular
and will be used basically for research. We want to analyze the
potential improvement available by tracking things like wind
fields and humidity in the atmosphere.

Susskind: Do you make any attempt to account for the effects of
water vapor on the brightness temperatures in those channels?

Toldalagi: Not for the moment. I do not have any equation but
Bourke[1] used several complex models for moisture.

Westwater: I think I noticed that at one of the upper levels
the rms errors and the retrievals using the ordinary linear
statistical inversion. What would you attribute this to?

Toldalagi: The problem probably is due partly to the fact that
we have very poor *a priori* climate information

---

[1]*See Ref. 7 in this paper.*

Westwater: But wouldn't the linear statistical inversion have the same amount of uncertainty due to the improper *a priori* statistics?

Toldalagi: Theoretically, you are right. We have, however, shown two different implementations for two different periods and climatologies. I could not answer with more precision. I think the problem is due to the *a priori* statistical assumptions.

Staelin: Could you talk about the advantages of this spectral model approach compared to the more traditional grid-point model?

Toldalagi: I probably rushed too much in this part of my presentation. The basic advantages of such a spectral model from a purely numerical forecasting standpoint are: one, the magnitude and spatial gradients of each parameter field (winds, temperatures, pressures) can be computed analytically with high precision up to the order of spectral truncation ($J = 15$); two, the polar regions do not introduce any computational singularity; and three, this type of model has very nice energy consideration properties inherently. On the other hand, the advantages from a remote sensing point of view are: one, the matrix $\underline{\underline{\Phi}}$ representing the dynamics of the temperature fields in the Kalman model is a direct by-product of the circulation model; two, this matrix $\underline{\underline{\Phi}}$ can also be used to modulate locally the *a priori* temperature covariance matrices; and finally, when the circulation model incorporates moisture equations, the observation matrix in the Kalman filter can be tuned locally, to account for the impact of water vapor on the radiances. The retrieval system is consequently "adapted" to the atmosphere being sensed.

TEMPERATURE RETRIEVALS FROM TIROS-N

J. Susskind
A. Rosenberg[1]

National Aeronautics and Space Administration
Goddard Space Flight Center
Greenbelt, Maryland

A nonstatistical, nonlinear iterative scheme, has been
developed, based on the methods of Chahine, to retrieve surface
temperature, vertical temperature profile and cloud distribution
from TIROS-N HIRS 2/MSU observations. Fundamental to the
technique is the ability to accurately calculate clear column
radiances given atmospheric and surface characteristics.
The cloud filtering algorithm, making simultaneous use of
IR 15 μm channels and microwave channels, allows accurate
infrared retrievals in the presence of up to 80 percent cloud
cover. Details for the iterative scheme are presented
together with preliminary results from analysis of TIROS-N
data.

[1]Present address: RCA-ASTRO, Advanced Mission Group,
Princeton, New Jersey

I.   INTRODUCTION

    TIROS-N, the current operational meteorological satellite,
contains three passive temperature sounding instruments: HIRS2
(a 20-channel infrared radiometer), MSU (a 4-channel microwave
radiometer), and SSU (a 3-channel pressure modulated infrared
radiometer sounding the upper stratosphere).   NOAA/NESS
operationally produces global atmospheric temperature soundings
by analysis of observations from these three instruments by using
statistical regression methods.   This paper describes a
fundamentally different approach to analysis of satellite
temperature sounding data, being developed and employed at NASA,
which relies more heavily on the ability to account for the
atmospheric physics giving rise to the observations rather than
on the statistical relationships of atmospheric properties to
satellite observations.

    The iterative scheme and cloud filtering methods used in
the inversion process are closely related to those of Chahine
(1,2).   Intrinsic in the use of the Chahine iterative retrieval
scheme is the ability to accurately solve the forward problem,
that is, given a guess set of atmospheric and surface parameters,
to accurately compute the corresponding observations as seen by
the satellite instruments.   The method involves iterative
modification of the atmospheric parameters, starting from an
initial guess, until sufficient agreement between computed and
observed radiances is reached.   First, considerations of the
forward problem are discussed, then the iterative scheme used,
and finally some preliminary results of TIROS-N retrievals are
compared with colocated radiosonde reports.

II. THE RADIATIVE TRANSFER EQUATION

Given atmospheric and surface conditions, the clear column radiances R_i observed by a sounding channel i can be expressed as

$$R_i = \varepsilon_i B_i [T_s] \tau_i (P_s) + (1 - \varepsilon_i) R_i \downarrow \tau_i (P_s)$$

$$+ \rho_i H_i \tau_{is} (P_s) + \int_{\ln P_s}^{\ln \bar{P}} B_i [T(P)] \frac{d\tau_i}{d \ln P} \, d \ln P \qquad (1)$$

where ε_i is the surface emissivity averaged over sounding channel i, $B_i [T]$ is the mean Planck black-body function, averaged over channel i, of the temperature T, $\tau_i (P)$ is the mean atmospheric transmittance from pressure P to the top of the atmosphere and evaluated at θ, the zenith angle of the observation, $R_i \downarrow$ is an effective atmospheric emission downward flux, $\rho_i H_i \tau_{is}$ is the reflected solar radiation in the direction of the satellite, and the subscript s refers to surface. The integral, taken from the surface to the satellite pressure P, represents the upwelling atmospheric emitted radiation, which is a mean value of the black body function of atmospheric temperature weighted by the channel weighting function $d\tau_i / d \ln P$. Table I shows the channels, centers, and peak of the weighting functions, or other relevant information, for the channels on MSU and HIRS. The current analysis does not employ the SSU observations.

The transmittance functions for the HIRS channels are taken to be a product of dry transmittance functions, parameterized as a function of temperature profile and zenith angle, and effective water vapor transmittances of the form

$$\tau_i (P, \theta) = e^{-a_i W \left(\frac{P}{P_s} \right)^{n_i}}$$

TABLE I.   HIRS2 and MSU Channels

Channel	$\nu(cm^{-1})$	Peak of $d\tau/d$ ln $P(mb)$	Peak of $B\ d\tau/d$ ln $P(mb)$
H1	668.40	30	20
H2	679.20	60	50
H3	691.10	100	100
H4	703.60	280	360
H5	716.10	475	575
H6	732.40	725	875
H7	748.30	Surface	Surface
H8	897.70	Window, sensitive to water vapor	
H9	1027.90	Window, sensitive to O_3	
H10	1217.10	Lower tropospheric water vapor	
H11	1363.70	Middle tropospheric water vapor	
H12	1484.40	Upper tropospheric water vapor	
H13	2190.40	Surface	Surface
H14	2212.60	650	Surface
H15	2240.10	340	675
H16	2276.30	170	425
H17	2310.70	15	2
H18	2512.00	Window, sensitive to solar radiation	
H19	2671.80	Window, sensitive to solar radiation	
M1	50.30^a	Window, sensitive to surface emissivity	
M2	53.74^a	500	
M3	54.96^a	300	
M4	57.95^a	70	

[a]Values in GHz.

where W is the column density of water vapor and a_i and n_i are
channel dependent constants (3,4). The infrared emissivity
ε_i is taken as 0.85 or 0.95 for land and water, respectively,
at wavelengths greater than 10 μm, and 0.95 or 0.98 for
wavelengths less than 5 μm (channels between 5 μm and 10 μm
were not used in the analysis). The effective downward flux,
$R_i\downarrow$, is calculated according to Kornfield and Susskind (5).
The solar radiation term is discussed later.

The microwave transmittances are taken to be products of O_2
transmittances, calculated as a function of temperature profile
and zenith angle according to Rosenkrantz and water vapor
transmittances having the same form as the infrared effective
water vapor transmittances (6). Water vapor retrievals using
the humidity sounding channels on HIRS2 are not done at this
time, and W, the water vapor column density, is estimated at
the 24-hour-lag analysis value, as is the surface pressure P_s.
The microwave emissivity is calculated from the 50.3 GHz
channel, as part of the iterative scheme, according to

$$\varepsilon = \frac{R_i - \int Td\tau - R_i\downarrow \tau_i(P_s)}{[T_s - R_i\downarrow]\, \tau_i(P_s)} \qquad (2)$$

where R_i is the 50.3 GHz observed brightness temperature, T_s is
the iterative surface temperature, and T(P) is the iterative
atmospheric temperature profile used in the calculation of the
upward and downward microwave fluxes emitted by the atmosphere.
The transmittance functions are, of course, corrected for
temperature, water vapor, and zenith angle as described
earlier, but possible effects of liquid water attenuation on
the 50.3 GHz channel are not accounted for. The emissivity
determined from the 50.3 GHz channel observation is used,
together with the iterative temperature profile, to calculate
brightness temperatures for the other MSU channels.

III.   ACCOUNTING FOR EFFECTS OF CLOUDS ON THE INFRARED
       OBSERVATIONS

The infrared radiance observed in an otherwise homogeneous
field of view, containing partial homogeneous cloud cover α,
is given, by a reasonable approximation, as

$$R_i = \alpha R_{i,CLD} + (1 - \alpha) R_{i,CLR} \qquad (3)$$

where $R_{i,CLD}$ and $R_{i,CLR}$ are the radiances which would have been
observed if the field of view were completely cloudy or clear,
respectively. Computation of clear column radiances $R_{i,CLR}$
can be done routinely as in Eq. (1), but computation of
$R_{i,CLD}$ requires accurate knowledge of the optical as well as
meteorological properties of the cloud. It is more advantageous
to be able to account for the effects of clouds indirectly than
to have to model their radiative transfer properties. A method
for doing this, as proposed by Chahine, using observations in
adjacent fields of view and using the assumption that both
fields of view are identical (up to an accountable effect of
zenith angle) was used (2). An estimate of the clear column
radiance, $R_{i,CLR}$, can be reconstructed from the observations
according to

$$R_{i,CLR} = R_{i,1} + \eta [R_{i,1} - R_{i,2}] \qquad (4)$$

where $R_{i,j}$ is the observation for channel i in field of view
j and η is given by $\alpha_1 / (\alpha_2 - \alpha_1)$, with $\alpha_2 > \alpha_1$. Since η is
dependent only on fractional cloud cover, η is independent of
channel and spectral region. Given η, clear column radiances
can be reconstructed from the observations by using Eq. (4)
and the effects of clouds are, in principle, accounted for.
These reconstructed clear column radiances are then used
in the iterative temperature scheme. It is seen from Eq. (4)
that large values of η will tend to amplify noise in the

observations and are, therefore, undesirable. As shown by
Chahine and Susskind *et al.*, η can be determined as part of an
iterative scheme according to

$$\eta^{(N)} = \frac{R_7^{(N)} - R_{7,1}}{R_{7,1} - R_{7,2}} \tag{5}$$

where $R_7^{(N)}$ is the computed clear column radiance for the 15 μm
surface channel, using the Nth iterative temperature profile
(2,7). The scheme will converge provided only 4.3 μm infrared
channels are used for temperature sounding in the lower troposphere. The rate of convergence increases with the difference
between the surface temperature and the cloud top temperature.
Under some high noise, low contrast conditions, divergent
solutions can occur in the sense that an overestimate of $\eta^{(N)}$
will cause an overestimate of the reconstructed 4.3 μm clear
column radiances which, in turn, will yield an increased lower
tropospheric temperature, produce an increased value of $R_7^{(N+1)}$,
and lead to an increased $\eta^{(N+1)}$, etc.

   This situation is greatly alleviated by incorporation of a
lower tropospheric microwave observation in the determination
of η. The error in $\eta^{(N)}$ is a result of either an error in
$R_7^{(N)}$, due to a wrong temperature profile or computational
uncertainties such as the effect of water vapor on the transmittance functions of channel 7, observational errors in $R_{7,i}$,
or errors in the assumption of only one degree of nonhomogeneity
in the combined fields of view. The error in $R_7^{(N)}$ due to a
wrong temperature profile can be well accounted for by adjusting
the computed brightness temperature for channel 7 by the difference in the observed and computed microwave brightness
temperatures for channel M2 according to

$$T_7 - T_7^{(N)} = T_{M2} - T_{M2}^{(N)} \tag{6}$$

where T_{M2} and $T_{M2}^{(N)}$ are the observed and calculated microwave brightness temperatures, $T_7^{(N)}$ is the calculated clear column brightness temperature for channel 7, and T_7 is the corrected clear column brightness temperature for channel 7. The corrected clear column radiance for channel 7, to be used in Eq. (5), is given by

$$R_7^{(N)} = B_7 \left[T_7^{(N)} + T_{2,M} - T_{2,M}^{(N)} \right] \tag{7}$$

This procedure not only speeds up convergence under all conditions, but stabilizes the solution in the sense that an increase in the iterative temperature profile in the lower troposphere will not, to a first approximation, cause an increase in η.

IV. DETERMINATION OF SURFACE TEMPERATURE

Given $\eta^{(N)}$, the clear column radiances for the three window channels 8, 18, and 19 are reconstructed according to Eq. (4). All three channels are relative atmospheric windows and are sensitive primarily to the surface (ground) temperature. The two 3.7 μm channels have the advantage of being more sensitive to surface temperature and less sensitive to uncertainties in surface emissivity and atmospheric water vapor than the 11 μm window channel. They have the disadvantage of being affected by solar radiation during the day, which must be accounted for before accurate surface temperatures can be calculated. At night, surface temperatures are taken to be the average of the surface temperature as determined from channels 18 and 19 where

$$T_{s,i}^{(N)} = B_i^{-1} \left[\frac{R_i^{(N)} - (1 - \varepsilon_i) \; R_i^{`(N)} \downarrow \tau_i^{(N)} (P_s) - \int_{\tau_i(P_s)}^{1} B_i(T^N) \; d\tau}{\varepsilon_i \; \tau_i^{(N)} P_s} \right] \quad (8)$$

In general, $T_{s,18}$ and $T_{s,19}$ agree with each other to $1°$, but differ by a larger amount from the surface temperature as determined from the 11 µm window channel 8, especially when the water vapor column density along the path of observation is greater than 3 gm/cm^2.

During the day, the effects of solar radiation on the 3.7 µm channels must be accounted for in obtaining accurate surface temperature retrievals from these channels. This can be done directly by subtracting $\rho_i H_i \tau_{is}(P_s)$ from $R_i^{(N)}$ and substituting the result into Eq. (8). $H_i \tau_{is}(P_s)$, the mean solar radiation across the channel, traversing the path from the sun to the earth and back to the satellite, can be well estimated as $2.16\pi \times 10^{-5} B_i[5600 \text{ K}] \cos \theta_H \tau_i(P_s, \theta_H + \theta)$ where θ_H is the solar zenith angle and the transmittance is computed at an effective zenith angle given by the sum of the solar and the satellite zenith angles.

The danger in such a procedure is the uncertainty in ρ_i. If the surface is Lambertian and the emissivity is known, ρ_i, the directional reflectance, is equal to $(1 - \varepsilon_i)/\pi$. Significant errors of up to a factor of 2 can be made in these estimations of ρ_i, which may produce errors of up to 10 K in retrieved surface temperature. These errors arise from uncertainties in ε_i and non-Lambertian character of the surface. The same uncertainties in ε_i, however, do not appreciably affect the calculated thermal radiation.

An error in ρ_i will produce different errors in the surface temperature as retrieved from the two 3.7 µm channels because $\frac{dT}{dB}$ in channel 19 is twice as great as that in channel 18.

Consequently, agreement to within $1°$ of $T_{s,18}$ and $T_{s,19}$, obtained by subtracting $(1 - \varepsilon_i)/\pi \, H_i \tau_i (P_s)$ from $R_i^{(N)}$ in Eq. (8), is taken as evidence of an accurate estimated value of ρ_i, and the average of $T_{s,18}$ and $T_{s,19}$ is taken as the surface temperature. If this agreement is not obtained, $T_{s,8}$, obtained from the 11 μm window channel, is taken as the ground temperature provided the water vapor column density along the observation path is less than 3 g/cm^2.

As a final alternative, an attempt to solve simultaneously for T_s and ρ_i is made by linearizing Eq. (1) about $T_{s,8}$, which is expected to be a reasonable approximation of T_s, according to

$$R_i = \varepsilon B_i [T_{s,8}] \, \tau_i (P_s) + \varepsilon \left(\frac{dB_i}{dT} \right)_{T_{s,8}} (T_s - T_{s,8})$$

$$+ \rho H_i \tau_{i,s} (P_s) + (1 - \varepsilon) \, R_i \!\downarrow \tau_i (P_s) + \int B_i \, d\tau \qquad (9)$$

Equation (9) gives two linear equations, one for channel 18 and one for channel 19, for the two unknowns, T_s and ρ, and can be solved in a straightforward manner.

V.   ITERATIVE PROCEDURE

Given an initial or Nth guess temperature profile $T^N(P)$, and surface temperature T_s^N, clear column radiances can be calculated from Eqs. (1) and (2) and algorithms to calculate transmittance functions and downward fluxes. Equations (4), (5), and (7) are then used to reconstruct clear column radiances $R_i^{(N)}$ for the infrared channels. T_s^{N+1} is obtained from Eq. (8) or (9). $T^{N+1}(P)$ is obtained in a relaxation manner analogous to that of Chahine. Chahine assigns a pressure P_i to each temperature sounding channel i, given approximately by the peak of the weighting function for channel i, and representing that

portion of the atmosphere in which local changes of temperature
will have the largest effect on the observed radiance for that
channel.   Chahine uses the iterative equation

$$\frac{B_i[T^{N+1}(P_i)]}{B_i[T^N(P_i)]} = \frac{\hat{R}_i^{(N)}}{R_i^{(N)}} \tag{10}$$

to determine a new estimate of temperature at each of the
pressures P_i.   Temperatures at other pressures can be obtained
by a variety of interpolation schemes or other constraints on
the solution.   Given $T^{N+1}(P)$ and T_s^{N+1}, the entire iterative
procedure can then be repeated.

   A more convenient form of the iterative equation for
temperatures can be written in terms of clear column brightness
temperatures T_i for each channel, that is, the temperature of
a black body with radiance R_i.   The analogous equation to
Eq. (10) becomes

$$T^{N+1}(P_i) = T^N(P_i) + \hat{T}_i^{(N)} - T_i^{(N)} \tag{11}$$

Equation (11) reflects the very good approximation that, given
a temperature profile and brightness temperature, if the entire
profile and surface temperatures were changed slightly by ΔT,
the brightness temperature will also change by ΔT regardless of
the frequency region or nature of the weighting function.

   The iterative scheme based on either Eq. (10) or,
equivalently, Eq. (11), sometimes produces divergent solutions
if the noise in adjacent channels is large and of opposite
sign.   It was found that stability was increased, with only
slight loss in speed of convergence, if the temperature at
pressure P_i was modified according to a weighted difference
of observed and calculated brightness temperatures for all
channels according to

$$T^{N+1}(P_i) = T^N(P_i) + \sum_j W_{ij}\left(\hat{T}_j^{(N)} - T_j^{(N)}\right) \Big/ \sum_j W_{ij} \tag{12}$$

where W_{ij} is the relative change in brightness temperature for channel j produced by a change in temperature at pressure P_j.

From Eq. (1), to a good approximation

$$\frac{dR_j}{dT(P_i)} = \left(\frac{dB_j}{dT}\right)_{T(P_i)} \left(\frac{d\tau_j}{d\ln P}\right)_{P_i} \tag{13}$$

W_{ij}, the relative change in brightness temperature for channel j produced by a change in temperature at P_i, is given by

$$W_{ij} = \frac{dT_j}{dT(P_i)} = \left(\frac{dT}{dB}\right)_{T_j} \frac{dR_j}{dT(P_i)}$$

$$= \left(\frac{dT}{dB}\right)_{T_j} \left(\frac{dB}{dT}\right)_{T(P_i)} \left(\frac{d\tau_j}{d\ln P}\right)_{P_i} \tag{14}$$

Using appropriate approximate forms for the black body function at microwave and infrared wavelengths, one gets.

$$W_{ij} = \left[\frac{T_j}{T(P_i)}\right]^2 \exp\left[1.439\ \nu_j\left(\frac{1}{T_j} - \frac{1}{T(P_i)}\right)\right]\left(\frac{d\tau_j}{d\ln P}\right)_{P_i} \tag{15a}$$

for infrared channels and

$$W_{ij} = \left(\frac{d\tau_j}{d\ln P}\right)_{P_i} \tag{15b}$$

for microwave channels.

Figure 1 shows the weighting functions as defined in Eq. (15) for the six TIROS-N channels used in the analysis of the data for a US standard atmosphere under nadir viewing. The observations are most sensitive to atmospheric temperature changes below 500 mb and between 150 mb and 50 mb, with relative

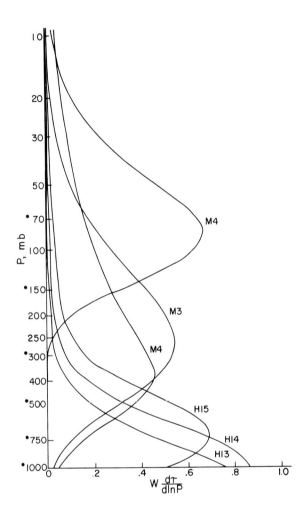

FIGURE 1.   Weighting functions for the six channels used to determine atmospheric temperature profile.   The * represents a pressure where iterative temperature adjustments are made.

weakness in between. The six characteristic pressures
used in the analysis are also shown in the figure.

VI. RESULTS

TIROS-N temperature soundings were produced for the period
January 3-8, 1979. Retrievals were run twice, by using either
a zonally averaged climatology initial guess temperature
profile, or one derived from a 24-hour-lag NMC analysis, to
show the dependence of the solution on the initial guess.
A shape preserved interpolation scheme, in which $T^{N+1}(P)$
- $T^N(P)$ was taken to be linear in log P, was used in both
cases (1). Figure 2 shows the root-mean-square (rms) errors
of mean temperature of layers bounded by the pressures 1000 mb,
850 mb, 700 mb, 500 mb, 400 mb, 300 mb, 200 mb, 100 mb, 70 mb,
and 50 mb for the temperatures retrieved by using both
initial guesses, and for the guesses themselves, as compared
with radiosonde reports colocated to 110 km in space, 6 hours
in time. The results shown are for colocated soundings over
the ocean between 18° N and 70° N latitudes.
In all the cases, a sounding was rejected if, after five
iterations, the rms difference of computed and reconstructed
brightness temperatures for the six temperature sounding
channels was greater than 1° or the computed and observed
brightness temperature for channel M2 differed by more than
1.5° . Approximately 30 percent of all soundings were rejected
by these criteria.
It is seen that considerable improvement is attainable over
the initial guess, and that moreover, in the regions of
maximum information below 500 mb and above 150 mb, the results
are relatively independent of the guess used. The increased
error beneath 700 mb is due, to a large extent, to low level

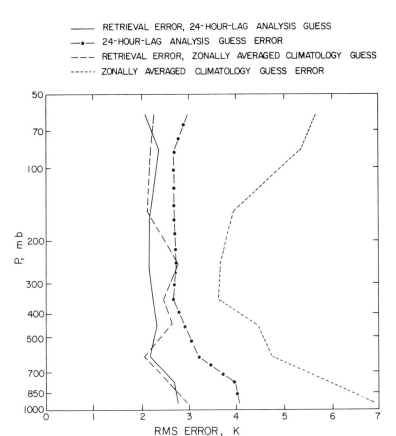

FIGURE 2. Root-mean-square errors of first guess and retrieved mean layer temperatures as compared with radiosondes. Results are shown for 6 hour, 110 km radiosonde colocations over oceans between 18° N and 70° N latitude for the period January 3-8, 1979. Mean layer errors greater than 7° are excluded from the statistics to guard against unrepresentative radiosonde reports. Inclusion of these errors causes slight degradation of the statistics.

inversions not resolved by the soundings and to larger spatial gradients in the lower tropospheric temperatures.

The following sample sounding and radiosonde comparison exemplify many of the concepts discussed in the text. The sounding was taken over the South Pacific at 28.3° S latitude, 142.9° W longitude on 2Z January 4, 1979. The colocated radiosonde was on Rapa Island 27.6° S, 144.5° W at 0Z January 4, 1979. The zenith angle of observation was 44.3° and the sounding was during the day. Columns 3 and 4 in Table II show the brightness temperatures observed in the two fields of view for each of the eight HIRS2 channels used in the analysis. Column 5 shows the reconstructed clear column brightness temperature for the HIRS2 channels as well as the observations for the MSU channels. Column 6 shows the brightness temperatures calculated for each channel using the initial guess temperature humidity profile and the ground temperature retrieved from the window channels using $T^{(0)}(P)$ and $\eta^{(0)}$.

The radiosonde report (at the mandatory levels) and the initial guess from a 24-hour-lag NMC analysis are shown in Table III, as well as the solution. It is observed that the guess follows the shape of the solution fairly well, that is, the vertical error structure is a smooth function and, in general, shows a warm bias ranging from 1° to 4° with maximum errors at 500 mb and 200 mb.

It is clear from the observations that the field of view 2 is considerably more cloudy than the field of view 1 with differences in brightness temperatures of 20° in channel 8, decreasing to 4° in channel 4. It appears from channel 7 that field of view 1 is almost, but not completely, clear, because the observed brightness temperature in field of view 1 is 2° colder than that calculated from the initial guess. If only channel 7 were used to construct $\eta^{(0)}$, a small positive value would result from Eq. (5). A comparison of observed and

TABLE II. Observed and Computed Brightness Temperatures (°K) for Channels Used in the Analysis of the Sounding.

Channel j	Pressure (mb) or function P_j	$T_{j,2}$	$T_{j,1}$	$\hat{T}_j(0)$	$T_j(0)$	$\hat{T}_j(0) - T_j(0) = \Delta_j$	$T^1(P_j) - T^0(P_j) = \Sigma w_{ji}\Delta_i$	$T(P_j) - T^0(P_j)$
4	300	222.0	226.0	226.0	228.3	-2.3	-2.4	-3.6
7	η	248.0	263.1	263.1	265.0	-1.9	–	–
8	Window	267.5	287.2	287.2	287.2	0	–	–
13	1000	263.6	273.3	273.3	274.1	-0.8	-1.3	-1.3
14	750	251.7	260.2	260.2	262.2	-2.0	-1.6	-2.0
15	500	240.4	246.7	246.7	248.7	-2.0	-1.9	-2.8
18	Window	283.3	292.2	292.2	290.1	-2.1	–	–
19	Window	288.4	294.9	294.9	290.7	-4.2	–	–
M1	Emissivity	–	–	244.6	244.6	0	–	–
M2	500, η	–	–	248.7	251.7	-3.0	–	–
M3	150	–	–	221.8	224.8	-3.0	-2.6	-3.7
M4	70	–	–	211.1	212.9	-1.8	-2.1	-1.9

TABLE III. *Comparison of Retrieved and Initial Guess Temperature Profiles to Colocated Radiosonde Reports*

Pressure (mb) P_i	$T_{RAD}(P_i)$	$T^O(P_i)$	$T^O(P_i) - T_{RAD}(P_i)$	$T(P_i)$	$T(P_i) - T_{RAD}(P_i)$
30	220.90	220.44	-0.46	218.90	-2.00
50	211.90	213.05	1.15	211.27	-0.63
70	205.30	207.92	2.62	205.99	0.69
100	202.90	204.13	1.23	202.90	-1.52
150	205.30	209.17	3.87	205.47	0.17
200	215.50	220.04	4.54	216.37	0.87
250	227.60	231.64	4.04	227.95	0.35
300	238.30	241.11	2.81	237.49	-0.81
400	254.10	256.01	1.91	252.86	-1.23
500	263.30	267.34	4.04	264.56	1.26
700	280.20	281.57	1.37	279.44	-0.76
850	290.60	289.91	-0.69	288.22	-2.37
1000	295.00	296.90	1.90	295.61	0.61

computed brightness temperatures for microwave channel M2 shows
the mean initial guess to be too warm by approximately 3° in the
lower troposphere. The microwave emissivity used in the calculation brightness temperatures was 0.51 as determined from Eq. (2).
When the microwave correction is made by using Eqs. (6) and (7),
η is estimated at a small negative number in Eq. (5), and is
set equal to 0, that is, field of view 1 is assumed to be clear
in the analysis.

The reconstructed clear column radiances shown in column 5
are equal to those observed in field of view 1, and correspond
to η = 0. By using these radiances in channels 8, 18, and 19,
ground temperatures of 292.26, 295.48, 298.00 were retrieved,
respectively, by using Eq. (8) and assuming no solar radiation.
When the solar correction was subtracted from the radiances of
channels 18 and 19, by using a directional reflectivity consistent with a Lambertian surface with emissivity of 0.95,
retrieved ground temperatures of 292.62 and 292.80 were
obtained. The sea-surface temperature was taken to be the
average of these two values because the agreement was within
1°.

The computed brightness temperatures shown in column 6 used
the computed ground temperature and the initial guess profile
but did not include solar radiation. Column 7 shows the
differences between these values and the reconstructed clear
column brightness temperatures. Column 8 shows the weighted
sum of these differences used to make the first iteration to
the guess profile by using Eq. (12). It is seen that all
changes are negative and follow the general error structure
of the guess, with a maximum correction applied at 150 mb. The
brightness temperature differences do not have the resolution to
pick the relative minimum error at 400 mb. Column 8 shows the
total difference of the final solution from the initial guess
at the six temperature levels. Most of the changes occurred

in the first iteration. The final ground temperature was
293.3°. The computed radiances agreed to with 0.2° of the
observations at the solution.

The retrieved temperatures of the mandatory levels and
comparisons with the radiosonde is included in Table III. The
general accuracy of the retrieved profile is attributable to
the smoothness of the error structure of the initial guess.
Significantly larger errors can occur in cases where structure
such as that due to low- or mid-level inversions or the
tropopause, is not well accounted for in the initial guess.

REFERENCES

1.   Chahine, M. T., *J. Atmos. Sci. 27,* 960-967 (1970).

2.   Chahine, M. T., *J. Atmos. Sci. 31,* 233-243 (1974).

3.   Mo, T., and Susskind, J., NASA Tech. Memo. 80253, 81-87
     (1978).

4.   Susskind, J., Halem, M., Edelmann, D., Tobenfeld, E.,
     Searl, J., Karn, R., Dilling, R., Sakal, D., Tung, L.,
     Carus, H., Rushfield, N., and Tsang, L., NASA Report
     X-1130-77-53, Appendix A (1977).

5.   Kornfield,J., and Susskind, J., *Mon. Wea. Rev. 105,*
     1605-1608 (1977).

6.   Rosenkranz, P. W., *IEEE Transactions on Antennas and
     Propagation, 23,* 498-506 (1975).

7.   Halem, M., Ghil, M., Atlas, R., Susskind, J., and Quirk, W. J.,
     NASA Tech. Memo. 78063, 2-28 to 2-33 (1978).

DISCUSSION

Westwater: Have you computed the statistics of your retrieval
results for TIROS-N data to see how they would do on the average
rather than on a few isolated cases?

Susskind: We are still in a very preliminary stage in our
analysis, so I wouldn't say that the statistics are particularly
meaningful at this time. We are taking a running track of
statistics as we go. We are getting some good solutions, some
poor ones. Typically, what we had to look at was the data for
the first week in January, and the mid-latitudes between about
$40°$ to the Tropics. The retrieval statistics are for slab-
average retrieved temperatures between 1000 and 850 mb, 850 and
700 mb, etc., versus radiosonde average temperature. The rms
errors there are of the order of 2 K or 2.1 K. They are considerably worse for latitudes above $40°$ especially in the lowest
altitudes between 700 and 850 mb and 850 and 1000 mb because
the profiles are categorized by very sharp temperature inversions.
We are working on a technique for determining these inversions
by independent measurements of ground and air temperatures, but
at this point we have not optimized it. The preliminary results
are at least a degree worse in the high latitudes. The number
may be different next week.

Green: I asked Dr. Abbas in private as to whether there is any
optimum wavelength one can use in this area. Are there any
general considerations as to whether one can go further into
the infrared or use higher resolution to optimize the sounding
ability?

Susskind: There are a number of considerations. However, there
are two main considerations. First is that you have to be looking
at where there is an absorbing gas in the atmosphere whose
distribution you know. So it is basically CO_2 and N_2O in the
infrared. The second one is limited by the signal of the
emitted atmospheric radiations. You don't want to go significantly shorter than say 2400 cm^{-1} that we are talking about.
And you like to be able to get the weighting functions as sharp
as you can. The considerations there are to, essentially, have
the onset of absorption occur as rapidly as possible as you go
through the atmosphere. The effects of that are that if the
absorption coefficient increases rapidly as you go down in
altitude, a very positive temperature dependence will sharpen
your weighting function; if the absorption coefficient is
increasing with the pressure so that if you are looking in the
wing of a line, that will sharpen your weighting function; and,
another possibility is that if you can use a gas whose mixing
ratio is increasing rapidly as you go down in altitude and you
know very accurately what that mixing ratio is, you can use

that--I don't believe anyone has attempted to do that yet. I
mean, if one knew exactly what the water vapor distribution was,
you could probably get very sharp temperature weighting functions
close to the ground because of the rapid increase of mixing ratio,
but you have a big noise problem there, because you do not know
accurately the distribution of the source of emission. But those
are the basic considerations. Of course there is the microwave
region, and that microwave region has been used with oxygen as
the absorbing gas.

Chahine: Are you finding any difference between day and night
retrievals with the 4.3 μm channels?

Susskind:[*] An embarrassing question! At present, on the basis
of very preliminary results, I must say we are definitely finding a difference. Once we get everything working correctly,
hopefully there will not be a big difference between the two.

_____

[*]*Editorial Footnote:* The following statement was submitted by
the authors soon after the Workshop--"Subsequent studies find no
appreciable difference in the quality of day and night time
soundings."

PERFORMANCE OF THE HIRS/2 INSTRUMENT
ON TIROS-N[1]

E. W. Koenig

ITT Aerospace
Optical Division
Fort Wayne, Indiana

The High Resolution Infrared Radiation Sounder (HIRS/2) was developed and flown on the TIROS-N satellite as one means of obtaining atmospheric vertical profile information. The HIRS/2 receives visible and infrared spectrum radiation through a single telescope and selects 20 narrow radiation channels by means of a rotating filter wheel. A passive radiant cooler provides an operating temperature of 106.7 K for the HgCdTe and InSb detectors while the visible detector operates at instrument frame temperature. Low noise amplifiers and digital processing provide 13 bit data for spacecraft data multiplexing and transmission. The qualities of system performance that determine sounding capability are the dynamic range of data collection, the noise equivalent radiance of the system, the registration of the air columns sampled in each channel and the ability to upgrade the calibration of the instrument to maintain the performance standard throughout life. The basic features, operating characteristics and performance of the instrument in test are described. Early orbital information from the TIROS-N launched on October 13, 1978 is given and some observations on system quality are made.

---

[1]*The HIRS/2 instrument was developed on Contract NAS5-23567 from NASA Goddard Space Flight Center.*

I. INTRODUCTION

The High Resolution Infrared Radiation Sounder (HIRS/2) was
developed and flown on the Television and Infrared Observation
Satellite, N Series, (TIROS-N) as one means of obtaining atmospheric vertical profile information. The HIRS/2 receives visible
and infrared spectrum radiation through a single telescope and
selects 20 narrow spectral channels by means of a rotating filter
wheel. A passive radiant cooler provides an operating temperature of 106.7 K for the HgCdTe and InSb detectors and the visible
detector operates at instrument frame temperature. Low noise
amplifiers and digital processing provide 13 bit data for spacecraft data multiplexing and transmission. The qualities of
system performance that determine sounding capability are the
dynamic range of data collection, the noise equivalent radiance
of the system, the registration of the air collumns sampled in
each channel, and the ability to upgrade the calibration of the
instrument to maintain the performance standard throughout life.

The basic performance of the instrument in test is described.
Early orbital information from the TIROS-N launched on October 13,
1978, are given and some observations on system quality are made.

II. THE HIRS/2 INSTRUMENT

The HIRS/2 program to provide an atmospheric sounding unit
is derived from the HIRS/1 instrument developed and flown on the
Nimbus 6 satellite (NASA Contract NAS5-21651). Results from
orbital data and system study showed promise of obtaining data
from which improved atmospheric soundings may be derived. The
basic design of the Nimbus HIRS system was modified to accommo-
date the TIROS spacecraft and orbital requirements. Several
changes in subsystem design were also made to improve the

sensor performance and the reliability of the filter wheel drive assembly. A protoflight instrument was designed and assembled and included some parts from the HIRS/1 program. Later models will use similar components with little change in design.

Multispectral data from one visible channel (0.69 µm), seven shortwave channels (3.7 to 4.6 µm) and twelve long-wave channels (6.7 to 15 µm) are obtained from a single tele-scope and a rotating filter wheel containing 20 individual filters. A mirror provides cross-track scanning of 56 increments of 1.8°. The mirror steps rapidly and then holds at each position while the filter segments are sampled. This action takes place each 0.1 second. The instantaneous field of view for each channel is approximately 1.2° which, from an altitude of 833 kilometers, is an area 17.45 kilometers diameter at nadir on the earth.

Three detectors are used to sense the radiation. A silicon cell detects the energy through the visible filter. An indium antimonide detector and a mercury cadmium telluride detector mounted on a passive radiator and operating at 107 K sense the shortwave and longwave energy. The silicon cell works at 288 K. The shortwave and visible detectors share a common field stop, whereas the longwave uses a different field stop, but one that is identical in configuration. The fields of view in all channels are determined by these field stops with a secondary effect from detector position. The proto-flight instrument has all channels registered within 3 percent of the field of view.

Calibration of the HIRS/2 is provided by programmed views of three radiometric targets: a warm target mounted to the instru-ment base, a cold target isolated from the instrument and operat-ing at near 265 K, and a view of space. Data from these views provide sensitivity calibrations for each channel at 256 second intervals if so desired. Internal electronic signals provide calibration of the amplifier chains at 6.4 second intervals.

Data from the instrument are multiplexed into a single data
stream controlled by the TIROS Information Processor (TIP) system
of the spacecraft. Information from the radiometric channels
and voltage telemetry are converted to 13-bit binary data.
Radiometric information is processed to produce the maximum
dynamic range so that instrument and digitizing noises are a
small portion of the signal output. Each channel is characterized by a noise equivalent radiance and a set of calibration
data that may be used to infer atmospheric temperatures and
probable errors.

The HIRS/2 instrument is a single package mounted on the
Instrument Mounting Platform of the TIROS-N spacecraft. The
unit is shown in Fig. 1. A thermal blanket encloses most outer
surfaces other than that of the radiating panel and door area.
The radiating surface views space and emits its heat to provide
passive cooling of the detectors to the 107 K temperature. An
earth shield prevents thermal input from that direction, and is
part of a door assembly that is closed during launch and for an
initial outgas period. The door is opened at the end of that
period to provide cooling. If indications of contamination
occur, the door remains open and heat is applied to bring both
stages of the radiative cooler to near 300 K.

Table I lists the general characteristics of the HIRS/2
Instrument. Table II lists the spectral channels and sensitivity requirement for the HIRS/2. Figure 2 shows the scan
pattern and sampled columns relative to the orbit and earth.

A. *System Description*

The HIRS/2 is a 20-channel scanning radiometric sounder
which utilizes a stepping mirror to accomplish crosstrack
scanning and directs the radiant energy from the earth to a
single, 15-cm (6-inch) diameter telescope assembly every tenth
of a second. Collected energy is separated by a beamsplitter

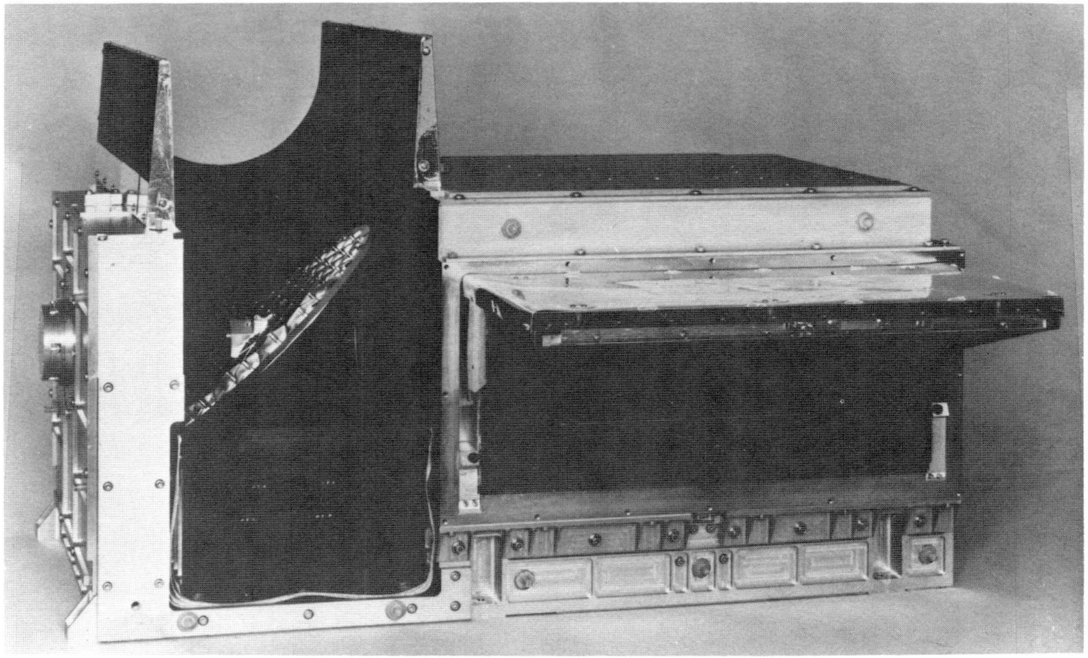

FIGURE 1. HIRS/2 Instrument

TABLE I. HIRS/2 System Characteristicts

Optical field of view, (typical)	1.22°
Included energy	97% within 1.80°
Channel to channel registration, LW	0.05 of FOV area, within band
Channel to channel registration, SW	0.02 of FOV area, within band
Earth scan angle	99.0°
Earth scan steps	56
Step and dwell time, (total)	100 ms
Retrace step time	0.8s
Total scan plus retrace time	6.4s
Earth swath coverage, 833 km orbit	2254 km
Earth field coverage	17.5 km (1.22° FOV)
Radiometric calibration	290 K black body,
	265 K black body, and
	space look
Frequency of radiometric calibration	256s, typical
Dwell time at calibration positions	5.6s (4.8s at space)
Longwave channels	12
Longwave detector	Mercury cadmium telluride
Shortwave channels	7
Shortwave detector	Indium antimonide
Visible detector	Silicon
Signal quantizing levels	8192 (13 bit coding)
Electronic calibration	32 equal levels each polarity
Frequency of electronic calibrations	One level each scan line (6.4s
Telescope aperture	15.0 cm (5.9 in)
Infrared detector temperature	107 K
Filter temperature	303 K
Instrument operating temperature	15° C nominal

TABLE II.   HIRS/2 Spectral Requirements

Channel	Channel Frequency (cm^{-1})	μm	Half-power bandwidth (cm^{-1})	Maximum Scene temperature (K)	Specified $NE\ \Delta\ N$ $mW\ m^{-2}\ st^{-1}\ cm$
1	669	14.95	3	280	0.75
2	680	14.71	10	265	0.25
3	690	14.49	12	240	0.25
4	703	14.22	16	250	0.20
5	716	13.97	16	265	0.20
6	733	13.64	16	280	0.20
7	749	13.35	16	290	0.20
8	900	11.11	35	330	0.10
9	1,030	9.71	25	270	0.15
10	1,225	8.16	60	290	0.15
11	1,365	7.33	40	275	0.20
12	1,488	6.72	80	260	0.10
13	2,190	4.57	23	300	0.002
14	2,210	4.52	23	290	0.002
15	2,240	4.46	23	280	0.002
16	2,270	4.40	23	260	0.002
17	2,360	4.24	23	280	0.002
18	2,515	4.00	35	340	0.002
19	2,660	3.76	100	340	0.001
20	14,500	0.69	1000	100%A	0.10%A

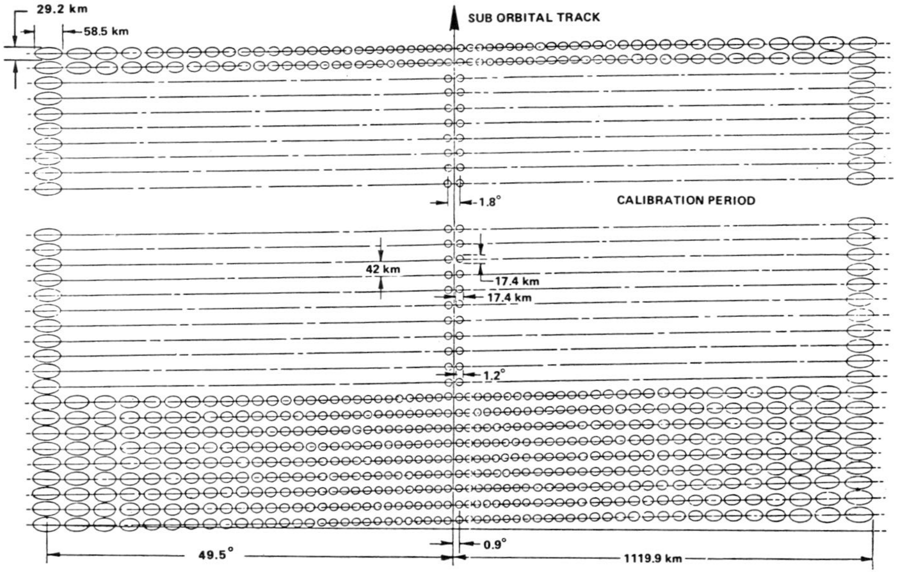

FIGURE 2. HIRS/2 Scan Pattern projected on earth.

into longwave (above 6.5 μm) and shortwave (visible to 4.6 μm)
energy which passes through field stops and through a rotating
filter wheel to cooled detectors. In the shortwave path a
second beamsplitter separates the visible channel to a silicon
detector.

The scan logic and control set the sequence of earth viewing
steps to provide a rapid scan mirror step motion to 56 fixed
positions for spectral sampling of each respective air column.
The filter wheel rotation is synchronized to this step-and-hold
sequence with approximately one-third of the wheel blank to
provide an 0.03-second step interval; the filters are positioned
for sampling only after the mirror has reached the hold position.
Registration of the optical fields for each channel to a given
column of air is dependent to some degree on spacecraft motion
and on the alignment of two field stops, which can be adjusted
to less than 1 percent of the field diameter.

Incident radiation energy is collected on cooled detectors
operating at a near optimum temperature of 107 K. A mercury
cadmium telluride detector and indium antimonide detector are
mounted on a two-stage radiant cooler. This assembly is large
enough to have reserve cooling capacity and thus permit active
thermal control to maintain the detectors at a fixed temperature.
This cooler and its housing are designed for contamination
prevention. Windows on the housing and first stage prevent
access of contaminant while controlling the heat input to the
detectors. Baffles and traps aid capture of water vapor in
other areas. A system of heating both stages is provided for
initial outgassing and for decontamination later if it should
be desired. The cold first-stage windows are heated several
degrees above their surroundings to reduce collection of con-
taminants on those elements.

Electronic circuits provide the functions of power conversion,
command, telemetry, and signal processing as shown in Fig. 3.
Amplification of the inherently weak signals from the infrared

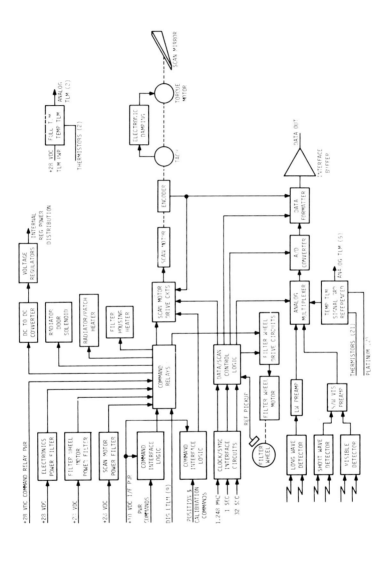

FIGURE 3. HIRS/2 System Block diagram.

(IR) detectors is accomplished in two low-noise amplifiers. The visible detector drives a separate preamplifier that joins the shortwave chain just after the low noise shortwave preamplifier. Radiance signals are fed through a base reference and memory processor, multiplexed, and A/D converted by a 13-bit range system. Once converted to digital format, the data are again multiplexed with HIRS/2 "housekeeping" data and provided as a serial data stream at the digital A output. Data from HIRS/2 are held in memory until called by the TIP request signals and clocked out of the instrument by the TIP clock.

Repetitive inclusion of electronic calibration signals and the periodic command to scan to space and two internal black-bodies provide the system with a complete set of data collection, calibration, and control that permits reliable operation in orbit.

B. *Power Input*

The HIRS/2 Instrument is powered from the Spacecraft +28 volt main bus, a +28 volt pulse load bus, a switched +28 volt analog telemetry bus, and a +10 volt interface bus. A separate input provides power to the base heater from the spacecraft Temperature Control Electronics (TCE) system.

All power systems are brought to the HIRS/2 system on separate leads. The power sources remain separate within the instrument, returns being maintained separate for each supply. A single chassis ground is brought to the output connector on an isolated pin.

The distribution of power is controlled primarily by the Instrument Power command. With this power switched on, there is power to the telemetry dc-dc converter, which provides output from analog telemetry as well as digital B telemetry even when other instrument functions are not turned on. The patch temperature control circuitry is activated by this command to provide overtemperature control during decontamination.

With instrument power ON, any combination of subsystem power may then be applied. This condition permits the operation of the scan motor, filter motor, filter housing heater, and door deployment independent of system electronics; thus, combinations of functions may be selected for launch, decontamination, standby, or other operating conditions.

The power required for system operation is obtained from all these sources. In normal operating (mission) mode, the average power demand is as shown in the following HIRS/2 Power Demand listing.

Peak current demands occur periodically during the scan cycle for short durations. At that time the power demand may reach 27.5 watts.

During system outgas an additional 20 wats may be required to maintain the cooler components at their prescribed temperatures. When the cooler door is closed at initial outgas, this peak load is intermittent and adds only 5 watts to the average load. The HIRS/2 power demand (mission mode) is as follows:

+28V Regulated	8.1 watts
+28V Pulse, Scan	8.2
+28V Pulse, F.W.	6.4
+10V Interface	0.1
+28V Command	0.
+28V Telemetry	0.03
Base Heater	0.
Total Average	22.8

C. *Commands*

The commands to the HIRS/2 instrument are all pulse commands, activating relays or directly controlling a circuit element. One feature of the HIRS/2 instrument is the automatic reset of many of the commands when instrument OFF is received. Many commands will reset to the OFF condition at that time.

This feature ensures that when the instrument is turned ON again, the system will be in a defined status, it also ensures a low power condition and provides a basis for a consistent turn-on sequence.

The operator has the option of operating the system in the repetitive calibration mode or disabling the autocalibration for continuous scan. After 37 lines of earth scanning in the repetitive mode, scan control moves the mirror to space and to the two internal targets for one-line time length only. He may also command the scan to dwell at either of the three calibration positions or at nadir. Other options available include patch control ON/OFF, filter motor High/Normal power, and filter housing heat ON/OFF.

D. Timing Signal Inputs

The HIRS/2 system is controlled by the spacecraft 1.248-MHz clock signal, the 32-second major frame signal and a 1-Hz clock signal (minor frame synch signal). From these inputs is generated the timing necessary to control the following: filter wheel rotation, scan mirror drive, data format and storage, power converter frequency, and signal collection and processing.

One of the features of the HIRS/2 timing system is that the scan system and filter wheel systems have independent timing circuitry for their individual use when the instrument electronics is OFF. When the electronics is ON, both systems are locked to the spacecraft clock system.

The filter wheel rotation starts independently of the clock system, but when the electronics ON command is given, the filter wheel is brought in phase with the 0.1-second minor frame synch signal to ensure data collection coincident with the data processing electronics. During each individual radiometric channel interval, the start and stop of the signal integration

is independent of the clock system, since it is controlled by a
timing disk on the filter wheel. This condition eliminates any
disturbance of signal timing by a slight variation in wheel
rotation during system operation.

The data from the HIRS/2 system is called from a storage
register by the TIP interrogation signal (A Select). The TIP
also provides an 8320-Hz clock to the instrument for clocking
out the Digital A data (C clock pulse).

In addition to the clock pulses, there is a Calibration
Start Pulse occurring every 256 seconds; this pulse activates
a sequence of scans that provide radiometric calibration of
the instrument. When the Calibration Disable command is in
effect, this calibration pulse will be ignored and the instrument will continue its normal scan sequence.

A 32-second major frame pulse is used to establish a
reference for the start of scan sequence and to synchronize the
retrace telemetry data to the TIP major frame so that electronic
calibration and telemetry data go into known data slots.

E. *Digital A Data Output*

The data from the HIRS/2 is provided to the TIP system from
a storage register. The TIP clock pulse (C_1) and Data Select
pulses determine the time at which data are called out. The
TIP formatter calls out groups of 8-bit words in a sequence
that multiplexes HIRS/2 data with that of other instruments.
Because of the large quantity of HIRS/2 data to be transmitted
and the use of 13-bit decoding of radiometric data, it was not
possible to format the HIRS/2 data into neat 8-bit segments.
The HIRS/2 data are therefore provided as a continuous stream
with 13-bit word lengths. During any minor frame there are 288
bits of data, each bit identified as to its purpose.

The data change during the scanning process, but the format
remains the same during the 56 earth scan time periods. During

retrace, the data format changes to provide measurement of the internal electronic calibration signals and to sample all the telemetry data.

Scan Element 0 describes the data at the time of viewing the first scan position. Scan Element 55 designates the last earth scan position. Scan Elements 56 to 63 occur during retrace when the system is in normal earth scanning mode. The same element number designations continue when the scan is commanded to a calibration target. Normally, the mirror motion between calibration targets takes place during the retrace interval. In the case of slew to the space-look position, the motion occurs during scan elements 0 to 7. Data reduction must take this into account as required.

F. *Telemetry Outputs*

The HIRS/2 system Digital B Telemetry output for command verification consists of single bit output from the TIP system at a sample rate of at least once every 32 seconds. In the HIRS/2 instrument, the Digital B outputs are discrete bilevel signals of 0 or 5V nominal relating to a Logic 1 or 0. The Digital B data are available at any time that instrument power is ON.

Selected temperature sensors in voltage divider circuits and supply voltages are monitored for analog telemetry, where the TIP system samples at 16 or 32 seconds and converts the output signal to a digital code independent of the HIRS/2. Sixteen positions are brought to the Analog TLM connector, two of which are full-time telemetry outputs.

A baseplate temperature sensor and a scan motor sensor are considered capable of defining the general system temperature under instrument OFF conditions and are powered by a separate voltage supply for full-time telemetry.

G. *Mechanical*

The HIRS/2 Instrument is mounted on the Instrument Mounting
Platform (IMP) of the TIROS-N Spacecraft. The instrument is
constructed in modules attached to a baseplate. The scan housing
containes the scan mirror, drive motor, encoder, and drive electronics. The mirror and housing are aligned to the optic axis
and baseplate to provide scan plane alignment to within $0.1°$.
The optic assembly contains the telescope, field stops, and
relay optics. The filter wheel housing is mounted to the base
and optic assembly, with detailed alignment for proper filter
positioning. The electronics housing, containing most of the
electronics, is another separable assembly, with its cover
plate tying the other modules together as a very rugged total
package. The cooler assembly mounts to the base, optics, and
cover plate and can be positioned for optical alignment, pinned,
and fastened in place.

Sun shields are added around the scan cavity to protect the
telescope and calibration targets from direct input and wide
temperature excursions within limitations on shield size and
placement.

H. *Thermal Interfaces*

Thermal characteristics of the HIRS/2 are based on a balance
providing independence of the instrument from thermal conduction
through the mounting surfaces. Temperature control is maintained
by the use of a thermal insulating blanket over much of the outer
surface of the instrument. The cooler housing openings, scan
cavity, and sun shields are exposed to space, earth, sun, and
spacecraft radiant inputs, depending on the view factors. Thermal
balance is maintained by radiation from the baseplate through
the temperature controlled louvers of the IMP. Because this
base radiant area is limited, some added radiant area is provided

to the scan housing (131 square cm) that has direct emission to space. With this design a constant 15 C temperature can be maintained by using only the vane control of the IMP system.

I. Filter Wheel Assembly

The separation of infrared channel information is provided by sequentially sampling individual spectral filters in a rotating filter wheel. The filters are selected for center wavenumber and bandwidth to provide an optimum sampling of the atmosphere for sounding. Twenty filters are mounted on a single wheel in two concentric patterns as shown in Fig. 4. The twelve longwave filters are positioned on the inner radius, whereas the seven shortwave filters and one visible filter are positioned on the outer radius. The channel locations were selected so that the reference channels 8 and 19 are centered in the grouping. This arrangement reduces registration error caused by satellite motion during the short time that the filters are passing through the optical path.

The filters are mounted in a wheel approximately 20 cm in diameter. The shape of the wheel is made to conform to the telescope optics and motor configuration. In addition to the filters mounted in the wheel, there are reference surfaces mounted to the wheel so that the detector sees a black surface through the filter as the first portion of a spectral sample. The blades having the black reference surfaces are on the telescope side. There is one reference surface for each filter; the blade is gold-coated on the telescope side and black on the filter side.

Within the central surface of the wheel there is a timing disc to permit control of the signals generated from the optical system. A slotted disc, opening a path between light emitting diodes and phototransistors, provides a timing pulse at the start of each spectral sampling interval.

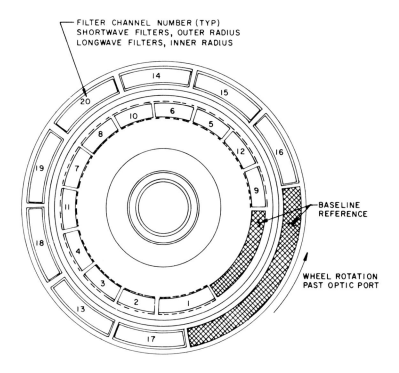

FILTER CHANNEL NUMBER (TYP)
SHORTWAVE FILTERS, OUTER RADIUS
LONGWAVE FILTERS, INNER RADIUS

BASELINE
REFERENCE

WHEEL ROTATION
PAST OPTIC PORT

FIGURE 4.   Filter wheel view from detector.

Because the filter wheel has a black reference for each
spectral channel, it becomes the blackbody reference for all
radiometric measurement. A constant uniform temperature is
therefore vital to system performance. To achieve this, the
wheel is gold-plated to reduce rapid thermal changes and is
mounted inside a temperature-controlled housing. The housing
has top and bottom cover plate blanket heaters and is painted
black to radiate heat to the wheel. A temperature sensor and
proportional control maintain a temperature of 303 K. The filter
wheel receives the radiant energy from the black-painted
enclosure to maintain its uniform temperature characteristic.
Mounting of the filter wheel to the drive motor by means of a
low conductivity material (synthane) maintains isolation of the
wheel from external thermal inputs.

The filter wheel is mounted to the rotor of a 40-pole
hysteresis synchronous motor. The rotor, having a cobalt
sleeve that becomes magnetized to provide the rotary poles,
is mounted on high-quality duplex bearings on the inner mount.
The filter wheel is mounted directly to the rotor; it is
driven at a rate of 10 revolutions per second. One feature of
the system is a dual-level drive system that permits operation
of the motor assembly at its most efficient power level while
providing a high-power backup mode.

J. *Radiant Cooler*

The HIRS/2 instrument performance is based on the use of
detectors cooled to the region of 105 K. A radiative cooling
system developed for the Advanced Very High Resolution Radiometer
(AVHRR) on Contract NAS5-21132 was adapted for the HIRS/2 system.
The cooler is a two-stage unit assembled in a housing that
mounts directly to the base of the instrument. Characteristics
of the cooler were determined by the available space, the
detector and optic positions, thermal loads of detector bias

and optics, interstage conduction and emission, and the
requirement for an earth shield that acts as a closed door
during launch and initial outgas. As the program developed, it
became necessary to modify the cooler assembly to change the
vibration characteristics as a means of noise reduction. This
change caused an increased thermal conduction between the first
and second stages (radiator and patch), increased the operating
temperature, and required a change from a nominal 105 K patch
temperature to one of 106.5 K. The cooler assembly has performed
well in system tests; thus, the contamination prevention methods
of blocking windows are serving their functions.

The radiative cooler is shown in Fig. 1 with the door (earth
shield) in the open position. The components of the assembly
may be better understood by reviewing the exploded view of Fig. 5.
The cooler housing is an aluminum shell, which is vacuum tight
when mounted in a test assembly, and contains the windows for
the longwave and shortwave optical paths. This housing is gold-
coated to reduce thermal emission to the cooled elements.

First-stage cooling is provided by the radiator, shown here
in two segments for ease of assembly and alignment. The space
facing surface is covered with honeycomb material and with
3M 401C flat black paint. Multilayer insulation (aluminized
mylar and polyester net) is placed between the radiator and
housing. The radiator also contains optical elements. A shortwave lens with a blocking filter is mounted with a loop of wire
to heat the lens approximately 8 K above the radiator temperature.
This heat, along with gold baffles, provides a trap for water
vapor that might collect on the lens surface. A longwave window
is mounted on the radiator with the same heat and baffle
arrangement.

The radiator is mounted to the cooler housing by eight
synthane support rods which provide sufficient strength to
withstand vibration and maintain alignment and yet transmit very

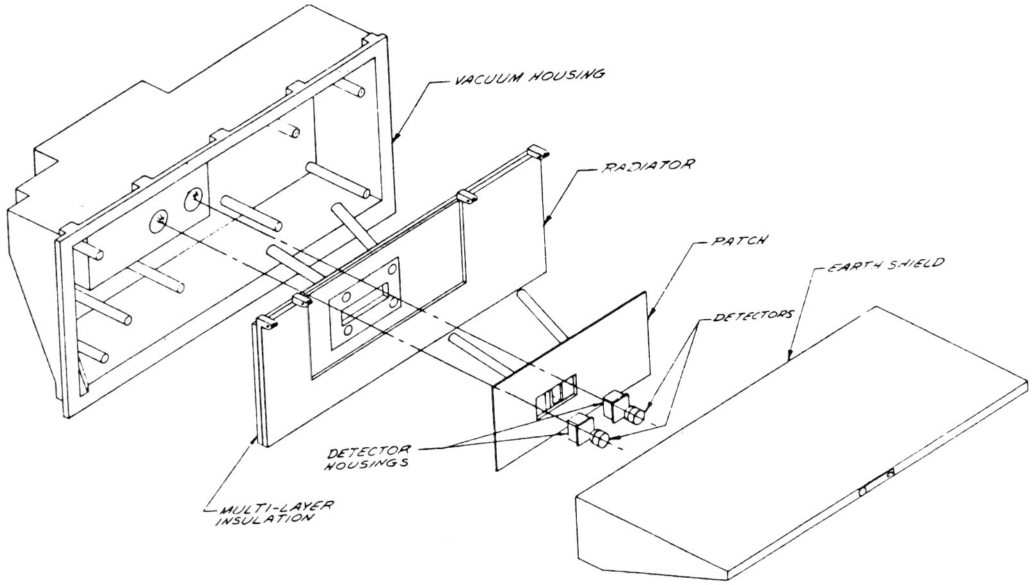

EARTH SHIELD

PATCH

DETECTORS

RADIATOR

VACUUM HOUSING

DETECTOR HOUSINGS

MULTI-LAYER INSULATION

FIGURE 5. Cooler components.

little heat from the cooler housing. The radiator will operate
at about 175 K in orbit and will provide an intermediate cold
condition for the second stage. For outgassing, the radiator
contains a 13-watt heating element and a platinum resistance
temperature sensor. The temperature of the radiator is
monitored by the data system.

The patch configuration, like the radiator, is basically a
flat plate with painted honeycomb for radiation to space. The
facing surfaces of the radiator and patch are gold-coated to
reduce coupling. Support for the patch is by means of four
synthane tubes mounted in wells in the radiator. This arrangement
provides ample stability while maintaining low levels of thermal
conductance. In addition to the detectors, the patch contains a
platinum resistance temperature sensor, a control heater, and an
outgas heater capable of 6 watts.

Each detector is mounted in a cylindrical housing having a
hemispherical aplanatic lens attached. After insertion, alignment,
and final wiring, a black cover is applied over the detectors to
provide a uniform black radiative surface for the patch.

The cooler has been checked in system tests in a vacuum
chamber and met the requirements with a narrow margin. In orbit,
the PFM system did not reach the cold temperatures anticipated
because of a malfunction during the launch sequence in which the
cooler cover opened prematurely. The patch stabilized at 107.0 K,
0.2 K above patch control point. Variation of patch temperature
was measured at less than 0.01 K; full operation of the system
was thus permitted and was nearly as stable as the control would
provide. Under these conditions the system is able to operate
accurately.

III.   PERFORMANCE CHARACTERISTICS

The ability to perform quality atmospheric soundings
requires a system having co-registration of all optical inputs,
a response capable of resolving the radiance quantities of
interest, and a low noise level that permits individual and
collective data to be interpreted within narrow uncertainties.

With a single telescope and field stop arrangement, the
HIRS/2 achieves co-registration of the seven channels within the
shortwave band to within 1 percent of the FOV width. The longwave channels are co-registered within 1.5 percent of the FOV.
The window channels (8 and 19) and the visible channel (20) are
co-registered within 1.5 percent.

All information passing through the HIRS/2 system is quantized
in a 13-bit analog-to-digital converter. Special circuits are
used to select the dynamic range of each channel to ensure
detection of the lowest levels (set by system and background
noise) to be at least one count out of the 8192 available.
The output of the system is determined as radiance per count
because the system is highly linear with respect to input
radiance. This factor, the slope of the radiance characteristic,
is an important indication of system quality. The responsivity
of the detectors, optical throughput, cooling capacity, and
system amplifier characteristics all contribute to this value.
A change in slope during operation would result from some effect
in these conditions and may indicate onset of component deterio-
ration or optical contamination.

The noise levels of the HIRS/2 are set by radiance effects
from instrument components in the optical path, background noise
in the detectors, amplifier noise and other disturbances. In the
protoflight system the predominant noise is from variable background inputs. Table III lists the characteristics of the proto-
flight unit in early orbital operation.

TABLE III. Orbital Performance

	PFM, TIROS-N in orbit		FM 1, NOAA-6 in orbit	
DATE	10/31/78		7/13/79	
PATCH TEMPERATURE, K	107.1		106.55	
RAD. TEMPERATURE, K	177.8		176.4	

Channel	SLOPE $mW/m^2\text{-}sr\text{-}cm^{-1}$	NE ΔN	SLOPE $mW/m^2\text{-}sr\text{-}cm^{-1}$	NE ΔN
1	.577	4.2	.2853	1.92
2	.272	.76	.1747	.506
3	.209	.61	.1497	.413
4	.150	.43	.1146	.311
5	.104	.31	.0793	.261
6	.111	.30	.0877	.295
7	.794E-1	.23	.567E-1	.174
8	.606E-1	.076	.334E-1	.062
9	.529E-1	.068	.404E-1	.071
10	.656E-1	.094	.619E-1	.137
11	.758E-1	.20	.321E-1	.191
12	.534E-1	.14	.494E-1	.150
13	.255E-2	.0053	.171E-2	.0025
14	.141E-2	.0042	.195E-2	.0032
15	.149E-2	.0033	.225E-2	.0037
16	.822E-3	.0015	.138E-2	.00207
17	.978E-3	.0020	.153E-2	.0019
18	.870E-2	.0022	.708E-2	.0009
19	.570E-3	.00065	.502E-3	.00049

The figure of merit of the radiometric performance of the HIRS/2 is defined as the Noise Equivalent Differential Radiance. This is the one sigma value of the radiance equivalent to the measured noise of the instrument. It is calculated from the slope (radiance/counts) and the system noise (counts).

The system noise is generally computed as the one sigma variation of the data when observing a black reference (space or one of the internal black bodies) for at least 48 sample times. This NE ΔN value can then be used as the system figure of merit for each instrument. In the HIRS/2 system, improvement of these values is attempted with each instrument assembled.

IV. CONCLUDING REMARKS

The HIRS/2 instrument is a compact highly sensitive instrument for measuring incident radiation in the infrared spectrum from the TIROS-N series satellites. The performance to date has shown it to be a rugged and highly reliable instrument. Performance in orbit has been consistent with test and analytical data and provides the meteorological community with an expanded capability for atmospheric soundings.

DISCUSSION

Chahine: What is the limit on your filters? How narrow in spectral band can you go? Can you go beyond the one percent filter?

Koenig: We could go below the one percent. We could go to a half percent.

Chahine: You have it on one of your channels which degraded a lot, this is channel number one.

Koenig: Yes, that is the one percent channel. We understand that a half percent of center wave number is about the best that we could expect. I don't think that this type of instrument is capable of going down to two wave-number bandwidths. But I believe that the usefulness of this instrument as it is now designed can be improved by improving its registration, maintaining its reliability, and reducing the noise values.

Chahine: Cooling would really help you a lot. Have you thought of going into a passive/active system?

Koenig: We have considered lower detector temperatures for future instruments, but as long as the spectral requirements do not extend beyond 15 microns the gain is not significicnat. Cooling from 107 to about 100 K would help on our longest wave channels. We are considering going down by this amount but not into the 70 to 90 K region.

Rosenberg: You said that you have some unwanted solar radiation coming into the instrument. Can you estimate the intensity of this radiation on the temperature sounding channels and is it corrected for the users?

Koenig: Other persons might be able to describe the effect on sounding better than I, but the solar radiation coming into the instrument disturbs the calibration black body temperature slightly. The calibration targets have been shielded as much as possible, but there are still some effects, not direct solar radiation, but mostly reflected off of other surfaces in the scan cavity. I cannot tell you how much these are corrected for at this time by the user. There are no corrections in the instrument data.

Planet: Just for information purposes, for perhaps the two people here who are interested in something other than temperature, the HIRS/2 and the MSU are also being used to obtain a measurement of total ozone globally by using a multichannel regression technique where we are regressing the total ozone from the Dobson stations or from selected Dobson stations and the radiances in several of the channels. That information will become hopefully an operational product eventually.

RETRIEVAL OF AEROSOL SIZE DISTRIBUTIONS FROM SCATTERING AND EXTINCTION MEASUREMENTS IN THE PRESENCE OF MULTIPLE SCATTERING[1]

A. Deepak
M. A. Box[2]
G. P. Box[2]

Institute for Atmospheric Optics and Remote Sensing
Hampton, Virginia

The paper deals with retrieval of aerosol size distribution from multispectral measurements of both scattered sky radiance and solar extinction by aerosol media. In general, the problem of making such retrievals from scattered radiance measurements is not only prohibitively expensive, but practically intractable, unless some simplifying approximations are made, and even then it remains quite a formidable task. The difficulty lies not just in the fact that the scattering by aerosol particles is a complicated process, which depends not on a single parameter but on at least four intermingled parameters--namely, size, real and imaginary parts of complex refractive index, and shape--but also on the fact that one must take into account multiple scattering in a spherical atmosphere. Essential to performing such retrievals are two sets of fast algorithms--one, for radiative transfer in the atmosphere; and the other, for retrieval of aerosol physical characteristics. In this paper, a fast technique is described based on an approximation to atmospheric radiative transfer in the region close to the sun, namely, the solar aureole. In this approximation, it is assumed that for a relatively clear day, the sky radiance is due to single scattering by molecules and aerosols

[1]*This work was supported by National Aeronautics and Space Administration Contract NAS1-15198 and NAS8-33135.*

[2]*Present address: Institute of Atmospheric Physics, University of Arizona, Tucson, Arizona.*

*plus multiple scattering by molecules alone; the contribution
from aerosol multiple scattering events being assumed negligible.
The accuracy of this approximation is discussed in connection
with inversion of scattered radiation data. In addition,
results for aerosol size distributions retrieved from multispectral solar extinction measurements by a numerical inversion
method are discussed.*

I. INTRODUCTION

At the 1976 Workshop on Inversion Methods in Atmospheric
Remote Sounding, the authors described the photographic solar
aureole measurement (PSAM) technique for making circumsolar
isophotes and measurements of angular distribution of almucantar
radiance, from which aerosol size distributions could be
determined using the single scattering (SS) approximation (1,2).
For references to other worker's work on solar aureole technique
using the SS approximation, see Refs. 1, 2 and 3. Since then
considerable progress has been made in the field of remote
sensing of aerosol size distributions using solar aureole
measurement technique. In this paper, the authors shall
discuss the retrieval of aerosol size distributions by inverting
(a) multispectral, multiangle, almucantar-radiance data using
the multiple scattering (MS) approximation, and (b) multispectral
extinction data for solar radiation traversing aerosol media
(4-8). In addition, the results of a sensitivity study of the
retrievals obtained from the solar aureole data is discussed.
    The problem of retrieving aerosol size distributions from
either multispectral multiangle scattered or direct attenuated
radiation is complicated by the fact that the scattering process
is dependent on the size, shape and complex refractive index of
the particles and the incident wavelength. It may further be
complicated by the fact that multiple scattering is also

present. Therefore, one must adopt strategies based on the
physics of the scattering phenomena to enable one to extract
information about the particular aerosol characteristic(s) of
interest. This is explained in the case of the solar aureole
technique described next.

II. SOLAR AUREOLE TECHNIQUE

    Since 1970, solar aureole method has been used successfully
to determine the altitude-integrated (or columnar) size distribution of atmospheric aerosols (1-3, 9,10). Solar aureole
is a region of enhanced brightness surrounding the sun's disk
within about 20° from it, due to the predominant forward
scattering by aerosols. For instance, the contribution to the
solar aureole, due to aerosols, is 10 to 10^3 times that due to
molecules. The advantage of solar aureole technique is that in
the forward direction, the scattered radiance is highly
sensitive to the particle size distribution $n(r)$ $[cm^{-3} \mu m^{-1}]$,
but is relatively less sensitive to the effects of particle
shape, refractive index, polarization and multiple scattering.
Molecular absorption effects can be neglected by working in
suitably selected spectral regions. The retrieval of aerosol
size distribution from multispectral solar aureole measurements
made photographically on May 6, 1977, as part of the University
of Arizona's Aerosol and Radiation Experiment (UA-ARE), is
discussed as follows (11).

A. *Measurements of Almucantar Radiance*

    The almucantar radiance measurements were obtained by
means of a photographic system, for which the experimental steps
and the data reduction procedures for making sky radiance
measurements are described in detail in Reference 2. The

digital values of sky radiance at various angular distances
from the sun along the almucantar for three wavelengths
(λ = 400, 500, and 600 nm) were then used as input data in the
retrieval scheme.

B. *Theory*

In much of the earlier work, the retrieval of aerosol size
distribution was made tractable by assuming that multiple
scattering (MS) is negligibly small compared to single scattering (SS), and could, therefore, be ignored (1-3, 9,10). In
the SS approximation, the sky radiance distribution (L_T^{SS}) along
the almucantar due to molecules and particulates is given by
the relation:

$$L_T^{SS} = L_M + L_P = \Phi_o \sec \theta \; e^{-\tau_T \sec \theta} \{F_{MC}(\psi) + F_{PC}(\psi)\} \qquad (1)$$

where $F_{MC}(\psi) = \tau_M P_M(\psi)$ is the columnar molecular scattering
function (sr^{-1}); $P_M = (3/16\pi)(1 + \cos^2 \psi)$, the molecular scattering
phase functions (sr^{-1}); $F_{PC}(\psi,\lambda)$, the columnar particulate
scattering function (sr^{-1}) is defined as

$$F_{PC}(\psi,\lambda) = \frac{1}{2k^2} \int_{r_1}^{r_2} (i_1 + i_2) \; N_C(r) \; dr \qquad (2)$$

with N_C as the columnar size distribution (cm^{-2}µm^{-1}); i_1 and i_2
are Mie intensity coefficients; r_1 and r_2 are lower and upper
limits of radii; Φ_o the incident solar flux; θ, the solar zenith
angle; subscripts M and P denote molecules and particles,
respectively; $\tau_T = \tau_M + \tau_P$, the total optical depth; and ψ, the
scattering angle given by the relation:

$$\cos \psi = \cos^2 \theta + \cos^2 \theta \; \cos \phi \qquad (3)$$

ϕ is the azimuth angle.

The molecular contribution can be determined by using the molecular density profile for the particular location; so that from Eq. (1) one can then experimentally determine $F_{PC}(\psi,\lambda)$ as a function of ψ and λ, by replacing L^{SS} by the L_T data as shown:

$$F_{PC}^{SS}(\psi,\lambda) = L_T \ \exp(\tau_T \ \sec \ \theta) \ \cos \ \theta/\Phi_0 \ - \ \tau_M P_M(\psi) \tag{4}$$

This SS approach is well discussed in Refs. 1 and 2. However, it has been shown by Box and Deepak that MS contribution to the solar aureole is not insignificant compared to SS, and that the mode radius of the size distributions retrieved on the basis of the SS alone could have errors up to 10 percent for $\tau \approx 0.4$ (10). It was also shown (10) that one can considerably improve the accuracy of the retrievals by using the MS approximation, which is a modification of the perturbation approach suggested by Deirmendjian (12) and Sekera (13). The perturbation approach assumes that the solar aureole radiance is essentially due to SS by molecules and aerosols and MS due to molecules alone, with all aerosol MS events being neglected. This approximation has the advantage that it permits one to retain the simplicity of the SS formulae, and, at the same time takes into consideration the effects of MS for non-zero ground albedo A. For the solar aureole region, the MS effects are taken into account by simply replacing $\tau_M P_M$ in Eq. (1) by an effective columnar scattering function $F_{ECM}(\psi)$ for the molecular plus aerosol atmosphere, so that the total radiance L_T^{MS} is then given, as in Ref. 14, by

$$L_T^{MS} = \Phi_0 \ \sec \ \theta \ \exp[-\tau_T \ \sec \ \theta]$$

$$[(\tau_M + \tau_{MS})P_M(\psi) \ + \ \tau_A P(0^\circ) \ + \ F_{PC}(\psi)] \tag{5}$$

where τ_{MS} and τ_A are the correction (optical depth) terms to take account of MS due to molecules and ground albedo A, respectively, and defined by the following formulae (14):

$$\tau_A = \frac{A\tau_2}{1 - A\tau_3}$$

$$\tau_{MS}(\equiv \tau_1) = 0.02\ \tau_{SS} + 1.2\ \tau_{SS}^2/\mu^{1/4}$$

$$\tau_2 = 1.34\ \tau_{SS}\ \mu(1 + 0.22(\tau_{SS}/\mu)^2)$$

$$\tau_3 = 0.9\ \tau_{SS} - 0.92\ \tau_{SS}^2 + 0.54\ \tau_{SS}^3 \qquad (6)$$

$$\tau_{SS} = \bar{\omega}\tau = \tau_M + \tau_{PS}$$

where τ_{PS} is the particulate optical depth due to scattering in contrast to particulate absorption; $\bar{\omega}$, the SS albedo (averaged); and $\mu = \cos\theta$. The experimental data for the scattering function is obtained by replacing L_T^{MS} in Eq. (5) by the L_T data and using the formula:

$$F_{PC}^{MS}(\psi,\lambda) = L_T\ \exp(\tau_T/\mu)\ \mu/\Phi_o$$

$$- [(\tau_M + \tau_{MS})\ P_M(\psi) + \tau_A P_M(0^o)] \qquad (7)$$

Since ground albedo A is usually unknown, it must be included among the parameters to be retrieved. Thus one employs the following model for the *total* columnar scattering function

$$F_{TC}^{MS}(\psi,\lambda) = \frac{1}{2k^2}\int_{r_1}^{r_2}(i_1 + i_2)\ N_C(r)\ dr$$

$$+ (\tau_M + \tau_{MS})\ P_M(\psi) + \tau_A P_M(0^o) \qquad (8)$$

This retrieval approach was applied to data obtained during the solar aureole experiment performed in Tucson on May 6, 1977. Almucantar radiance measurements were made for three wavelengths (λ = 400, 500 and 600 nm). From these measurements experimental data was obtained for $F_P(\psi)/F_P(3^o)$, (Fig. 1) which were inverted by using a two-term size distribution model, each term being a Haze M distribution

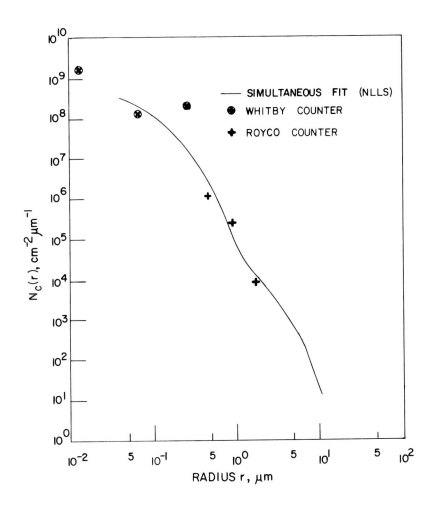

FIGURE 1. *Plots of experimental data for $F_p(\psi)/F_p(3°)$ and NLLS best fits to the data.*

$$N_C(r) = p_1 \, r \, [\exp(-\, p_2 \sqrt{r}) + p_3 \, \exp(-\, p_4 \sqrt{r})] \tag{9}$$

A nonlinear least squares (NLLS) program described in Section IV
was used to invert the spectral data. Initial estimates of the
parameters p_i for use in the NLLS program were obtained with the
help of a parameterized graphical catalog of phase function plots
corresponding to different size distribution models (15). The
details are given in Ref. 10.

C. *Discussion of Results*

The size distribution model used to fit the columnar
scattering function data for λ = 400, 500 and 600 nm was a
sum of two Haze M terms. Sensitivity studies were performed to
understand the effects on retrieved results due to different
factors, such as, aerosol refractive indices, limits r_1 and r_2
of integration, and ignoring multiple scattering. For example,
three refractive indices were used: 1.45 - i(0.00), 1.50 -
i(0.00), and 1.50 - i(0.01). Essentially no difference was
found between the fits obtained for the different refractive
indices although the retrieved parameters p_i were slightly
different. In addition, the decrease of r_1 from 0.04 μm to
0.01 μm made no difference in the final retrieved $N_C(r)$
results. For scattering angles greater than 12°, it was found
that the effects of multiple scattering needed to be taken
into account. Retrieved size distributions were relatively
more sensitive to r_2 and surface albedo A.

The retrieval size distribution obtained by assuming m = 1.55
- i(0) is shown in Fig. 2. Preliminary ground-truth SD data,
obtained by airborne Whitby (⊠) and Royco (+) counters, and
plotted in Fig. 2, were provided by Prof. J. Reagan, University
of Arizona. Even though the ground-truth data is sparse and

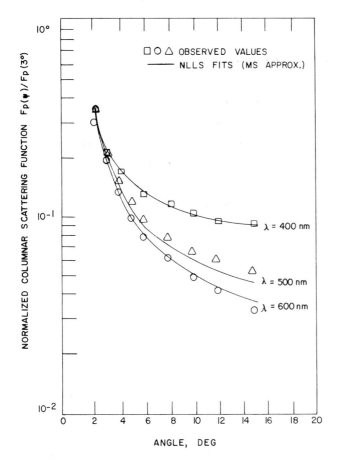

FIGURE 2. Plots of retrieved and measurement data for columnar size distribution $N_C(r)$.

is for May 7, 1977, the day following the day (May 6) on which
solar aureole measurements were made, they are reasonably close
to the retrieved columnar size distribution results to give
confidence in the latter.

The solar aureole technique is a simple and accurate method
for determining the columnar size distribution of tropospheric
aerosols. Use of forward scattered solar radiance (i.e., aureole) measured from the ground, as shown here, or from a satellite-
borne radiometer scanning up to 10° on either side (horizontal)
of setting/rising sun, is, perhaps, the most accurate technique
for measuring aerosol size distributions in the atmosphere. The
deployment of this aureole scanning capacity on satellite radiometers such as SAGE II is, therefore, strongly recommended.

III. MULTISPECTRAL OPTICAL EXTINCTION TECHNIQUE

Multispectral optical extinction technique, also referred to
as solar radiometry or laser beam transmissometry, essentially
involves the measurements of attenuated irradiance $I(\lambda)$ of solar
or laser radiation scattered and absorption during transit through
length L(m) of the medium, and the use of Bouguer-Beer law

$$I(\lambda) = I_0(\lambda) \; e^{-\tau_T(\lambda)}$$

(10)

where $I_0(\lambda)$ is the unattenuated irradiance. For solar radiometer
measurements, the total optical depth $\tau_T(\lambda)$ can be obtained by
the well-known Langley plot method. With the help of tables and
measurements of surface pressure, the molecular and ozone optical
depths $\tau_M(\lambda)$ and $\tau_{O_3}(\lambda)$, respectively, can be computed, so that
the particulate optical depth $\tau_P(\lambda)$ can be obtained from the
relation

$$\tau_P(\lambda) = \tau_T(\lambda) - \{\tau_M(\lambda) + \tau_{O_3}(\lambda)\}$$

(11)

τ_p is related to $N_C(r)$ by the relation

$$\tau_p(\lambda) = \int_0^\infty \pi r^2 Q_{EXT}(x,m)\ N_C(r)\ dr \tag{12}$$

where Q_{EXT} is the extinction efficiency factor. It is therefore possible to retrieve aerosol size distribution information from multispectral $\tau_p(\lambda)$ measurements. Several workers have employed this technique to determine aerosol size distributions by adopting different retrieval approaches; e.g., modified Twomey techniques (16,17), well explained by Herman (4), have been used by some researchers (6,18), and approximate methods (7,19) and table search methods (8) have been used by others. Here, an example of retrieving aerosol columnar size distribution from optical depth $\tau_p(\lambda)$ measurements by using the NLLS method, described in Section IV, will be given.

Measurements of $\tau_p(\lambda)$ made by means of a calibrated, seven-channel solar radiometer during a University of Arizona-Aerosol and Radiation Experiment, May 6-19, 1977, were provided to us by Professor John A. Reagan, University of Arizona. Three sets of $\tau_p(\lambda)$ measurements, obtained by using Eq. (11) are shown in Table I. The assumed columnar size distribution model was the Haze H model, vis.,

$$N_C(r) = \frac{1}{2}\ ab^3 r^2 e^{-br} \tag{13}$$

for which the parameters p_1 and p_2 retrieved by the NLLS from $\tau_p(\lambda)$ data (Table I) are shown in Table II. For details, see Ref. 7.

TABLE I. *Multispectral Measurements for the Particulate*
Optical Depths

Wavelength λ (μm)	Particulate optical depth $\tau_p(\lambda)$		
0.4400	0.0852	0.0603	0.0850
0.5217	0.0756	0.0527	0.0829
0.5556	0.0676	0.0507	0.0687
0.6120	0.0649	0.0490	0.0728
0.6708	0.0629	0.0459	0.0658
0.7797	0.0613	0.0425	0.0683
0.8717	0.0648	0.0364	0.0577

TABLE II. *Retrieved Size Distribution Parameter a and b by*
Means of NLLS Program

Observation data sets	b (μm^{-1})	$\pm\Delta b$ (μm^{-1})	a	$\pm\Delta a$
I	14.60	1.690	212.00	133.00
II	16.50	0.844	298.00	85.30
III	15.10	1.260	271.00	123.00

IV. NONLINEAR LEAST SQUARES (NLLS) METHOD

The nonlinear least squares (NLLS) code is based on the linearization method for solving the nonlinear problem (20,21). This is explained as follows:

Consider the model

$$g = f(x,p) + \varepsilon \tag{14}$$

where

g = a vector of observed values

$f(x,p)$ = vector of predicted values

p = vector of parameters

x = vector of independent variables

ε = random error vector

To estimate the parameters p, minimize the sum of squares of the errors

$$S = \{g - f(x,p)\}^T \, W^{-1} \, \{g - f(x,p)\} \tag{15}$$

where W is a weighting matrix.

The estimates of p are given by the solution of

$$\frac{\partial S}{\partial p} = 0$$

i.e.,

$$\left\{\frac{\partial}{\partial p} f(x,p)^T\right\} W^{-1} \{g - f(x,p)\} = 0 \tag{16}$$

or, letting

$$H = \frac{\partial}{\partial p} \{f(x,p)\} \tag{17}$$

one obtains,

$$H^T \, W^{-1} \, \{g - f(x,p)\} = 0 \tag{18}$$

However, $f(x,p)$ is nonlinear in p, so in order to linearize
Eq. (18), carry out a first order Taylor series expansion of f
about an initial parameter estimate p_O; i.e.,

$$f(x,p) = f(x,\hat{p}_O) + \left[\frac{\partial}{\partial p} f(x,p)\right]_{p=p_O} (p - p_O) \tag{19}$$

substituting Eq. (19) in Eq. (18) gives

$$-H^T W^{-1}\{g - f(x,\hat{p}_O)\} + H^T W^{-1} H(p - p_O) = 0 \tag{20}$$

Eq. (20) is linear in $(p - p_O)$ and the solution is given by

$$p - p_O = (H^T W^{-1} H)^{-1} H^T W^{-1} \{g - f(x,p_O)\} \tag{21}$$

or

$$p = p_O + (H^T W^{-1} H)^{-1} H^T W^{-1} \{g - f(x,p_O)\} \tag{22}$$

In practice, solution of nonlinear least squares system of
equations is an iterative process, which continues until the
solution converges. The new estimate, p, at the end of each
iteration, becomes the initial estimate p_O for the next iteration.

The convergence criteria used in this method were: (1) that
the variances for the $(n - 1)^{th}$, and n^{th} and estimated variance
for the $(n + 1)^{th}$ iteration not differ from one another by more
than 1 percent; (2) that the change in each parameter between the
$(n - 1)^{th}$ and n^{th} iteration be no more than 0.1 percent.

In preceding sections, the equations being solved are Eqs.
(8) and (12). A schematic diagram of the NLLS inversion scheme
is shown in Fig. 3.

V.   CONCLUDING REMARKS

From the above discussions, it is quite evident that the solar
aureole technique is a simple and relatively inexpensive method
for determining the columnar size distribution of aerosols. The
use of the NLLS inversion scheme yields retrievals that are

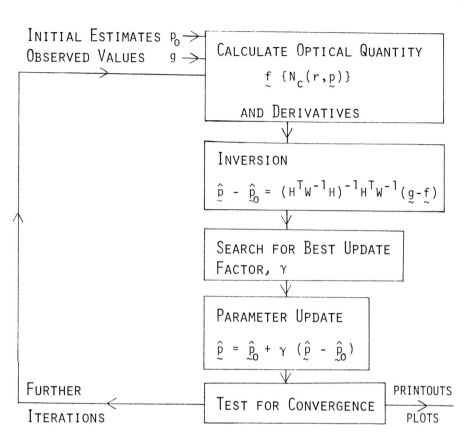

INITIAL ESTIMATES $p_0 \rightarrow$
OBSERVED VALUES $g \rightarrow$

CALCULATE OPTICAL QUANTITY

$$\underset{\sim}{f} \{N_c(r, \underset{\sim}{p})\}$$

AND DERIVATIVES

INVERSION

$$\underset{\sim}{\hat{p}} - \underset{\sim}{\hat{p}}_0 = (H^T W^{-1} H)^{-1} H^T W^{-1} (\underset{\sim}{g} - \underset{\sim}{f})$$

SEARCH FOR BEST UPDATE FACTOR, γ

PARAMETER UPDATE

$$\underset{\sim}{\hat{p}} = \underset{\sim}{\hat{p}}_0 + \gamma (\underset{\sim}{\hat{p}} - \underset{\sim}{\hat{p}}_0)$$

FURTHER ITERATIONS

TEST FOR CONVERGENCE

PRINTOUTS

PLOTS

FIGURE 3. Schematic diagram of the NLLS inversion scheme.

accurate when checked against ground truth measurements of $N_C(r)$. These retrievals were performed by taking into account the multiple scattering contributions to the sky radiance within the region of the solar aureole by the use of a MS approximation. The retrievals are sensitive to the upper limit of integration over radius and the surface albedo on which further work is in progress.

As shown earlier, the NLLS method can also be used to retrieve $N_C(r)$ parameters from multispectral optical depth measurements. Even though the theoretical procedures are in hand, an experimental validation still needs to be done in retrieving $N_C(r)$ information from the combined measurements of F_{PC} and τ_p, which must be taken with the same instrument. The data for F_{PC} and τ_p presented in the earlier sections were taken with different instruments having different calibration constants.

In addition, work is in progress in developing fast retrieval techniques (8), which can be used to develop automated inversion systems.

ACKNOWLEDGMENTS

It is a pleasure to acknowledge the valuable assistance and cooperation of Professor John A. Reagan, University of Arizona, in enabling us to participate in the UA-ARE experiment and providing us with both $\tau_p(\lambda)$ and ground truth-size distribution measurements. The valuable assistance of R. R. Adams and B. Poole, NASA-Langley Research Center, in taking photographs of the solar aureole and data reduction, is gratefully acknowledged.

REFERENCES

1. Green, A. E. S., Deepak, A., and Lipofsky, B. J., *Appl. Opt.* *10*, 1263 (1971).

2. Deepak, A., *in* "Inversion Methods in Atmospheric Remote Sounding" (A. Deepak, ed.), p. 265. Academic Press, New York (1977).

3. Twitty, J. T., *J. Atmos. Sci.* *32*, 584 (1975).

4. Herman, B. M., *in* "Inversion Methods in Atmospheric Remote Sounding" (A. Deepak, ed.), p. 469. Academic Press, New York (1977).

5. King, M. D., Byrne, D. M., Herman, B. M., and Reagan, J. A., *J. Atmos. Sci.* *35*, 2153 (1978).

6. Chu, W. P., *in* "Inversion Methods in Atmospheric Remote Sounding" (A. Deepak, ed.), p. 505. Academic Press, New York (1977).

7. Box, G. P., Box, M. A., and Deepak, A., Spectral Sensitivity of Approximate Method for Retrieving Aerosol Size Distributions from Multispectral Solar Extinction Measurements, IFAORS Report No. 126. (Available from IFAORS, P. O. Box P, Hampton, VA 23666.)

8. Box, G. P., and Deepak, A., Fast Table Search Method for Retrieval of Aerosol Size Distribution from Multispectral Optical Depth Measurements, NASA Report (1980). (Available from NTIS, Springfield, VA., 22161).

9. Box, M. A., and Deepak, A., *Appl. Opt.* *17*, 3794 (1978).

10. Box, M. A., and Deepak, A., *Appl. Opt.* *18*, 1376 (1979).

11. Reagan, J. A., Herman, B. M., Byrne, D. M., and King, M. D., Third Conference on Atmospheric Radiation, June 28-30, 1978, Davis, Calif., p. 241. American Meteorological Society, 45 Beacon Street, Boston, Mass. 02108.

12. Deirmendjian, D., Use of Scattering Techniques in Cloud Microphysics Research. I. The Aureole Method, Rand Corp. Report R-590-PR (1970).

13. Sekera, Z., *Adv. Geophys. 3*, 43 (1956).

14. Box, M. A., and Deepak, A., A Multiple Scattering Approximation to Radiative Transfer in a Turbid Atmosphere: Almucantar Radiance Formulation, IFAORS Report No. 134. (Available from IFAORS, P. O. Box P, Hampton, VA 23666.)

15. Deepak, A., and Box, G. P., Analytic Modeling of Aerosol Size Distribution, NASA-CR 159170 (1979).

16. Twomey, S., *J. Assoc. Comput. Mach. 10*, 97 (1963).

17. Twomey, S., *J. Franklin Inst. 279*, 95 (1965).

18. Shaw, G. E., Reagan, J. A., and Herman, B. M., *J. Appl. Meteorol. 12*, 374 (1973).

19. Box, M. A., and Lo, S. Y., *J. Appl. Meteorol. 15*, 1068 (1976).

20. Draper, N., and Smith, H., "Applied Regression Analysis." John Wiley and Sons, Inc., New York (1966).

21. Deutsch, R., "Estimation Theory." Prentice-Hall, Inc., Englewood Cliffs, New Jersey (1965).

DISCUSSION

Lenoble: Are you assuming the particles to be spherical?

Deepak: That is correct. So far, we have been assuming only spherical particles.

Zardecki: In your solar aureole method, you said that the aureole is due to scattering, so it means that these aerosol particles are much larger than the wavelengths. Is it right?

Deepak: These particles are sizes comparable to the visible wavelengths. We used Mie scattering in all of these calculations.

Zardecki: Does it mean that when you consider multiple scattering you just take the molecules into account and not the aerosols?

Deepak: We find that the multiple scattering effect of molecules is much stronger than that of aerosols. Because aerosols scatter predominantly in the forward direction, the major contribution to aerosol multiple scattering is due to single scattering, under normal sky conditions. On the other hand, for molecules, the complete multiple scattering needs to be taken into account. Thus, for optical depths of up to 0.6, the multiply scattered contribution due to scattering events involving only aerosols is much smaller than the single scattering due to aerosols and single plus multiple scattering due to molecules, and have therefore been ignored. The retrieval problem, using such an approximation, becomes practically as simple as, but much more accurate than, with the use of the single scattering approximation.

ATMOSPHERIC EFFECTS IN REMOTE SENSING
OF GROUND AND OCEAN REFLECTANCES

P. Y. Deschamps
M. Herman
J. Lenoble
D. Tanre
M. Viollier

Laboratoire D'Optique Atmospherique, E.R.A. 466
Universite Des Sciences Et Techniques De Lille

An analytical expression of the measured reflectance is
established for the general case of a nonlambertian and nonuniform ground. The signal is nearly linear in function of the
intrinsic atmospheric reflectance, the actual target reflectance,
and two average reflectances, angular and spatial; the relative
importance of these contributions is discussed for various cases
of turbidity. The method is applied in order to study the
reduction of contrast for targets of various size including
bidirectional properties. Particular attention is paid to the
contribution of the target environment to the apparent reflectance of a lambertian ground. Another application concerns the
ocean diffuse reflectance and a correction algorithm for such
measurements is proposed; the possibility of remote sensing of
chlorophyll content is discussed.

I. INTRODUCTION

The terrestrial atmosphere perturbs the measurement of ground reflectance from space mainly by molecular and aerosol scattering (1-3). In some channels, the absorption by ozone or water vapor has to be taken into account, but this effect is rather simple to introduce and has not been considered here. The purpose was to separate the main mechanisms of interactions between ground and atmosphere and to compare their relative importance.

In the next section, the signal measured from space is analyzed. Starting from the formalism for a uniform and lambertian ground leads to the definition of important average reflectances in the general case. In the third and fourth sections, the relative importance of the different terms and the average reflectances are discussed for three atmospheric models. The fifth section deals with the problem of contrast reduction and the last section deals with ocean remote sensing and retrieval of chlorophyll content.

II. GENERAL ANALYSIS

The signal received from a uniform lambertian ground of albedo ρ is analyzed first based on Tanre *et al.* (4). The apparent albedo measured in a direction (μ,ϕ) above an atmosphere of optical thickness τ illuminated by a solar flux f from the direction $(\mu_o \phi_o)$ is

$$\rho^*(\mu,\phi;\mu'\phi') = \pi I^+(o;\mu,\phi)/\mu_o f \qquad (1)$$

and it can be written as

$$\rho^*(\mu,\phi;\mu_o',\phi_o') = \pi S(\mu,\phi;\mu_o,\phi_o) + e^{-\tau/\mu_o} \rho \, e^{-\tau/\mu}$$

$$+ T(\mu_o) \, \rho \, e^{-\tau/\mu} + e^{-\tau/\mu_o} \rho \, T^*(\mu)$$

$$+ T(\mu_o) \, \rho \, T^*(\mu) + \left(e^{-\tau/\mu_o} + T(\mu_o) \right)$$

$$\rho \left[e^{-\tau/\mu} + T^*(\mu) \right] \frac{\rho \, \bar{S}^*}{1 - \rho \, \bar{S}^*} \tag{2}$$

Here $S(\mu,\phi;\mu_o,\phi_o)$ and $T(\mu,\phi;\mu_o,\phi_o)$ are the scattering and transmission functions of the atmosphere alone, defined by

$$I_{atm}^+(o;\mu,\phi) = \mu_o f S(\mu,\phi;\mu_o,\phi_o)$$

$$I_{atm}^-(\tau;\mu,\phi) = \mu_o f T(\mu,\phi;\mu_o,\phi_o) \tag{3}$$

In a similar way S^* and T^* are defined for the atmosphere
illuminated at the bottom. Let us also define

$$T(\mu) = \int_0^{2\pi} \int_0^{+1} T(\mu,\phi;\mu',\phi') \, \mu' \, d\mu' \, d\phi'$$

$$T(\mu) = \int_0^{2\pi} \int_0^{+1} T(\mu',\phi';\mu,\phi) \, \mu' \, d\mu' \, d\phi' \tag{4}$$

$$\bar{S} = \frac{1}{\pi} \int_0^{2\pi} \int_0^{+1} \int_0^{2\pi} \int_0^{+1} S(\mu,\phi;\mu',\phi') \, \mu \, d\mu \, d\phi \, \mu' \, d\mu' \, d\phi'$$

$$\bar{S} = 2 \int_0^{+1} S(\mu) \mu \, d\mu \tag{5}$$

and similarly for the other functions.

In the right-hand side of Eq. (2), the first term is the contribution of the atmosphere alone (Fig. 1a), the second term is due

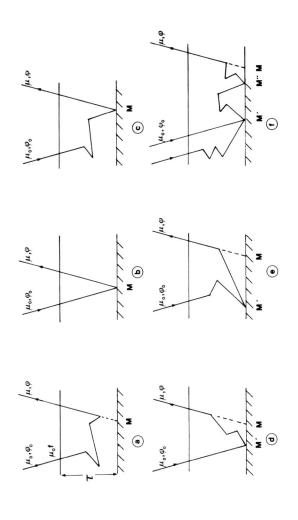

FIGURE 1. Successive orders of radiation interactions in the ground-atmosphere system.

to photons directly transmitted to the ground and after reflection (Fig. 1b) and the third term to photons diffusely transmitted to the ground and directly transmitted after reflection
(Fig. 1c). Terms 4 and 5 correspond respectively to photons
directly (Fig. 1d) or diffusely (Fig. 1e) transmitted to the
ground and diffusely transmitted after reflection. In terms 4
and 5, the portion of the ground intervening is not that
directly looked at by the detector. Finally, the last term

$$\frac{\rho \, \bar{S}^*}{1 - \rho \, \bar{S}^*} = \sum_{n=1}^{\infty} (\rho \, \bar{S}^*)^n \tag{6}$$

holds for multiple interactions between the ground and the
atmosphere. (See Fig. 1f for the double interaction.) Of
course, Eq. (2) can be simplified as

$$\rho^* (\mu, \phi; \mu_o, \phi_o) = \pi S(\mu, \phi; \mu_o, \phi_o) + \left(e^{-\tau/\mu_o} + T(\mu_o) \right)$$
$$\left[e^{-\tau/\mu} + T^*(\mu) \right] \rho / (1 - \rho \, \bar{S}^*)$$

In the case of a nonuniform, nonlambertian ground, a
similar formulation can be used

$$\rho^* (\mu, \phi; \mu_o, \phi_o) = \pi S(\mu, \phi; \mu_o, \phi_o) + e^{-\tau/\mu_o} \rho(\mu, \phi; \mu_o, \phi_o) \, e^{-\tau/\mu}$$

$$+ T(\mu_o) \, \bar{\rho}(\mu, \phi; \mu_o, \phi_o) \, e^{-\tau/\mu}$$

$$+ e^{-\tau/\mu_o} < \rho > (\mu, \phi; \mu_o, \phi_o) \, T^*(\mu)$$

$$+ T(\mu_o) << \rho >> (\mu, \phi; \mu_o, \phi_o) \, T^*(\mu)$$

$$+ \text{(multiple interactions)} \tag{7}$$

where $\frac{1}{\pi} \rho(\mu, \phi; \mu_o, \phi_o)$ is the actual directional reflectance of
the observed target, and

$$\bar{\rho}(\mu,\phi;\mu_o,\phi_o) = \frac{1}{T(\mu_o)} \int_0^{2\pi} \int_0^{+1} T(\mu',\phi';\mu_o,\phi_o)$$

$$\rho(\mu,\phi;\mu',\phi') \ \mu' \ d\mu' \ d\phi' \qquad (8)$$

is a directional average reflectance of the target.

The spatial average reflectances are defined by

$$<\rho>(\mu,\phi;\mu_o,\phi_o) = \frac{1}{T^*(\mu)} \int_0^\infty \int_0^{2\pi} \int_0^{2\pi} \int_0^{+1} \rho(r,\psi;\mu',\phi';\mu_o,\phi_o)$$

$$T^*(r,\psi;\mu,\phi;\mu',\phi') \ dr \ d\psi\mu' \ d\mu' \ d\phi' \qquad (9)$$

$$<<\rho>>(\mu,\phi;\mu_o,\phi_o) = \frac{1}{T^*(\mu)} \int_0^\infty \int_0^{2\pi} \int_0^{2\pi} \int_0^{+1} \bar{\rho}(r,\psi;\mu',\phi';\mu_o,\phi_o)$$

$$T^*(r,\psi;\mu,\phi;\mu',\phi') \ dr \ d\psi\mu' \ d\mu' \ d\phi' \qquad (10)$$

where $\rho(r,\psi;\mu',\phi';\mu_o,\phi_o)$ is the reflectance at a point of coordinate (r,ψ), the target being at the origin; and $T^*(r,\psi;\mu,\phi;\mu',\phi')$ refers to the contribution of point (r,ψ) to the transmission function $T^*(\mu,\phi;\mu',\phi')$ at the observation point.

The multiple interactions term involves even more complex spatial average reflectances. In what follows, it is assumed that $<<\rho>> \simeq <\rho>$, and the same value given by Eq. (9) can be used in the last term of Eq. (7).

Therefore, Eq. (7) can be written as

$$\rho^* = \pi S + A\rho + B\bar{\rho} + C<\rho>$$

$$+ \frac{<\rho> \ \bar{S}^*}{1 - <\rho> \ \bar{S}^*} \left[(A + B) \ \bar{\rho} + C<\rho> \right] \qquad (11)$$

where the variables $(\mu,\phi;\mu_o,\phi_o)$ have been deleted

$$A = e^{-\tau/\mu} \, e^{-\tau/\mu_o}$$

$$B = e^{-\tau/\mu} \, T(\mu_o)$$

$$C = \left[e^{-\tau/\mu_o} + T(\mu_o) \right] \, T^*(\mu) \tag{12}$$

III. RELATIVE IMPORTANCE OF CORRECTION TERMS

Only the second term on the right-hand side of Eq. (11) carries the information on the actual reflectance of the observed target, the other terms being error terms, the importance of which depends on the atmospheric properties, on the illumination conditions, and the ground itself.

Three atmospheric models have been chosen: pure molecular atmosphere, molecular atmosphere with aerosols corresponding to the model defined by McClatchey et al. for ground visibilities of 5 km and 23 km. The necessary atmospheric functions have been computed for wavelengths of 450 nm, 550 nm, and 850 nm; the corresponding optical thicknesses are given in Table I.

TABLE I. Optical Thickness for Molecules (τ^R) and Aerosols (τ^P) for Two Visibilities, $V = 23$ km and $V = 5$ km.

λ (μm)	0.4500	0.5500	0.8500
τ^R	0.2157	0.0948	0.0163
τ^P (V = 23 km)	0.2801	0.2348	0.1550
τ^P (V = 5 km)	0.9305	0.7801	0.5151

TABLE II. Function $\bar{S}^*$ for Four Wavelengths and the Three
Atmosphere Models

Wavelength (nm)	Molecular Atmosphere	Turbid Atmosphere (V = 23 km)	Turbid Atmosphere (V = 5 km)
450	0.1605	0.2128	0.3080
550	0.0807	0.1403	0.2432
650	0.0438	0.1038	0.2056
850	0.0157	0.0698	0.1606

Table II shows the values of $\bar{S}^*$, which increases with
turbidity and toward short wavelength due to Rayleigh scattering.
As $\bar{S}^*$ appears multiplied by $<\rho>$ in the multiple interactions
term, it is concluded that over ocean or low reflecting ground
($<\rho> < 0.05$), the corresponding contribution is of the order of
1 percent in the worst cases and that it can be completely
neglected. Over high reflecting ground ($<\rho> \simeq 0.50$), the
contribution reaches 15 percent. But in remote sensing of
terrestrial sites, such high reflectances are generally found
for vegetation only in the near infrared; therefore, the interaction term is only about 7 percent for the worst visibility
cases.

This last term will therefore be deleted and Eq. (11) is
written as

$$\rho^* = \pi S + A\rho + B\bar{\rho} + C<\rho> \qquad (13)$$

The atmosphere contribution πS is very variable with (μ, ϕ) and
(μ_o, ϕ_o) and is shown for some cases in Table III. It is mainly
important for low reflecting sites and when absolute values of
ρ and not only contrasts are sought. Therefore, the discussion
will be reported to the section concerning ocean reflectance.

TABLE III. Atmospheric Contribution $\pi S(\mu,\phi;\mu_o,\phi_o)$ for Vertical Observation $(\mu = 1)$, $\theta_o = $ Arc cos μ_o.

λ_{mn}	θ_o^o	Rayleigh	$V = 23$ km	$V = 5$ km
450	15	0.0838	0.1050	0.1603
550	15	0.0367	0.0567	0.1071
650	15	0.0184	0.0366	0.0815
850	15	0.0061	0.0208	0.0559
450	60	0.0988	0.1281	0.2027
550	60	0.0448	0.0708	0.1420
650	60	0.0228	0.0454	0.1096
850	60	0.0077	0.0250	0.0747

Figure 2 shows the coefficients A, B and C for $\mu = 1$ (vertical observation) as functions of μ_o for three visibilities and three wavelengths 450 nm, 550 nm, and 850 nm. The variation with μ_o is slow except for large incidence angles, but the correcting terms become more important when the incidence increases. It appears that in many cases the contribution of the three terms ρ, $\bar{\rho}$, and $<\rho>$ are about of the same order. The relative contribution of $\bar{\rho}$ and $<\rho>$ decreases for clearer atmospheres and longer wavelengths. But, even for a very clear atmosphere $(V = 30$ km) and $\lambda = 850$ nm, the contribution of $\bar{\rho}$ and $<\rho>$ remain about 10 percent of the contribution of the actual reflectance ρ to be measured.

IV. EVALUATION OF THE AVERAGE REFLECTANCES

The direction average $\bar{\rho}$ has been computed by Eq. (8) from the exprimental values of the reflectance $\rho(\mu,\phi;\mu_o,\phi_o)$ of the savanna given by Kriebel (6). The computation has been

124

P. Y. DESCHAMPS *et al.*

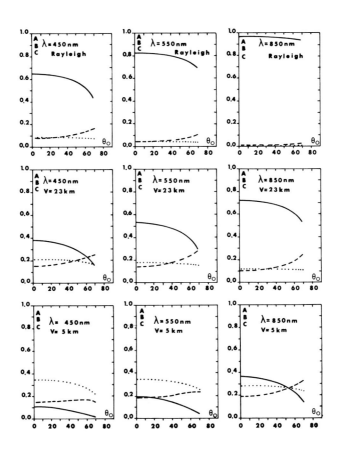

FIGURE 2. Relative contributions A, B, and C of reflectances ρ, ρ̄, and <ρ> as functions of the solar zenith angle, for a vertical observation (lines A, solid; B, dashed; C, dotted).

performed for the three atmosphere models and for two wavelengths 450 nm and 850 nm in each case for different incidence and viewing angles. The result is that $\bar{\rho}$ depends very little on (μ,ϕ) and (μ_o,ϕ_o) and that this dependence can be included in the dependence on ρ. Then $\bar{\rho}$ can be expressed by

$$\bar{\rho} = \bar{\rho}_o + a\rho \tag{14}$$

where $\bar{\rho}_o$ and a depend slightly on λ and on atmospheric conditions for a given ground. Figure 3 shows $\bar{\rho}$ as function of ρ for the savanna. It may be seen that the coefficient a is small and increases slightly with turbidity; this condition is probably due to the forward peak of aerosol scattering.

The spatial average $<\rho>$ is much more difficult to evaluate. As the atmosphere optical thickness is at the most of the order of 1, the first orders of scattering play an important part. The contribution of the photons scattered once and of the photons scattered once or twice to $T^*(\mu)$ have been computed for the two visibilities 5 km and 23 km and for two wavelengths 450 nm and 850 nm with $\mu \simeq 1$; a contribution varying from 89 percent $(V = 23$ km; $\lambda = 850$ nm) to 51 percent $(V = 5$ km; $\lambda = 450$ nm) was found for one scattering. For two scatterings, the corresponding contributions are 98 percent and 75 percent. Moreover, the photons scattered twice have a mean path rather similar to those scattered once because of the strong forward scattering. Therefore, it seems reasonable to express $T^*(r,\psi;\mu,\phi;\mu',\phi')/T^*(\mu,\phi:\mu',\phi')$ in the single scattering approximation (7).

Also, the evaluation of $<\rho>$ will be limited to a lambertian ground and a vertical observation ($\mu = 1$). Then, Eq. (9) can be rewritten as

$$<\rho> = \int_0^\infty \int_0^{2\pi} p(r)\ \rho(r\psi)\ r\ dr\ d\psi \tag{15}$$

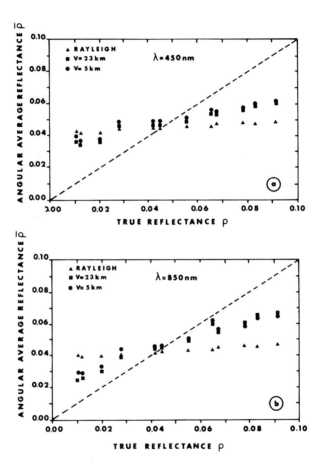

FIGURE 3. Angular average reflectance $\bar{\rho}$ for the incident
diffuse radiation as function of reflectance for three atmosphere
models at $\lambda = 450$ nm and $\lambda = 850$ nm. Each point corresponds to
different geometric conditions (solar incidence and observation
angle).

where

$$p(r) = T^{*}(r,\psi;\mu = 1)/T^{*}(\mu = 1) \qquad (16)$$

The relative contribution rp(r) of a ground element of a
distance r from the target is shown in Fig. 4 for the three
atmospheric models and the three wavelengths. Figure 5 shows
for the same cases the integrated contribution of the environment
within a distance r from the target

$$F(r) = 2\pi \int_{0}^{r} r'\, p(r')\, dr' \qquad (17)$$

The contribution of the environment is mainly due to points
near to the target for strong turbidity; but, when the turbidity
decreases, the immediate neighborhood contributes less and
farther areas contribute more (80 percent within 1 km for a
visibility of 5 km and within 4 km for a visibility of 23 km).
For a pure Rayleigh atmosphere, the contribution remains
important to distances larger than 10 km. On the other hand,
the variation with wavelength is not very important and is
mainly due to Rayleigh scattering, as it disappears when the
turbidity is large.

By varying the models, it has been found that p(r) is rather
insensitive to the aerosol phase function, that is, to their
exact size distribution and refractive index, but it is very
sensitive to the aerosol vertical profile. This is due to the
fact that a limited and well-located atmospheric layer is
mainly responsible for the contribution of a given area of the
environment. (See Fig. 6.) The relative contribution to rp(r)
of a layer at an altitude z is plotted against z for r = 50 m,
r = 500 m and r = 5000 m for the two visibilities and two
different aerosol profiles. It is seen that the points near the
target contribute to the average reflectance through the low
atmospheric layers and the points far from the target through the
high atmospheric layers.

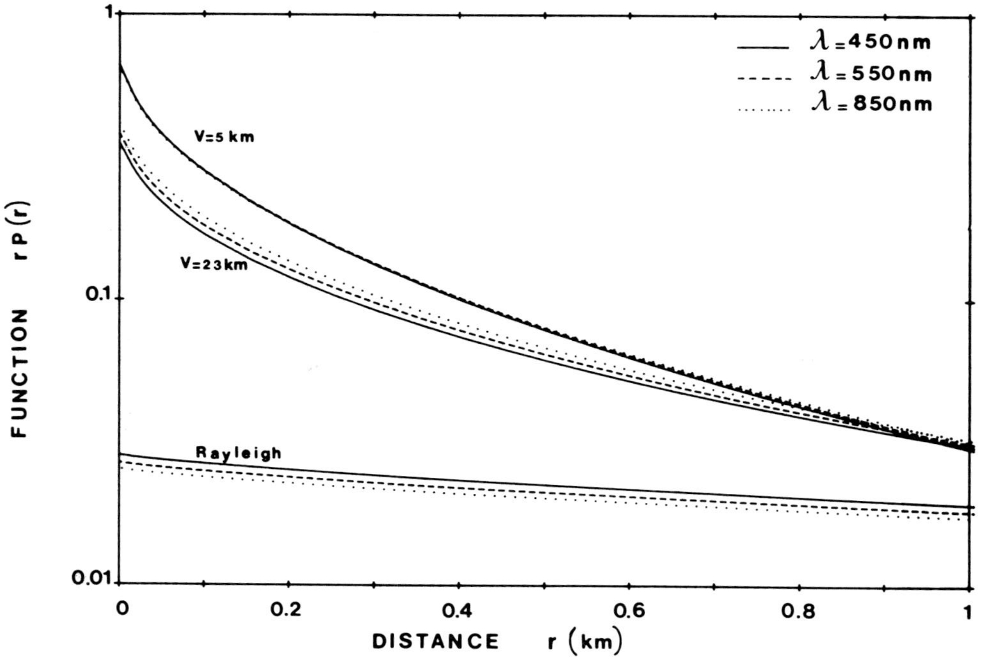

FIGURE 4. Weighting function of environment.

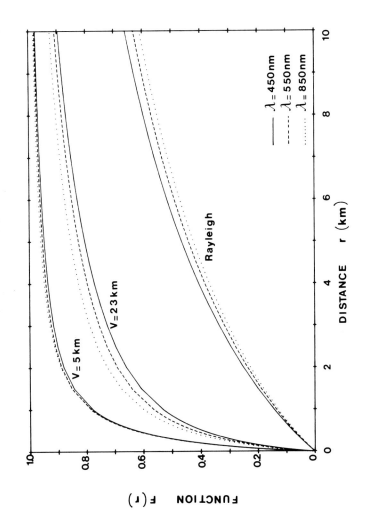

FIGURE 5. Integrated contribution of the environment.

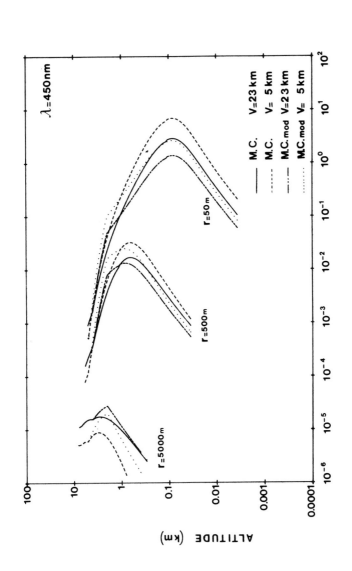

FIGURE 6. Contribution of the various atmosphere layers to the environment weighting function $rP(r)$. M.C. refers to McClatchey's model, M.C. mod. refers to the same model with a modified altitude profile.

V. APPLICATION TO GROUND REMOTE SENSING--CONTRAST REDUCTION

In ground remote sensing, attention is focused on the
reduction of contrast introduced by the atmosphere. If ρ_1 and
ρ_2 are the actual reflectances in two different measurement
conditions, ρ_1^* and ρ_2^*, the measured apparent reflectances, the
contrast reduction is expressed by

$$R = \frac{\rho_1^* - \rho_2^*}{\rho_1 - \rho_2} \tag{18}$$

First, consider the spatial contrast of two neighboring targets
for which it can be assumed that the direct atmosphere contribution is identical $(S_1 = S_2)$.
  Then, from Eq. (13)

$$R = A + B \frac{\bar{\rho}_1 - \bar{\rho}_2}{\rho_1 - \rho_2} + C \frac{<\rho>_1 - <\rho>_2}{\rho_1 - \rho_2} \tag{19}$$

For Lambertian targets $(\bar{\rho} = \rho)$, Eq. (19) reduces to

$$R = A + B + C \frac{<\rho>_1 - <\rho>_2}{\rho_1 - \rho_2} \tag{20}$$

Two limiting cases are:
  1. Very small targets within the same environment, such that
$<\rho>_1 = <\rho>_2$, and

$$R = A + B = e^{-\tau/\mu}\left[e^{-\tau/\mu_o} + T(\mu_o)\right] \tag{21}$$

  2. Very large targets, observed far from the borders, such
that $<\rho>_1 = \rho_1$ and $<\rho>_2 = \rho_2$, and

$$R = A + B + C = \left[e^{-\tau/\mu} + T^*(\mu)\right]\left[e^{-\tau/\mu_o} + T(\mu_o)\right] \tag{22}$$

The general case is between these two limits with $<\rho>_1$ and $<\rho>_2$ computed by Eq. (9) or Eq. (15) from the actual ground reflectance.

The results for these two cases are shown in Fig. 7, where the contrast reduction is plotted against λ for the three atmospheric models. The degradation is, as expected, stronger for small targets than for large targets and for short wavelengths than for large wavelengths.

It is possible to have an idea of the range of validity of these approximations, by considering a circular target of radius $\bar{r}$ and albedo ρ in a uniform environment of albedo ρ_B. For a small target, assuming $<\rho> = \rho_B$ in Eq. (13) introduces on ρ an error

$$\Delta\rho = \rho^{mes} - \rho = \frac{C F(\bar{r})}{A + B + C\ F(\bar{r})} \ (\rho^{mes} - \rho_B)$$

$$\Delta\rho = E_S(\bar{r}) \ (\rho^{mes} - \rho_B) \qquad\qquad (23)$$

Similarly, for a large target, assuming $<\rho> = \rho$ in Eq. (13) gives an error

$$\Delta\rho = \rho^{mes} - \rho = - \frac{C(1 - F(\bar{r}))}{A + B + C\ F(\bar{r})} \ (\rho^{mes} - \rho_B)$$

$$\Delta\rho = - E_\ell(\bar{r}) \ (\rho^{mes} - \rho_B) \qquad\qquad (24)$$

The functions $E_S(\bar{r})$ and $E_\ell(\bar{r})$ are given in Figs. 8 and 9. For the worst case $\rho^{mes} - \rho_B \simeq 1$, an error $\Delta\rho \geq 5\%$ is obtained by the approximations

1. For $V = 23$ km: in blue if $100\ m < \bar{r} < 6$ km

    in IR if $300\ m < \bar{r} < 2$ km

2. For $V = 5$ km: in blue if $10\ m < \bar{r} < 4$ km

    in IR if $50\ m < \bar{r} < 2$ km

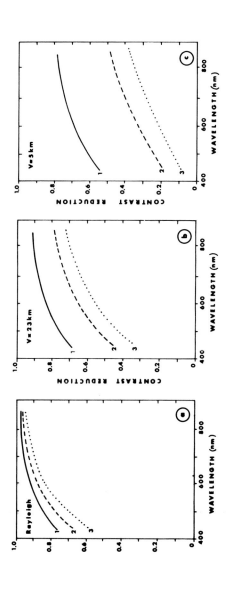

FIGURE 7. Contrast reduction as function of wavelength for a molecular atmosphere (a) and turbid atmospheres for V = 23 km (b) and V = 5 km (c). (μ = 0.866, μ₀ = 0.750). Large lambertic targets designated by 1; small lambertian targets, by 2; and small target with the directional reflectance of the savanna, by 3.

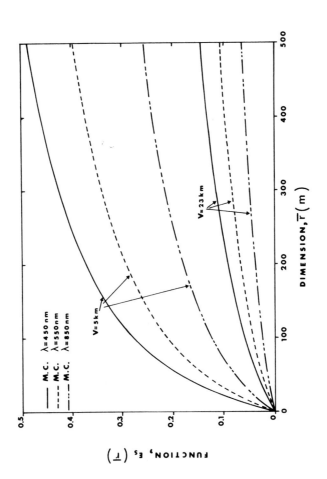

FIGURE 8. Error due to the small target approximation for a circular target of radius r̄.

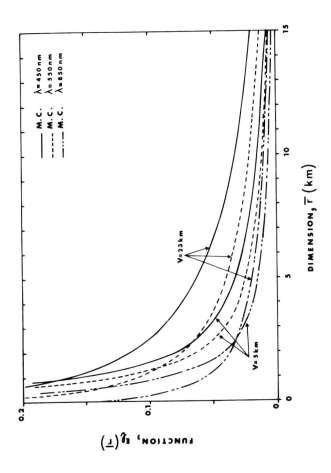

FIGURE 9. Error due to the large target approximation for a circular target of radius r̄.

135

The size interval in which neither of the approximations applies is larger for the clearest atmosphere, that is, in the case where the reduction of contrast is smaller and where both approximations give nearer results.

Let us now consider the same nonlambertian target observed in two different directions; a directional contrast reduction can be defined from Eq. (18) and Eq. (13) by

$$R_d = \pi \frac{(S_1 - S_2)}{\rho_1 - \rho_2} + A + B \frac{\bar{\rho}_1 - \bar{\rho}_2}{\rho_1 - \rho_2} + C \frac{<\rho>_1 - <\rho>_2}{\rho_1 - \rho_2} \qquad (25)$$

The main difficulty for a complete discussion of Eq. (25) is due to the first term on the left-hand side, that is, the direct atmospheric contribution, which is very variable with the direction of observation.

The very particular case of a small highly reflecting target within a uniform Lambertian environment will only be considered here. Therefore, the term S can be neglected in ρ^*, and $<\rho>_1 = <\rho>_2$ as it is due to the Lambertian environment. Moreover, $\bar{\rho}$ can be expressed by Eq. (14). Then Eq. (25) becomes

$$R_d = A + a\, B = e^{-\tau/\mu} \left[e^{-\tau/\mu_o} + aT(\mu_o) \right] \qquad (26)$$

The corresponding results are shown in Fig. 7 (curve 3), with the value of a found for the savanna.

The fact of neglecting $\vec{S}^*$ really limits the use of this result to the recognition of vegetation in infrared with low turbidity. A further consideration should be devoted to this problem of directional contrast.

VI. APPLICATION TO OCEAN REMOTE SENSING

Over the ocean reflectance varies slowly only for large
distances, and there is no more interest in contrast observation,
but there is interest in the measurement of the radiation
backscattered by the sea which contains information on the sea
water.

Nevertheless, for coastal waters, the term $<\rho>$ in Eq. (13)
is dependent on the neighboring coastal ground. To get an idea
to what extent this ground contribution is nonnegligible, $<\rho>$
has been computed in the case of two adjacent half planes with
extreme reflectances $\rho_1 = 0$ and $\rho_2 = 1$ (Lambertian). Figure 10
shows $<\rho>$ as function of the distance of the observed point to
the border for the two turbidities and $\lambda = 450$ nm. The coastal
effect is important to a distance of a few kilometers and has to
be taken into account for any retrieval concerning coastal
waters. Figure 11 shows for the same case the measured signal
ρ^* as obtained from Eq. (13) for a solar incidence of 60°.

The ocean reflectance can be considered as the addition of a
diffuse reflectance ρ resulting from backscattering from the
water mass and assumed to be nearly Lambertian and a specular
surface reflectance ρ' which obeys Fresnel's laws. In remote
sensing, the interest is focused on ρ in order to get information
on the water content.

Because of the waves the direct solar beam is not reflected
only in the direction of specular reflection by a plane surface,
but rather it is distributed around this direction more as the
wind speed is larger. This constitutes the "glitter," which is
not considered here. In the following discussion it is assumed
that the detector does not receive the glitter radiation.
Therefore, the specular reflectance ρ' plays a role only in the
interactions with the atmosphere and it can be included in the
term πS.

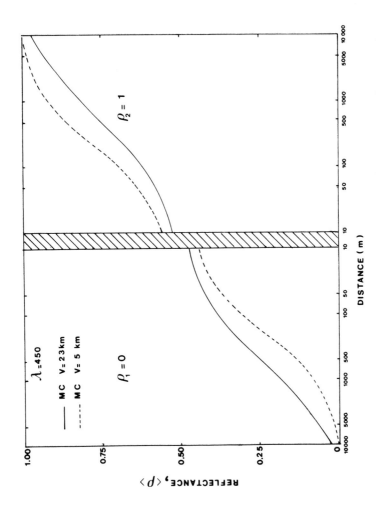

FIGURE 10. Average reflectance of environment as function of the distance between the target and the separation of the two half planes ($\rho_1 = 0$ and $\rho_2 = 1$) for two McClatchey's models ($V = 23$ km and $V = 5$ km) at $\lambda = 450$ nm.

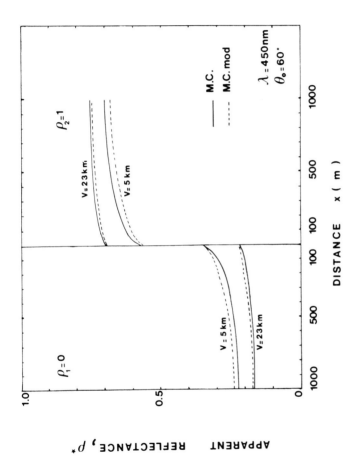

FIGURE 11. Variations of the apparent reflectance as function of the distance between the target and the separation of the two half-planes ($\rho_1 = 0$, $\rho_2 = 1$) for the four models (M.C.--McClatchey; M.C. mod.--McClatchey modified) at λ = 450 nm for a zenithal solar angle θ_O = 60°.

It is important now to consider more carefully this atmospheric contribution πS since ρ is small and absolute values of ρ are sought. As said previously, this term is a very variable function of (μ,ϕ) and (μ_o,ϕ_o) and has to be computed from the equation of transfer for each atmospheric structure. Therefore, for practical remote sensing problems, it is necessary to look for simpler approximate expressions.

Table IV shows for Rayleigh scattering the accuracy of a primary scattering and small optical thickness approximation given by

$$\pi S^R(\mu,\phi;\mu_o,\phi_o) = \frac{P^R(\mu,\phi;\mu_o,\phi_o)}{4\mu\,\mu_o}\,\tau^R \tag{27}$$

where P^R is the Rayleigh phase function. Also, on Table IV the effect of specular reflectance on S^R is shown.

For observation near to the vertical and usual aerosol optical thickness $(\tau^P < 0.7)$, it has been shown that the molecular and aerosol contributions to the atmospheric functions can be separated; moreover, the aerosol contribution is roughly proportional to the aerosol optical thickness and independent of the aerosol profile. This is illustrated in Fig. 12 for the S function, which can be written as (8)

$$\pi S(\mu,\phi;\mu_o,\phi_o) = \pi S^R(\mu,\phi;\mu_o,\phi_o) + b(\mu,\phi;\mu_o,\phi_o)\,\tau^P \tag{28}$$

Therefore, if the law of the spectral variation of the aerosol optical thickness $\tau^P(\lambda)$ is known (from independent measurements or from statistics) the following practical procedure can be suggested

1. Measure the apparent reflectance $\rho_{\lambda_1}^* = \pi S_{\lambda_1}$ at a wavelength λ_1 in the near infrared, where the ocean is nearly black

TABLE IV. Atmosphere Reflectance πS for Rayleigh Scattering (1--Exact without Polarization; 2,4--Exact with Polarization; 3,5--Approximated by Eq. (27); θ = Arc cos μ--Direction of Observation; and θ_o = Arc cos μ_o--Solar Zenith Angle).

	λ (nm)	Without specular reflection			With specular reflection	
		1	2	3	4	5
θ = 0						
θ_o = 15	450	0.0791	0.0838	0.0809	0.0884	0.0842
	550	0.0355	0.0367	0.0356	0.0391	0.0370
	650	0.0181	0.0184	0.0180	0.0193	0.0187
θ_o = 60	450	0.1009	0.0988	0.1011	0.1096	0.1091
	550	0.0453	0.0448	0.0444	0.0506	0.0479
	650	0.0229	0.0228	0.0225	0.0257	0.0243
θ = 30						
θ_o = 15	450	0.0816	0.0846	0.0822	0.0903	0.0855
	550	0.0364	0.0373	0.0361	0.0397	0.0375
	650	0.0185	0.0187	0.0183	0.0199	0.0190
θ_o = 60	450	0.1119	0.1098	0.1109	0.1209	0.1197
	550	0.0501	0.0496	0.0487	0.0555	0.0526
	650	0.0253	0.0252	0.0247	0.0279	0.0267

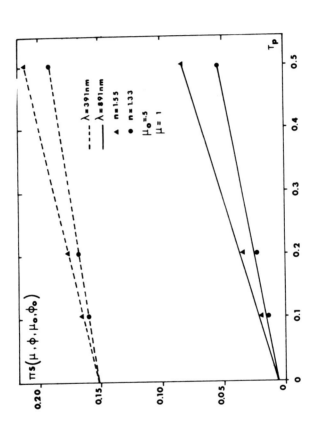

FIGURE 12. Atmosphere reflectance as function of aerosol optical thickness.

2. Then deduce

$$b\tau^P(\lambda_1) = (\rho^*_{\lambda_1} - \pi S^R_{\lambda_1})$$

3. At any visible wavelength λ_2, the atmospheric contribution is obtained by

$$\pi S_{\lambda_2} = \pi S^R_{\lambda_2} + b\tau^P(\lambda_1) \; \frac{\tau^P(\lambda_2)}{\tau^P(\lambda_1)}$$

where the Rayleigh contribution is computed from Eq. (27). Similar approximate treatment can be applied to the transmission function.

A simulation has been done of the retrieval of chlorophyll content by using this procedure (9). At sea level a relation has been found from aircraft and sea truth measurements between the chlorophyll concentration and the difference of the reflectance at 525 nm and 466 nm. This is illustrated in Fig. 13, which also shows results of a theoretical modeling.

The atmospheric effect has been computed for four chlorophyll concentrations between 0.1 and 3 mg/m^3 and two turbidities, low $\tau^P_{700} = 0.18$ and high $\tau^P_{700} = 0.93$. The spectral variation of τ^P has been taken from statistics in Azores Islands in the form

$$\frac{\tau^P(\lambda_1)}{\tau^P(\lambda_2)} = \left(\frac{\lambda_1}{\lambda_2}\right)^{-\alpha}$$

the annual average of α is $\bar{\alpha} = 0.93$ and its variance $\Delta\alpha = 0.3$.

The main cause of error in the retrieval is due to the uncertainty of $\tau(\lambda)$; it has been checked by using the two values of $\bar{\alpha} - \Delta\alpha$ and $\bar{\alpha} + \Delta\alpha$. The extreme retrieved values are given in Table V, in comparison with the initial values.

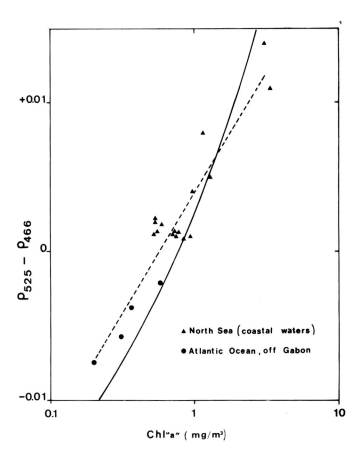

FIGURE 13.   *Difference of reflectances* ρ *at sea level at*
525 nm and 466 nm compared with chlorophyll concentrations.
Dashed line indicates experimental regression; solid line,
theoretical modeling.

TABLE V. *Initial and Retrieval Values of Chlorophyll Concentration* (mg/m^{-3})

Initial	Retrieval	Disparity in log interval unit
	Low turbidity	
0.10	0.07 to 0.15	0.18
0.31	0.23 to 0.41	0.14
1.00	0.84 to 1.22	0.09
3.13	2.75 to 3.66	0.06
	High turbidity	
0.10	0.03 to 0.25	0.50
0.31	0.17 to 0.61	0.29
1.00	0.65 to 1.61	0.21
3.13	2.24 to 4.49	0.15

VII. CONCLUDING REMARKS

It has been shown that the apparent reflectance measured from space is a nearly linear function of the intrinsic atmospheric reflectance, the actual target reflectance, and two average reflectances. Higher orders of interaction between atmosphere and ground can be taken into account roughly by a simple correction factor, but they are generally negligible, except for high reflectances and short wavelength observation.

The atmospheric reflectance is very variable with incidence and observation directions. It is important when absolute values of low reflectances are sought (ocean problem).

It has been shown that molecular and aerosol contribution can be separated, the aerosol effect being roughly proportional to the aerosol optical thickness.

The effects of actual reflectance and the two average reflectances are often of about the same order.

The effect of environment is due to low atmospheric layers for points near to the target and to high atmospheric layers for points far from the target. It is, therefore, very sensitive to the aerosol profile.

The contrast reduction of two targets is influenced by the environment, but approximate formulas are given for small and large targets. The critical size range is between a few hundred meters and a few kilometers.

For ocean remote sensing, the perturbation due to the coast contribution extends to a few kilometers. In open sea, the possibility of obtaining chlorophyll content is discussed and a simple correction procedure has been proposed, which uses a simultaneous reflectance measurement in the near infrared.

It clearly appears that the major effort for improving corrections to remotely sensed reflectances concerns simultaneous information on aerosols.

REFERENCES

1. Herman, B. M., Browing, S. R., and Curran, R. J.   *J. Atm. Sci. 28,* 419-428 (1971).

2. Ueno, S., Haba, Y., Kawata, Y., Kusaka, T., and Terashita, Y., *in* "Remote Sensing of the Atmosphere: Inversion Methods and Applications" (A. L. Fymat and V. E. Zuev, eds.), pp. 305-319. Elsevier Scientific Publishing Company, New York (1978).

3. Gordon, H. R., *Appl. Opt. 17,* 1631-1636 (1978).

4. Tanŕe, D., Herman, M., Deschamps, P. Y., and Deleffe, A., *Appl. Opt. 18,* 3581-3596 (1979).

5. McClatchey, R. A., Fenn, R. A., Selby, J. E. A., Voltz, F. E.,
    and Garine, J. S., "Optical Properties of the Atmosphere,"
    AFCRL 71-0279, Envir. Res. Papers n$^{\circ}$ 354, Bedford,
    Massachusetts (1971).

6. Kriebel, K. T., *Appl. Opt. 17*, 253-259 (1978).

7. Tanre, D., and Herman, M., Influence of the Background
    Contribution upon Space Measurements of the Ground
    Reflectances, *Appl. Opt.* (1980).

8. Tanre, D., and Herman, M., "Correction de l'Effet de
    Diffusion Atmosphérique pour les Données de Télédétection,"
    Proc. of Int. Conf. on Earth Observation Toulouse, ESA--
    SP 134, 355.360., ESA Space Documentation Service, Paris,
    France.

9. Viollier, M., Tanre, D., and Deschamps, P. Y., "An Algorithm
    for Remote Sensing of Water Color from Space," IUCRM Colloq.
    on Passive Radiometry of the Ocean, Sidney, B. C. Canada,
    June 14-21 (1978). (Also in *Boundary Layer Meteorology*
    to appear 1980.)

DISCUSSION

Zardecki: The multiple interactions term has the form of a
geometric series. How do you arrive at this term?

Lenoble: You mean, the last term of Equation (11). Of course,
it is a geometrical series in $<\rho>\bar{S}$ and we get it by summing from
two reflections at the ground to an infinite number of reflections.

Zardecki: Does it mean that you iterate the radiative transfer
equation.

Lenoble: No when we have this expression, we can just use it
like that, computing $\bar{S}$ by any radiative transfer method. We
do not use iteration. We use mainly spherical harmonics method
to get $\bar{S}$.

Gordon: Just one comment. We do not, in general, believe the
aerosol optical thickness is independent of wavelength. We just
use that in the initial attempt to correct the images. What we
plan to do in the future is to use clear water that is in the
scene to derive the correction factors and use those same
correction factors in the coastal aeras. And, when we do that,
we find that the aerosol optical thickness indeed does depend
on wavelength. We also find that it doesn't satisfy Angstrom's
law. We are not quite sure why. And the other comment is you
did not indicate whether the transmittances that you have in
these equations are direct or diffuse transmittances. In my
slides, I was multiplying the water radiances by a quantity
small t which I called the transmittance and never really said
anything else about it. And the reason I did not say much about
it was that I do not really know whether to put the diffuse or
the direct transmittance there, but, it turns out that if you
use an algorithm that involves taking the ratios of two bands,
it really does not matter if you use the diffuse or the direct
transmittance because the aerosol does not have a very large
effect on the diffuse transmittance and as long as the Angstrom
exponent is fairly close to zero, it does not have any effect
on the diffuse transmittance either when you are taking the
ratios.

Lenoble: We use both the direct and diffuse transmittance in
Equations (12) and (13), which are supposed to be exact if we
neglect the multiple interaction term. I agree with the advantage you mentioned for using a ratio. We used the difference in
order to eliminate the specular reflectance, which is merely a
constant with wavelengths. That was the purpose. And, as it
seemed to work well, we kept this method.

ANALYSIS AND INTERPRETATION OF LIDAR
OBSERVATIONS OF THE STRATOSPHERIC
AEROSOL

P. Hamill, T. J. Swissler, and M. Osborn
Systems and Applied Sciences Corporation
Hampton, Virginia

M. P. McCormick
NASA Langley Research Center
Hampton, Virginia

*Remote sensing of the stratosphere with lidar is an important
source of information on the characteristics and properties of
the stratospheric aerosol layer. Data from the NASA Langley
48 inch telescope lidar system are compared with results obtained
by using a one-dimensional stratospheric aerosol model to analyze
various microphysical processes which influence the formation of
this aerosol. In particular, how lidar data can help determine
the composition of the aerosol particles, and how the layer
responds to variations in the temperature profile are considered.*

*Lidar measurements taken at Hampton, Virginia, during the
period 1974 to 1979 show seasonal variations in the aerosol
layer as well as a gradual decrease in stratospheric loading
from the primarily volcanogenic aerosol following the eruption
of Volcán de Fuego to present-day background levels.*

*Comparisons of lidar observations with theoretical calcula-
tions using the aerosol model show that the backscatter profile
is consistent with the composition of the particles being
sulfuric acid and water, but is not consistent with an ammonium
sulfate composition. It is shown that the backscatter ratio is
not sensitive to the composition or stratospheric loading of
condensation nuclei such as meteoritic debris or the residue
from evaporating cloud droplets. The possible benefits of multiwavelength lidar studies of the stratospheric aerosol are
considered.*

I. INTRODUCTION

One of the most important methods presently used for studying
the stratospheric aerosol layer is the analysis of the backscatter of laser light from stratospheric particles. In this
paper the type of information obtained with lidar studies of the
stratosphere and how this data may be used to evaluate various
physical properties and characteristics of the layer and to
extract information on the microphysical processes responsible
for the formation and maintenance of the aerosol are considered.
    Although lidar is a relatively new tool for remote sensing
of the aerosol layer, fortunately there is a fairly comprehen-
sive lidar record of the aerosol layer for the years 1974 to
1979 taken with the NASA Langley 48 inch telescope lidar
system.
    A lidar system measures the intensity of light backscattered
from atmospheric particles and gases. The data can be used to
obtain a relationship between the backscattered light and the
altitude at which the scattering occurred. Thus, one can construct vertical profiles of backscattered ratio which yield a
great deal of information on the characteristics and properties
of the aerosol layer. However, one can obtain much more insight
into the nature of the aerosol if these experimental results are
coupled to a theoretical analysis of the aerosol layer, such as
is afforded by using a theoretical model of the stratospheric
aerosol. In this way, by balancing theoretical studies with
experimental results, it should be possible to extract a
maximum amount of information about the system. Recently, such
an analysis has been started. This paper is intended primarily
to describe the methodology and to present some preliminary
results of the work.

II. THE 48 INCH LIDAR SYSTEM

The NASA Langley lidar system uses a ruby laser as a source of 6943 Å wavelength coherent light. The beam is directed vertically upward. Light which strikes particulate matter in the atmosphere will be scattered in all directions; some of it will be scattered back down where it will be collected by the 48 inch Cassegranian telescope.

For a single laser pulse, the backscattered light arriving at the telescope will be received with a time delay proportional to the altitude from which the scattering occurred. This fact allows one to determine the backscatter as a function of altitude.

It should be kept in mind that the laser light pulse will also be scattered by molecules (Rayleigh scattering) so the total backscattering cross section per unit volume (F) is a sum of the backscattering due to aerosols (F_A), and the Rayleigh backscattering (F_M) due to molecules. Thus,

$$F(\lambda,Z) = F_A(\lambda,Z) + F_M(\lambda,Z) \tag{1}$$

The total backscattering cross section for a given altitude can be determined from the received signal by using the lidar equation

$$N_S(\lambda,Z) = \frac{K(\lambda)}{Z^2} Q^2(\lambda,Z) \ F(\lambda,Z) \tag{2}$$

where $N_S(\lambda,Z)$ is the number of charge carriers generated by the detector, $K(\lambda)$ is a calibration factor which depends on the characteristics of the lidar system (such as collector area, system optical efficiency, etc.) and $Q(\lambda,Z)$ is the transmission, a term which describes the two-way attenuation of radiant energy from the transmitter to altitude Z and back again. The denominator in Eq. (2) describes the $1/r^2$ spreading of the beam of scattered light.

The signal depends, then, on the backscattering cross section (F) and the transmission (Q). Physically significant information on the aerosol scatterers is contained in F_A. Equation (2) can be solved for F if an expression for Q such as can be obtained from models of atmospheric extinction is available. Then, if the molecular backscatter (F_M) is known from measurements of atmospheric density or by use of a model, $F_A(\lambda, Z)$ can be obtained from Eq. (1). The information on the aerosol backscatter of laser light is most frequently presented in terms of the backscatter ratio $R(\lambda, Z)$ defined as

$$R(\lambda, Z) = \frac{F_M(\lambda, Z) + (F_A(\lambda, Z))}{F_M(\lambda, Z)} = 1 + \frac{F_A(\lambda, Z)}{F_M(\lambda, Z)} \qquad (3)$$

The analysis will, of course, introduce systematic and random errors into the result (cf. Ref. 1). At this point it is merely mentioned that the error in $F_A(\lambda, Z)$ is proportional to F_M/F_A and increases as F_A decreases at higher altitudes. Consequently, information obtained from 30-km altitude will contain greater uncertainties than that from an altitude of 20 km where most of the aerosol particles are found.

III.   THE LIDAR RECORD

The lidar record during the period 1974 to 1979, as given in Fig. 1, shows that the peak value of the backscatter ratio has decreased significantly during the last several years. The high values of R in late 1974 and early 1975 are clearly due to the effect of Volcán de Fuego which erupted in October 1974 and injected a large amount of material into the stratosphere. As pointed out by McCormick *et al.* the 1/e decay time of this stratospheric volcanic dust is about 1 year (2). An inspection of Fig. 1 suggests that the low values of R observed during the past several years is a constant "background" quantity. If the

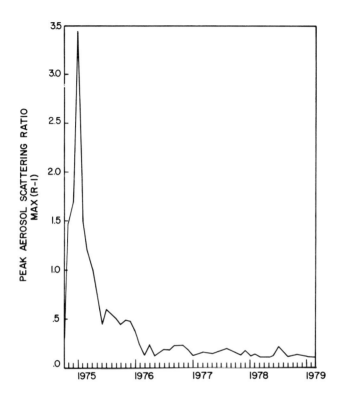

FIGURE 1.   Peak value of R - 1 plotted as a function of time.
Note large increase in backscatter ratio due to Volcán de Fuego,
and initial rapid decrease until early 1976.   Post-1976 lidar
record appears to indicate a steady-state aerosol.

layer has not decayed during the recent volcanically quiescent
period, there must be some mechanism at work maintaining the
aerosol at a certain minimum value. (The climatological impli-
cations of this fact are, of course, quite interesting.)

In the one-dimensional aerosol model described below, it is
assumed that the nonvolcanic component of the layer is maintained
by the upward diffusion of tropospheric OCS. The OCS presumably
is photodissociated to SO_2 which is then transformed into gaseous
H_2SO_4. The H_2SO_4 participates in the formation and growth of
sulfuric acid-water solution droplets. The droplets fall under
the action of gravity but they can also be mixed vertically by
eddy diffusion. Particles mixed upward into regions of warmer
temperatures will evaporate.

If this picture of the stratospheric aerosol is correct, then
the lidar record would be expected to exhibit an evaporation
level, i.e., a temperature above which the aerosol particles are
not found, or expressed in a somewhat different manner, it should
be possible to determine a temperature corresponding to the top
of the layer. However, analysis of the lidar record does not
show a clear-cut top to the aerosol layer. The standard version
of the one-dimensional aerosol model described below predicts an
evaporation layer at 33 km for a spring/fall standard atmosphere
(3). Recent measurements of the vapor pressure of sulfuric acid
give a lower value than previous estimates and would change the
predicted aerosol evaporation level to 37 km for spring/fall
standard atmosphere. These measurements correspond to temperatures of -41° C and -31° C. The lidar data does not extend to
such high altitudes and/or temperatures. Nevertheless, it is
interesting to note that in the 21- to 33-km altitude range,
the aerosol scattering function F_A is falling off rapidly. As
shown in Figs. 2 and 3, the drop off in F_A is correlated with
temperature with a correlation coefficient of 0.802 and with

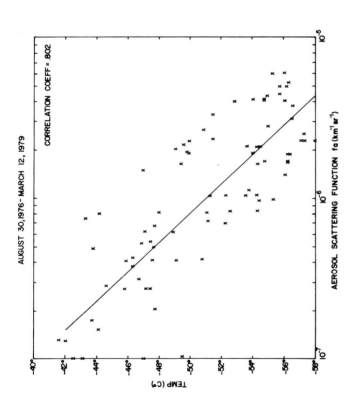

FIGURE 2. Plot of aerosol scattering function and temperature. Correlation coefficient is 0.802.

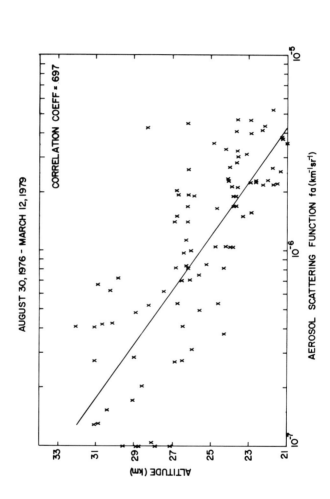

FIGURE 3. Plot of aerosol scattering function and altitude. Correlation coefficient is 0.697.

altitude with a correlation coefficient of 0.697. Since altitude and temperature are highly correlated, it is not clear that the drop off in aerosol loading is physically a function of temperature rather than altitude.

The temperature profile and the backscatter ratio can also be studied together to see whether there is any correlation between the point of minimum temperature and the aerosol properties. It is found, from model calculations for standard atmospheric conditions, that the number density of particles with radius greater than 0.15 μm peaks at 16 km and the number density of particles with radius greater than 0.25 μm peaks at 18 km. This is for a temperature profile with tropopause at 14 km. In all the model calculations, it is found that the maximum value for the large particle mixing ratio occurs well above the tropopause, usually some 10 km higher. Comparisons with lidar studies indicate that the peak in scattering ratio R is generally found a few kilometers above the minimum temperature. As illustrated in Fig. 4, 60 percent of the time the peak is within 5 km above the minimum temperature; thus, general agreement with model results is indicated. The data in Fig. 4 indicates that 82 percent of the time the peak value in scattering ratio is at altitudes between 16 and 21 km.

In Fig. 5, an average lidar backscatter ratio and average temperature profile is illustrated. Note the clear inverse relationship and the upward shift of the maximum in backscatter ratio. In Fig. 6 the data are presented on a seasonal basis. It is interesting to note that the winter profiles of both backscatter ratio and temperature are shifted upward from the summer profiles.

It should be mentioned that characteristic times for the evaporation of H_2SO_4 solution aerosols at high altitudes depend on the water vapor content of the environment. At 30 km an

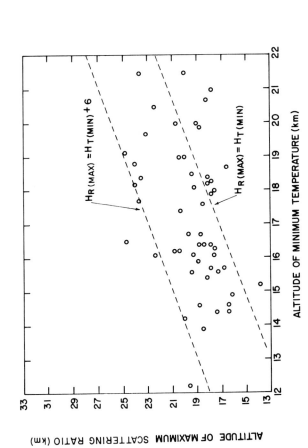

FIGURE 4. Plot of altitude of maximum scattering ratio (R) and altitude at which minimum temperature occurs. Note that the maximum scattering ratio generally occurs a few kilometers above the point of lowest temperature.

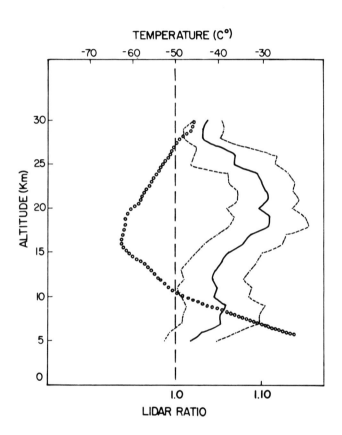

FIGURE 5.   *Average backscatter ratio profile (solid curve).*
Dot-dash curves represent one standard deviation from average
curve.   Open circles give average temperature profile.   Note
that maximum average backscatter ratio is a few kilometers
above altitude of minimum temperature.

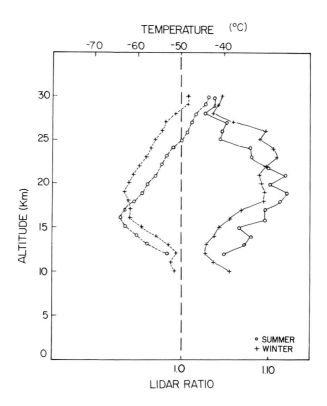

FIGURE 6. *Seasonal averages of backscatter profile and tem-*
perature profile for summer 1978 and winter 1978-1979.

0.5-μm particle can evaporate in a time ranging from 1 day to about 3 weeks. Therefore, particle characteristics can lag behind temperature changes. This is a point which deserves further study.

IV. THE AEROSOL MODEL

The one-dimensional stratospheric aerosol model utilized in the analysis has been described in detail elsewhere (4-7). Here it will suffice to delineate a few of the pertinent characteristics of the model.

The model is intended to simulate a background (nonvolcanic) stratospheric aerosol layer. As mentioned, it assumes that the primary source of sulfur-bearing compounds in the stratosphere is the release of OCS at ground level (8). This tropospheric OCS diffuses into the stratosphere where it is photodissociated into SO_2, and leads, finally, to the production of gas phase sulfuric acid. Particle formation is assumed to be due to the heteromolecular-heterogeneous nucleation of gaseous sulfuric acid and water vapor onto pre-existing solid particles of tropospheric origin. The aerosol particles are assumed to be spherical, liquid, H_2SO_4- H_2O solution droplets with solid cores. These particles are subjected to the physical processes of condensation, evaporation, coagulation, sedimentation, and vertical eddy mixing (4). The model incorporates these chemical and physical mechanisms and predicts altitude profiles for all the important aerosol parameters, including size distribution, number concentration, and weight percentage sulfuric acid in the drops.

The model predicts stratospheric aerosol concentrations and compositions; these are then incorporated into a Mie model and the backscattered radiation of 6943 Å light is calculated for

each altitude. In Fig. 7 the model predicted backscatter ratio
R is compared with the backscatter ratio obtained by averaging
16 lidar observations taken by P. B. Russell and colleagues (9).
Note that agreement between model and observation is quite good
and certainly within the indicated range of measurement variation.

V. NUMERICAL EXPERIMENTS

By using the model and comparing theoretical results with the
experimentally obtained lidar record, it is possible to carry out
a series of numerical experiments to obtain information on the
properties of the aerosol layer.

Thus, one can vary the index of refraction of the aerosol
particles in the model. Figure 8 shows the effect of using 1.3,
1.4, and 1.5 for the real part of the index of refraction. By
comparing with observational results (Fig. 7) it is noted that an
index of refraction of 1.5 is outside the expected range of
values. This is significant because the index of refraction of
ammonium sulfate is 1.5. Thus, the lidar record indicates that
the stratospheric particles are probably not composed of
ammonium sulfate, as was previously believed.

The variations in the backscatter ratio obtained by changing
the stratospheric temperature profile are indicated in Fig. 9
where model results using a warm stratospheric profile (July,
75° North) are compared with the model results using a spring/
fall 45° (mid-latitude) temperature profile. The theoretical
result shows the same trend as the measurements illustrated in
Fig. 6; that is, the profile which has higher temperatures in
the 15 km to 30 km range is associated with a lower cut-off in
the aerosol layer. However, further theoretical work on this
problem is required before any hard conclusions can be drawn on
the relationship between the temperature profile and the lidar
sounding of the aerosol layer.

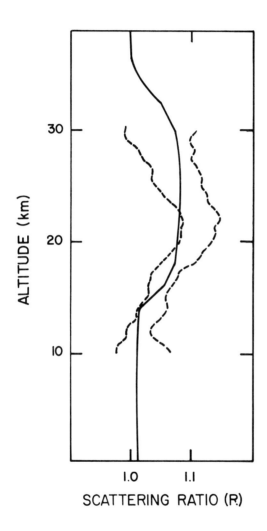

FIGURE 7. *Scattering ratio profile as predicted by one-dimensional aerosol model (solid line). Dashed lines represent one standard deviation from average of 16 measured backscatter ratio profiles.*

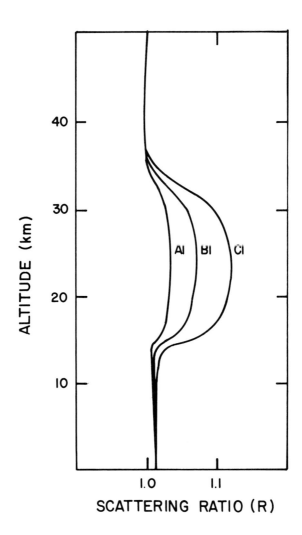

FIGURE 8. *Predicted backscatter ratio profiles using one-dimensional model for different indices of refraction of the aerosol particles. Curve Al is for n = 1.3, curve Bl is for n = 1.4 and curve Cl is for n = 1.5.*

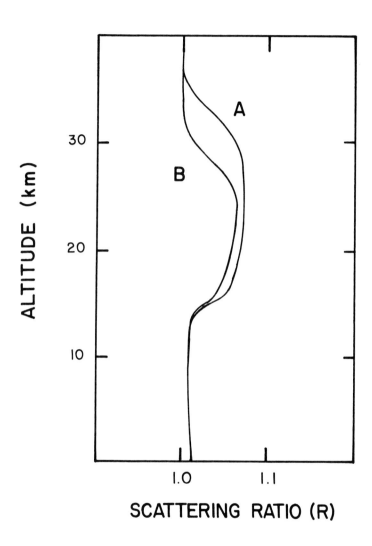

FIGURE 9.   *Model predicted backscatter ratio for a standard atmosphere mid-latitude temperature profile (curve A) and for a 60° N July temperature profile (curve B).*

The model distinguishes between sulfate aerosol particles
(assumed to be liquid sulfuric acid-water solution droplets) and
condensation nuclei which are assumed to be small solid particles
which serve as nuclei for the formation of the stratospheric
aerosol droplets. A series of numerical experiments were made
with the model and it was determined that the lidar profile is
not sensitive to the number density or the composition of these
condensation nuclei, assuming that they are small and obey an
$1/r^4$ size distribution. Therefore, meteoritic debris or evaporating cloud droplet residue may be important in the formation of
the layer, but lidar studies cannot be expected to yield much
information in this regard.

Finally, the model can be used to evaluate the expected backscatter ratio profiles to be obtained with a lidar system using
a different wavelength. Thus, for example, several simulations
were made for $\lambda = 1060$ Å and it was determined that scattering
ratios are consistently higher but do not seem to contain much
more information on index of refraction than the 6943 Å return.
However, the ratio of the two can be used to evaluate the
particle mean size as a function of altitude. In Fig. 10 are
presented model predictions for the ratio $\beta(0.6943)/\beta(1.06)$ as a
function of altitude. The bars represent the results of measurements made by Iwaska (10). The apparent discrepancy may be due
to experimental error, since this is a very new technique. It
will be interesting to consider future experimental results in
light of the prediction made in Fig. 10.

VI. CONCLUDING REMARKS

An analysis of the lidar record in conjunction with theoretical modeling allows one to evaluate physical properties of the
aerosol particles and to isolate important processes which affect
the formation and evolution of the aerosol layer.

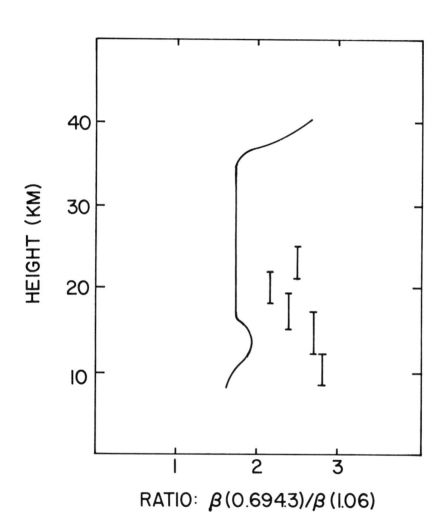

FIGURE 10. Model predicted ratio of 0.6943 μm wavelength
backscatter ratio to 1.06 μm backscatter ratio (solid line).
Bars represent data obtained with a two-color lidar system.

REFERENCES

1. Russell, P. B., Swissler, T. J., and McCormick, M. P.,
    Appl. Optics, 19(22), 3783 (1979).

2. McCormick, M. P., Swissler, T. J., Chu, W. P., and Fuller,
    W. H., *J. Atmos. Sci. 35,* 1296 (1978).

3. COESA, United States Standard Atmosphere, NOAA-S/T 76-1562.
    U.S. Government Printing Office, Washington, DC (1976).

4. Turco, R. P., Hamill, P., Toon, O. B., Whitten, R. C., and
    Kiang, C. S., *J. Atmos. Sci. 36,* 699 (1979).

5. Toon, O. B., Turco, R. P., Hamill, P., Kiang, C. S., and
    Whitten, R. C., *J. Atmos. Sci. 36,* 718 (1979).

6. Hamill, P., Toon, O. B., and Kiang, C. S., *J. Atmos. Sci.*
    34, 1104 (1977).

7. Hamill, P., Swissler, T. J., Turco, R. P., and Toon, O. B.,
    Nature, 278, 149 (1979).

8. Critzen, P., *Geophys. Res. Lett. 3,* 73 (1976).

9. Russell, P. B., Viezee, W., Hake, R. D., and Collis,
    R. T. H., *Q. J. Roy. Met. Soc. 102,* 675 (1976).

10. Iwasaka, Y., The Vertical Change of the Stratospheric
    Aerosol Size Distribution Measured by Two Color Lidar.
    Water Res. Inst., Nagoya U., Chikusa-Ku, Nagoya 464,
    Japan (1979).   (Unpublished manuscript.)

DISCUSSION

Staelin: To what extent can direct chemical measurements of the aerosol composition test your hypothesis?

Hamill: That is a real problem. There have been no real tests of the composition of the aerosol *in situ*. If you take a sulfuric acid water solution droplet collected up there, for example on a wire impactor, and bring it down to your laboratory, open up the box and look in it, breathe on it, and so forth, you may get ammonium sulfate. Furthermore, if you bring it down, even if you are very careful about not letting it be contaminated with anything else, you are going to change the temperature. All of these things make it very difficult to determine the composition of the particles. However, Jim Rosen, quite a few years ago, using a balloon-borne instrument, measured the boiling point of the aerosol particles and showed it was consistent with a 75% sulfuric acid water solution. People generally believe that nowadays. Before that measurement they believed it was ammonium sulfate.

Green: Didn't the recent volcanic eruption--within the last year--perturb your baseline?

Hamill: The most recent volcano was La Soufriere which occurred about 5 weeks ago, on April 13. My record does not include such recent measurements. There was a volcano, Mt. Augustine in Alaska, which erupted in 1976 or late 1975, which was believed to inject material into the statosphere, but there was little evidence of new stratospheric particles after that one. I am not going to claim there is no volcanic material up there, but as far as we know, it looks as if there is none.

REMOTE SENSING OF OZONE IN THE MIDDLE ULTRAVIOLET[1]

A. E. S. Green
James D. Talman[2]

Interdisciplinary Center for Aeronomy and (other)
Atmospheric Sciences (ICAAS)
University of Florida
Gainesville, Florida

Earlier studies of the remote sensing of the ozone distribution from satellites in the middle ultraviolet are extended,
giving particular attention to the analytic model method. After
a brief introduction, the historical background of this subject
is presented. This is followed by a summary and review of early
work based upon the analytic model. Then a linearized approach
to the problem and a linearized equation for a density correction
are discussed. These indicate some of the general limitations
of the inversion method. Finally the parametrized model approach
is reexamined by using both simulated data and some actual
satellite data.

[1]This work was supported in part by National Science
Foundation Grant ATM75-21962 and National Aeronautics and
Space Administration Grant NAS522908.
[2]On leave from Applied Mathematics Department, University
of Western Ontario, London, Canada.

I.   INTRODUCTION

   The middle ultraviolet spectral region from 170 nm to 340 nm
has a number of important aspects from the viewpoint of remote
sensing and solar-terrestrial relations. For example, the depth
to which 1/e of the incident solar intensity penetrates for an
overhead sun is about 100 km at 170 nm. It descends rapidly and
irregularly to 35 km at 200 nm due to absorption in the Schumann-
Runge bands of O_2, and rises to 45 km at 250 nm at the peak of
the Hartley absorption continuum of O_3. Proceeding toward longer
wavelengths in the Hartley continuum and Huggins bands the 1/e
depth descends rapidly through the stratosphere and troposphere
reaching the ground at about 310 nm. The weakly absorbing features of ozone influences the solar radiation reaching the ground
out to about 340 nm. Thus, depending upon the specific wavelength chosen within this spectral octave, one is involved with
mesospheric, stratospheric, or tropospheric phenomena.

   The next section gives a brief historical introduction to
ozone sounding from satellites. Then an earlier work with an
analytical profile technique is described. A linearized approach
to the ozone sounding problem and a linearized equation for a
density correction which indicates some of the limitations of the
inversion method are then discussed. Finally the parametrized
model approach using simulated data and some actual satellite
data is reexamined.

II. HISTORICAL BACKGROUND

   It appears that in 1957 Singer and Wentworth (1) made the
first suggestion as to the possibility of determining the vertical
ozone distribution in the stratosphere by measuring backscattered
solar radiation in the 300 nm region. Their specific discussion
dealt with the potential information content of backscattered
light at 280 nm and 300 nm at various solar zenith angles. 1961
was an important year for ozone sounding. In that year Kaplan (2)

suggested using a spectroscope as a tool for atmospheric sounding
by satellite with specific but not detailed reference to the
ultraviolet (UV)-ozone sounding. Sekera and Dave discussed the
determination of the vertical distribution of oxone from measurement of diffusely reflected UV solar radiation (3). Finally
Twomey made a concrete contribution which has had considerable
impact by using the single scattering approximation to show that
when the constituent absorbs differently, at different wavelengths, the spectral distribution of reflected energy depends
on the vertical distribution of the absorbing constituent (4).
In particular, if the vertical distribution of the absorber is
represented by expressing the pressure as an explicit function
of the mass of the absorber above the atmospheric pressure level,
the spectral energy distribution can be related to the Laplace
transform of the former function. This result can be utilized
to deduce the approximate vertical distribution of the absorbing
constituent.

     In 1962 Green also used the single scattering approximation
to develop an approximate analytic expression for the scattered
solar irradiance seen by a satellite in terms of the wavelength,
the look angle, the sun angle, the scattering angle, and three
parameters which characterize a useful analytic form which nicely
describes the overall ozone distribution (5,6). An approximate
analytic formula makes it possible to exhibit the influences of
the major variables quite conveniently.

     These results were utilized in interpreting the early
experimental measurments of Mayfield, Friedman, Rawcliffe and
Meloy and Friedman, Rawcliffe and Meloy (7,8). Twomey and
Howell next discussed the instability problems of indirect
sounding (9). They discussed and applied a method which allows
a stable, smooth solution to be obtained in certain cases and
gave one of the earliest discussions of the limitations of
indirect sounding.

Early experimental observations by Barth, Hennis et al., and
Friedman et al. (10, 11, 8) greatly reduced the range of
uncertainty of the backscattered radiance to be expected from a
satellite when looking down upon the earth's atmosphere in the
middle ultraviolet (MUV). Work in the USSR by Yakoleva et al. on
the ozone layer paralleled that in the USA during this period (12).
A monograph on the Middle Ultraviolet, Its Science and Technology
by Green et al. assembled together much of the physical information related to the atmospheric ozone problem and the ultraviolet
radiance of the earth in this middle ultraviolet spectral region.
In Chapter 5 on "The Radiance of the Earth and the Middle
Ultraviolet" the earlier 1962 analytic approach of Green was
refined and a perturbation treatment was developed to help to
go beyond the simple three parameter description of the ozone
distribution (14). Chapter 6 by Dowling and Green examines the
magnitude of second order scattering correction (16). In
Chapter 7 approximate corrections for the spherical nature of
the earth's atmosphere are given (15).

    In Chapter 10 of this monograph Barth considers the ultra-
violet spectra of the earth from the viewpoint of the MUV dayglow
and gives an example of the earth's albedo as measured with an
altitude controlled rocket at 168 kilometers (17). The year
1966 might thus be regarded as the date when the exploratory
phase of remote sensing of the ozone layer ended and the detailed
work began.

    The ultraviolet airglow experiment on Ogo by Barth and Mackey,
launched on July 28, 1967, has provided one of the earliest
"production" sets of UV irradiance data (18, 19). This data is
still being subjected to analyses and useful systematics on the
global distribution of stratospheric ozone are being extracted
by London (20). Work during this era in the USSR also advanced
beyond the exploratory stage (21).

Measurements made with the Backscatter Ultraviolet (BUV) instrument launched in April, 1970 by Heath, Mateer and Krueger, presently constitute the most massive body of data now available (22,23). Evaluation of the data from BUV led to a number of design features of the SBUV/TOMS experiment aboard the Nimbus 7 satellite which was launched in October, 1978. These new instruments are now providing data which undoubtedly will improve the knowledge of the global stratospheric ozone distribution. The SBUV subsystem is similar in design to the BUV. It features a state-of-the-art double monochromator that maximizes optical throughput while providing good wavelength resolution, accuracy and a very high stray-light rejection. The TOMS subsystem uses a single monochromator that is a close match of the first mono-chromator of the SBUV. It uses six wavelengths from 312.5 to 380 nm to accomplish total ozone mapping.

The remainder of this report will be concerned with the interpretation of SBUV-type data giving particular attention to the analytic method.

III.   THE ANALYTIC METHOD

The analytic approach, which exhibits explicitly the dependence of radiance observations upon the parameters of the ozone distribution and the observational degrees of freedom, can serve as an intuitive way of introducing this subject. Typical ozone distributions are fitted quite well by a generalized distribution function (GDF) shown in Fig. 1. This "standard" ozone distribution is shown fitted by the differential form

$$n(y) = \frac{n_o (1 + \alpha)^2 e^{y/h}}{[\alpha + e^{y/h}]^2} \tag{3.1}$$

with the parameters $n_o = 3.07 \times 10^{-4}$ cm/km, $\alpha = 152$, and $h = 4.63$ km.

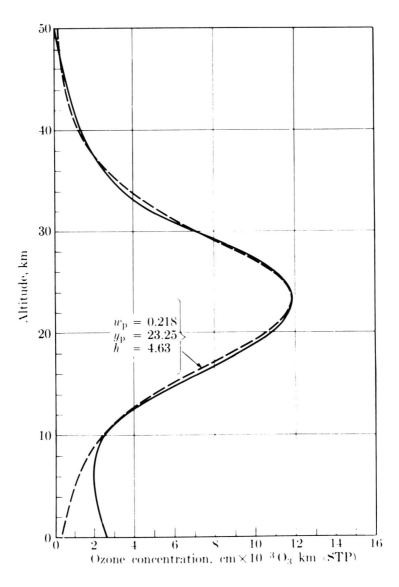

FIGURE 1. A "standard" density distribution for ozone. The solid curve shows a "standard" density for ozone. It corresponds to 0.229 atm-cm of ozone in a vertical column. The dashed curve represents the analytical fit with the parameter w_p = 0.218, y_p = 23.25, and h = 4.63. (From Ref. 6.)

In the range from 0 km to 50 km, the vertical air density distribution of the earth may be approximately characterized by the same generalized distribution function with $n_o = 1$, $\alpha = 0.312$ km, and h = 6.42 km. The value n_o corresponds to 2.55×10^{19} molecules/cm^3. The integrated distribution corresponds to a total vertical air mass N(O) = 8.00 km. The integral distribution corresponding to Eq. (3.1) may be expressed in the form

$$w = w_p \frac{1}{1 + \exp[(y - y_p)/h]}$$ (3.2)

where w_p, y_p, and h are adjusted constants. The negative of the derivative of this function which is proportional to the ozone density (compare with Eq. (3.1)) is then given by

$$\rho = -\frac{dw}{dy} = \frac{w_p}{h} \frac{\exp[(y - y_p)/h]}{[1 + \exp\{(y - y_p)/h\}]^2}$$ (3.3)

Apart from the region close to the ground, this relation has the general layer characteristic of the "standard" ozone distribution (see Fig. 1).

For the distribution shown in Fig. 1, these parameters are $w_p = 0.218$ cm, $y_p = 23.2$ km, h = 4.63 km. To approximately correct the vertical path for slant angles away from the nadir, one simply multiplies w in each case by the secant of the angle from the zenith. This procedure, together with the Lambert-Beer's law, leads to the transmission function.

$$T = \exp\left[-\frac{k(\lambda)w_p \sec\theta}{1 + \exp\{(y - y_p)/h\}}\right]$$ (3.4)

where $k(\lambda)$ is the absorption coefficient for ozone.

The absorption coefficient of ozone has a very large variation in the wavelength interval of interest. For an overall
representation, this curve may be fitted approximately by the
expression (see Fig. 2)

$$k(\lambda) = 300 \exp\left(\frac{\lambda - 254}{25}\right)^2 \text{cm}^{-1} \tag{3.5}$$

As a smooth overall approximation for the solar radiance in this
range, use

$$H(\lambda) = 0.1 \exp\left(\frac{\lambda - 254}{25}\right) \frac{\text{watts}}{\text{m}^2\text{nm}} \tag{3.6}$$

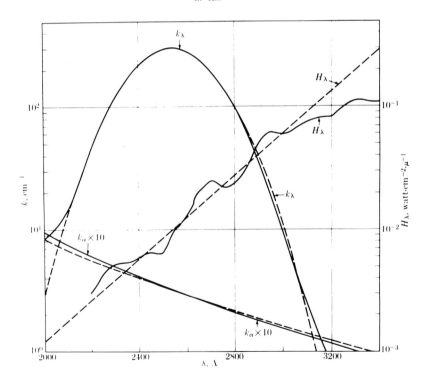

FIGURE 2. The solid curve k_λ is the absorption coefficient
of O_3 at $P = 760$ mm and $\theta = 18\,^\circ C$. The other curves are discussed in text (from Ref. 6).

For the Rayleigh scattering coefficient, use

$$k_a(\lambda) = 0.315 \left(\frac{254}{\lambda}\right)^4 \; km^{-1} \tag{3.7}$$

where the entire atmosphere is equivalent to 8.00 km. These analytical curves are shown in relationship to the experimental values in Fig. 2.

To convert the effective Rayleigh absorption coefficient to an appropriate form for this application, the equivalent atmospheric thickness in the exponential form is used

$$w_a = w_{ap} \exp\left(-\frac{y - y_p}{h_a}\right) = w_{ao} \exp\left(- y/h_a\right) \tag{3.8}$$

where $w_{ao} = w_{ap}\delta$, $\delta = \exp(y_p/h)$, and in the neighborhood of 23 km, $w_{ap} \approx 0.20$, $y_p = 23$, $h_a \approx 7.00$, and all numbers are in kilometers. The attenuation above 23 km due to Rayleigh scattering of the incident beam is small compared with the attenuation due to ozone. For this reason, the attenuation due to Rayleigh scattering is ignored initially in calculating the radiances of the earth.

In Fig. 3, the geometry for the calculation is given. For simplicity, the diagram assumes a flat earth and assumes that the sun line, zenith line, and satellite line lie in a plane.

The various analytical functions that have been developed may now be used to calculate the earth's radiance over the spectral range of interest. It is assumed, as an approximation, that the attenuation of the incident and scattered radiation is due to ozone absorption alone. For the assumed geometric arrangement and an exponential air thickness function, the effective scattering mass is

$$\Delta M = \frac{w_a}{h_a} R^2 \; \Delta\Omega \; \Delta y \; \sec \phi \tag{3.9}$$

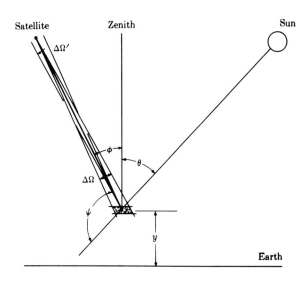

FIGURE 3. Diagram illustrating the angles involved in albedo calculation, where θ is zenith angle of sun, φ is zenith angle of satellite, and ψ is the angle of scattering (from Ref. 6).

where R is the range to the satellite, and $\Delta\Omega$ is the solid angle accepted by a detector element in the optical system. By using Rayleigh scattering theory for the differential cross section for scattering into a solid angle $\Delta\Omega'$, it follows directly that the total scattering power into an area of the detecting system defined by $R^2 \Delta\Omega'$ per unit wavelength interval per unit solid angle is

$$B = \frac{3}{16\pi} (1 + \cos^2\psi) \, H(\lambda) \, k_a(\lambda) \, w_{ao} \, T \sec\phi \qquad (3.10)$$

where ψ is the scattering angle and

$$T = \frac{1}{h_a} \int_0^\infty \exp\left(-\frac{y}{h_a}\right) \exp\left(-\frac{\alpha}{1 + \delta e^{y/h}}\right) dy \qquad (3.11)$$

and

$$\alpha = w_p k(\lambda) \; (\sec \theta + \sec \phi) \tag{3.12}$$

When α is large compared with 1, it can be shown that

$$T = \frac{\Gamma(v + 1)}{\alpha^v} + \frac{v\Gamma(v + 2)}{\alpha^{v + 1}} + \cdots \tag{3.13}$$

where Γ is the gamma function and

$$v = h/h_a \tag{3.14}$$

Under most circumstances, the leading term greatly exceeds all the others. Accordingly, to a good approximation

$$B = 3.75 \times 10^{-5} \left[\frac{100 \; w_{ap} \; \Gamma(v + 1)}{(600 w_p)^v} \right] f\beta(\lambda) \; \beta(\theta,\phi,\psi),$$

$$\frac{watts}{m^2 nm \; sterad} \tag{3.15}$$

where

$$\beta(\lambda) = \frac{\exp\left(\dfrac{\lambda - 254}{25}\right)}{\left[\exp\left\{ -\left[\dfrac{\lambda - 254}{25} \right]^2 \right\} \right]^v} \; \frac{254}{\lambda}^4 \tag{3.16}$$

$$\beta(\theta,\phi,\psi) = \frac{1 + \cos^2 \psi}{2^{1-v}(\sec \theta + \sec \phi)^v} \; \sec \phi \tag{3.17}$$

and f is a factor that represents the departures of $\beta(\lambda)$ from the smooth function given by Eq. (3.10), largely because of the irregularities in $H(\lambda)$. Apart from the numerical coefficient, all the factors in Eq. (3.15) are now of the order of magnitude of unity, and $\beta(\lambda) = 1$ at $\lambda = 254$, and $\beta(\theta,\phi,\psi) = 1$ at $\theta = \phi = 0$. The irradiance function associated with this distribution

is shown in Fig. 4 along with early unpublished calculations
(24,25) and several early satellite measurements (7,8,10,11).
Shown in Fig. 4 are the predicted results for several ozone
distributions characterized by the parameters shown.

According to these calculations, the scattered radiation
has a minimum intensity at about 240 nm and increases as the
wavelength changes in either direction. The calculation becomes
suspect as $\alpha(\lambda)$ in Eq. (3.12) approaches unity, since several
approximations break down. Accordingly, one must use the
results longward of 300 nm purely as a guide.

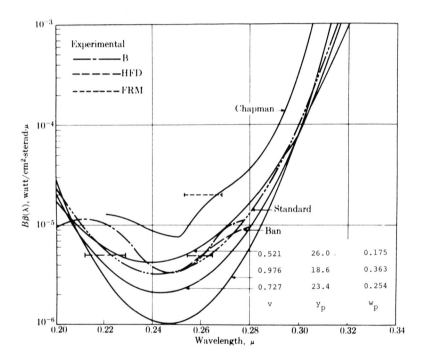

FIGURE 4. *Theoretical calculations of scattered solar
irradiance plotted against wavelength as viewed from above the
atmosphere looking toward the nadir with the sun at the zenith.
Early experimental values are shown (from Ref. 6).*

The factor $\beta(\phi,\theta,\psi)$ given by Eq. (3.17) indicates the dependence of the radiance upon angular factors. The results for the extreme angles $\phi = 90°$ and $\theta = 90°$ must be discounted, since the secant approximation breaks down. Nevertheless, this function shows that significant changes in albedo might be noted, depending upon the look angle, the sun angle, and the scattering angle. Figure 5 illustrates the expected results.

The functional form of $\beta(\theta,\phi,\psi)$ shows an interesting aspect of radiative transfer problems. If $v = h/h_a < 1$, as in the cases illustrated, the dependence upon sec ϕ in the denominator is overwhelmed by that in the numerator. Thus, it would be expected to have limb brightening as the horizon is approached. On the other hand, if $v > 1$, limb darkening would be expected. The distributions of the absorbing and scattering components are of importance in assessing the variation of radiance in the neighborhood of a limb.

The results of Eqs. (3.15 to 3.17) suggest that the angle dependence and wavelength dependence primarily determine $v = h/h_a$. Since h_a is known, this fixes the scale height h which characterizes the high-altitude ozone distribution. From absolute radiance measurement, one could use Eq. (3.15) to estimate w_p which is approximately the total ozone column thickness.

A more refined calculation in the spirit of the analytic method was carried out by Green by considering O_2 and O_3 absorption and utilizing thickness functions as the variables (14).

In this calculation for transmission $T(\theta,y)$ of the incident radiation to a particular height interval with air thickness $dw_a(y)$ is followed by the scattering by a mass of air

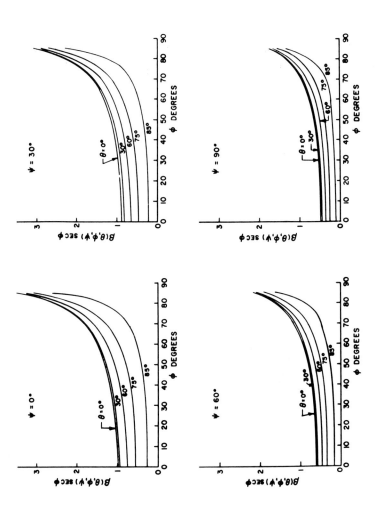

FIGURE 5. The angular function β(θ, φ, ψ) for a series of scattering angle values and for the standard ozone distribution (ν = 0.666). The results near θ = 90° and φ = 90° must be viewed with care since the secant function fails (from Ref. 6).

$$dm = r^2 d\Omega \sec \phi \left(- \frac{dw_a(y)}{dy} \right) dy \qquad (3.18)$$

and the transmission $T(\phi,y)$ to the satellite altitude. Finally, all the scattered radiation from all elements accepted by the "look" cone are summed. The air scattering (a) and attenuation due to ozone (3) and oxygen (2) are characterized by the optical thicknesses

$$\tau_a(y) = k_a(\lambda) \, w_a(y) \, (\sec \theta + \sec \phi) \qquad (3.19)$$

$$\tau_2(y) = k_2(\lambda) \, w_2(y) \, (\sec \theta + \sec \phi) \qquad (3.20)$$

$$\tau_3(y) = k_3(\lambda) \, w_3(y) \, (\sec \theta + \sec \phi) \qquad (3.21)$$

where transmission and optical depths are related by

$$T = e^{-\tau} \qquad (3.22)$$

The radiance of the atmosphere due to first-order scattering may be placed in the form of Eq. (3.10) where $W = w_{ao}\tau_a$, the transmission weighted effective air thickness, is now given by

$$W(\theta,\phi,\lambda) = \int_0^\infty e^{-\tau(y,\theta,\phi,\lambda)} \left[- \frac{dw_a(y)}{dy} \right] dy \qquad (3.23)$$

Here, the $\tau(y,\theta,\phi,\lambda)$ function represents the sum of all the optical depth functions.

The first-order scattering calculation thus depends on the determination of the effective thickness of air. This depends on the geometry through the angles θ and ϕ in a way which is determined by the absorption coefficients $k_a(\lambda)$, $k_2(\lambda)$, and $k_3(\lambda)$.

The evaluation of the effective air thickness integral can be carried out in a number of ways. For some purposes, it is convenient to integrate by parts to give

$$W = w_a(0) \; e^{-\tau(0)} + \int_0^{\tau(0)} w_a[y(\tau)] e^{-\tau} \, d\tau \qquad (3.24)$$

where the variable of integration has been simultaneously
transformed to the optical depth itself. Here $w_a(0)$ is the
total air mass from sea level upward and $\tau(0)$ is the total
optical depth to and from this level along the slant path. To
illustrate the use of this result, treat the case previously
studied in which ozone and air are represented by simple
analytic functions, and only the attenuating influence of ozone
is considered.

If the attenuating influence of the air scattering is
ignored, then the effective air thickness is

$$W = w_{ao} \; T[\alpha(\theta,\phi),v,\delta] \qquad (3.25)$$

where T is given by Eq. (3.11). The optical depth may then be
transformed as the variable of integration by using

$$y(\tau) = h \, \ln\left[\frac{\alpha - \tau}{\delta\tau}\right] \qquad (3.26)$$

Then T, the net transmission factor is given by

$$T(\alpha,v,\delta) = e^{-\tau/(1+\delta)} + \int_0^{\alpha/(1+\delta)} \left(\frac{\delta\tau}{\alpha - \tau}\right)^v e^{-\tau} d\tau \qquad (3.27)$$

If the integrand is expanded and integrated term by term, the
series

$$T = e^{-\alpha/(1+\delta)} + \frac{\delta^v}{\alpha^v}\left[\gamma\left(v + 1, \frac{\alpha}{1+\delta}\right) + \frac{v}{\alpha}\gamma\left(v + 2, \frac{\alpha}{1+\delta}\right)\right.$$
$$\left. + \frac{v(v+1)}{1 \cdot 2\alpha^2}\gamma\left(v + 2, \frac{\alpha}{1+\delta}\right) + \cdots\right] \qquad (3.28)$$

is obtained where $\gamma(a,x)$ is the incomplete gamma function. For
large α the first term is small, and the incomplete gamma
function goes over into the usual gamma function giving

$$T \approx \frac{\delta^v}{\alpha^v}\Gamma(v + 1) \qquad\qquad (3.29)$$

Equation (3.29), together with Eqs. (3.10) and (3.25), is
equivalent to the approximate formula obtained previously
(Eq. (3.13) or Refs. 5 and 6).

The transmission factor $T(\alpha,v,\delta)$ is a well-behaved function
of the parameters a, δ, and v, even where Eq. (3.29) breaks
down. A numerical program may be used for its evaluation for
various values of the parameters α and v and for a fixed value
of δ. In this way, it is possible to explore the predictions for
small values of α which arise in wavelength regions where the
absorption coefficient is small. Values of the integral, as a
function of α, for a number of values of v and a fixed value of
δ are shown in Fig. 6.

The effective thickness can now be calculated under the same
assumptions as before by using the exact form of the integral.
The results are given by the curve labeled 0 in Fig. 7. Values
of $k_o(\lambda)$, $k_a(\lambda)$, and $H(\lambda)$ used in calculating the results are
given in Ref. 15 (page 126).

Figure 7 also shows the results of a perturbation calculation
in which some of the simplifying assumptions previously made were
removed. Each correction term was treated by the identical
mathematical procedure as the main term. The results of the
equivalent air thickness associated with these three corrections
are shown in Fig. 7 for $\theta = 0°$ and $\phi = 0°$. Also shown in Fig. 7
is the total air thickness including the three corrections.
Corrections for the spherical nature of the earth's atmosphere
have also been considered.

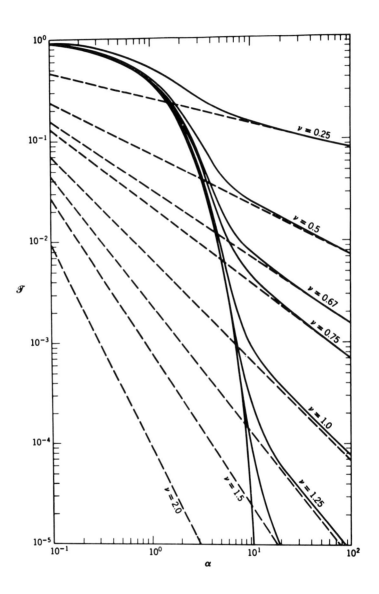

FIGURE 6.   Transmission integral given as a function of α for
standard δ(0.006594) and for various values of ν.   The dotted
curves represent the asymptotic limit given by Eq. (3.21)
(from Ref. 14).

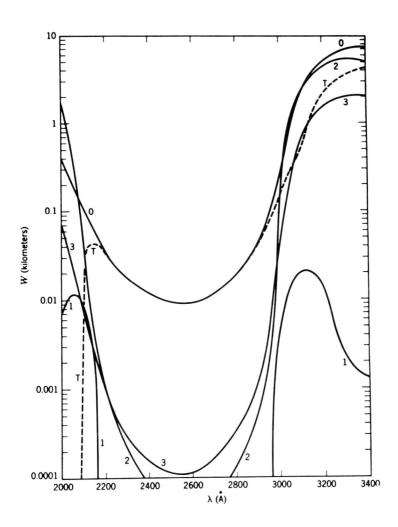

FIGURE 7. *Effective air thicknesses as function of wave-length. (1) represents standard unperturbed thickness correction due to air attenuation; (2) correction due to second ozone component; (3) correction due to second air component; (T) total effective air thickness (from Ref. 14).*

Thus far, the discussion has largely consisted of developing
an approximate analytic expression to relate the ultraviolet backscattered radiance as functions of the variables λ, θ, ϕ, to
the parameters which characterize the ozone distribution. To
invert the ozone distribution from the radiance observations,
one utilizes numerical search routines to optimize the fit of the
calculated distribution to the observed radiance. Some tests
of this procedure were carried out by Brinkman *et al.* by using
rocket data (26).

The assessment made in 1965 based upon the analytic method
was that it is unlikely that more detailed information as to
the ozone distribution could be gathered from satellite or
rocket observations than the fixing of two, three, or possibly
four parameters of a suitably chosen ozone distribution function.
It is still a matter of debate as to whether this assessment
now needs to be greatly modified.

Second-order scattering contributions to the UV backscattered
radiance have been estimated by Dowling and Green by using
variations of the analytic approach (16). Their results are
illustrated in Figs. 8 and 9. It is seen that for $\lambda > 300$ the
second order becomes significant to varying percentages
depending upon the detailed geometric parameters.

IV. THE LINEARIZED PROBLEM

In the single scattering approximation for the SBUV and
ground-based remote sensing of ozone, the observed quantity may
be expressed in the form

$$F(D) = \frac{1}{w_1(0)} \int_0^\infty e^{-[\alpha w_1(y) + \beta w_2(y)]D} \frac{dw_1}{dy} \, dy \qquad (4.1)$$

(cf., Eq. (3.23)). Here $\rho_1(y)$ and $\rho_2(y)$ denote the density of
air and ozone, respectively, and

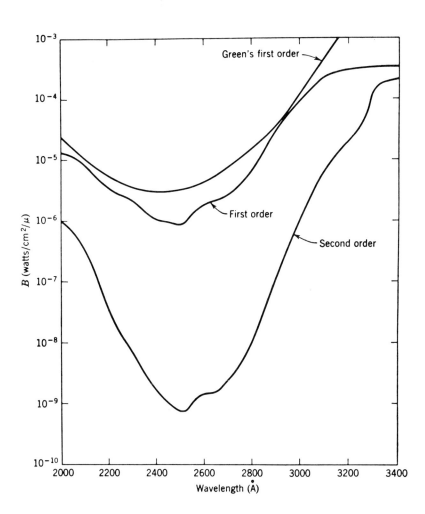

FIGURE 8.   Results of calculation for sun and satellite
overhead (from Ref. 16).

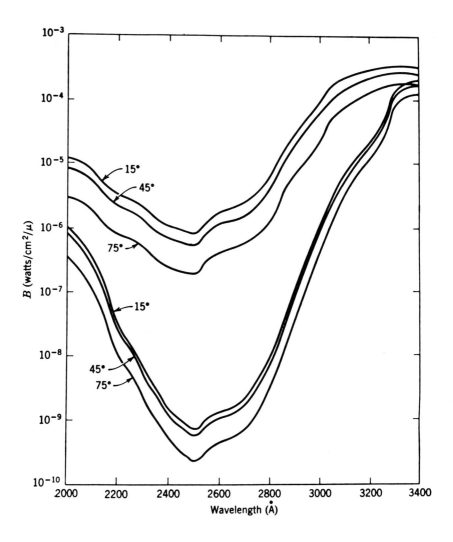

FIGURE 9.   Results for satellite overhead and sun angles
$\theta = 15°$, $45°$, and $75°$ (from Ref. 16).

$$w_i(y) = \int_y^\infty \rho_i(y')dy' \tag{4.2}$$

Also $D = \sec\theta \pm \sec\phi$ (negative for ground-based measurements).
It is convenient to absorb the factor $w_1(0)$ into $w_1(y)$, and
thereby assume $w_1(0) = 1$. Here $\alpha = k_a(\lambda)\,w_1(0)$ is the optical
depth from the ground of air and β is the ozone absorption
coefficient.

Note that $F(D)$ can also be written

$$F(D) = \int_0^\infty e^{-[\alpha w_1 + \beta w_2(w_1)]D}\,dw_1 \tag{4.3}$$

If the observed intensity data are given as a function $F(D)$
for fixed α and β in Eq. (4.1), it is possible by a change of
variables to linearize the problem. Usually the data are not
given in this form, but it is of interest to discuss the
linearized problem to illustrate the difficulties involved.
The problem is linearized by introducing a new variable of
integration

$$z = \alpha w_1(y) + \beta w_2(y) \tag{4.4}$$

in Eq. (4.1). This is a valid transformation since $w_1(y)$ and
$w_2(y)$ are monotonically decreasing. Equation (4.1) then
becomes

$$F(D) = \int_0^b e^{-Dz}\,g(z)\,dz \tag{4.5}$$

where $b = \alpha w_1(0) + \beta w_2(0)$, and

$$g(z) = \frac{dw_1}{dz}(z) \tag{4.6}$$

Equation (4.5) is a linear integral equation for $g(z)$. If
it can be solved, $w_2(y)$ can be found by calculating

$$G(z) = \int_0^x g(z')\,dz' + w_1(0)$$

$$= w_1(z) \tag{4.7}$$

This relationship can then be inverted to find

$$z = G^{-1}(w_1)$$

and from Eq. (4.4)

$$w_2 = [G^{-1}(w_1) - \alpha w_1]/\beta \tag{4.8}$$

It is reasonable to assume that the limit b in Eq. (4.5) is
known from other observations.

A difficulty with this procedure is that the integral
Eq. (4.5) is ill-posed in the sense of Hadamard in that the
inverse of the integral operator is unbounded. This means that
a small error in F(D) can give rise to arbitrarily large errors
in g(z).

The difficulty in solving the problem can be illustrated as
follows. Equation (4.5) can be symmetrized by considering
instead the problem of minimizing

$$S = \int_{D_1}^{D_2} \left[F(D) - \int_0^b e^{-Dz} g(z)\,dz \right]^2 dD \tag{4.9}$$

where $[D_1, D_2]$ is the interval on which measurements of F(D) are
made. The minimization problem leads to a symmetric integral
equation on the interval [0,b]:

$$\int_0^b A(y,z)\,g(z)\,dz = h(y) \tag{4.10}$$

where

$$A(y,z) = \int_{D_1}^{D_2} e^{-D(y+z)} dD$$

$$= \frac{1}{y+z} \left[e^{-D_1(y+z)} - e^{-D_2(y+z)} \right] \qquad (4.11)$$

$$h(y) = \int_{D_1}^{D_2} e^{-Dy} F(D) \, dD \qquad (4.12)$$

Equation (4.10) can also be derived by multiplying Eq. (4.5) by e^{-Dy} and integrating over $[D_1, D_2]$.

Equation (4.10) can be solved, in principle, by expanding $A(y,z)$, $g(z)$, and $h(y)$ in the orthonormal eigenfunctions of the kernel; these satisfy

$$\int_0^b A(x,y) \, \phi_n(y) \, dy = \lambda_n \phi_n(x) \qquad (4.13)$$

$$\int_0^b \phi_m(y) \, \phi_n(y) \, dy = \delta_{mn} \qquad (4.14)$$

Then

$$A(x,y) = \sum_n \lambda_n \phi_n(x) \, \phi_n(y) \qquad (4.15)$$

If $g(z)$ and $h(y)$ are expanded as

$$g(z) = \sum_n \alpha_n \phi_n(z) \qquad (4.16)$$

$$h(y) = \sum_n \mu_n \phi_n(y) \qquad (4.17)$$

it follows that

$$\lambda_n \alpha_n = \mu_n \qquad (4.18)$$

which is immediately solved for α_n and hence $g(z)$.

The problem is that the eigenvalues λ_n approach zero very rapidly as n increases and therefore the coefficients μ_n of the observed function $h(z)$ also approach zero rapidly and are

quickly lost in the noise. For the kernel of Eq. (4.13), it has
been shown by Hille and Tamarkin that λ_n approaches zero more
rapidly than $n^{-n/d}$ where d is any number greater than 4 (27).
Therefore, a slight error in μ_n can be magnified greatly when
α_n is calculated. Explicitly, in a mathematical sense, there
is no degree of accuracy in h(y) (the observations) that will
ensure that a prescribed degree of accuracy in g(z) will be met.

It has been frequently suggested that this problem can be
avoided to some extent if some constraint for g(y) is known.
This approach has been analyzed in detail by Miller (28). As an
example, it could be supposed that g(z) satisfied a constraint
of the form

$$\int_0^b g(z)^2 dz < E \qquad (4.19)$$

Then Miller suggests that rather than minimize S,

$$S' = S + \frac{\varepsilon}{E} \int_0^b g(z)^2 dz \qquad (4.20)$$

should be minimized. Here ε is an upper bound on S, which is
assumed to be known, that is, by estimating the experimental
errors. It is not difficult to show that

$$S = \sum_n (\lambda_n \alpha_n - \mu_n)^2$$

and that

$$S' = \sum_n (\lambda_n \alpha_n - \mu_n)^2 + (\varepsilon/E)\alpha_n^2 \qquad (4.21)$$

This relation is minimized by

$$\alpha_n = \lambda_n \mu_n / (\lambda_n^2 + \varepsilon/E) \qquad (4.22)$$

For $\lambda_n^2 \gg \varepsilon/E$ this is the same as Eq. (4.18), but the
instability as $\lambda_n \to 0$ is damped out.

Miller has shown that for a fairly general class of constraints, this method is stable; that is, as $\varepsilon \to 0$, $g(z)$
approaches the true $g(z)$ at least weakly. To achieve strong
convergence requires a more stringent constraint. For example,
the constraint of Eq. (4.19) would not guarantee strong convergence. In the present problem, it is not, in fact, clear
that there are any global physical constraints that can be
imposed.

It has, moreover, recently been shown by Bertero, De Mol,
and Viano that for the problem of object restoration from a
diffraction limited image, this convergence is very slow for a
fairly general class of constraints (29). The problem in this
case is to solve the integral equation

$$y(t) = \int_{-1}^{1} \frac{\sin[c(t-s)]}{\pi(t-s)} x(s) \, ds \qquad (4.23)$$

It was shown that the error in the restored object is of the
order of $|\ln(\varepsilon)|^{-1}$ where ε is an estimate of the error in the
determination of the image. The argument requires rather
precise estimates of the eigenvalues of the kernel in Eq. (4.23)
and may not be applicable in the present case. It indicates,
however, that even with some global constraint to stabilize the
inversion, a small error in the observed data can give rise to
large errors in the $g(z)$ inferred.

In practice, the function $F(D)$ is not known for all D values,
but only for a finite subset D_1, D_2, ..., D_N. This problem can
be treated in the same way, by making a least-squares fit to
$F(D_i)$ by minimizing an S as in Eq. (4.9). In this case, the
integration is replaced by a summation over the observed points
and Eqs. (4.11) and (4.12) become

$$A_1(x,y) = \sum_i e^{-D_i(x+y)} \qquad (4.24)$$

$$h_1(y) = \sum_i e^{-D_i y} F(D_i)$$ (4.25)

The kernel $A_1(x,y)$ has only a finite number N of nonzero eigenvalues. The integral Eq. (4.10) is then singular and it is possible to add to a solution $g(z)$ any eigenfunction $\phi_n(z)$ corresponding to a zero eigenvalue. It is natural then to solve Eq. (4.10) in the subspace corresponding to the nonzero eigenvalues.

Even this device is not very satisfactory, however, because the eigenvalues of $A_1(x,y)$ decrease very rapidly. In the case $b = 9$, and D_i values of 0.1, 0.2, ..., 1.9, and 2.0, the eigenvalues of A_1 in decreasing order are

        14.5, 2.26, 0.301, 0.0121, 0.00052, ...

It is seen from this, and Eq. (4.18), that the coefficients μ_n in the expansion of h_1 decrease very rapidly, and if the experimental data are valid to say 1 percent accuracy, probably only the first three values μ_1, μ_2, and μ_3 can have any significance. The problem is compounded by the fact that the quantity of interest is

$$\rho_2(y) = -\frac{dw_2}{dy}$$ (4.26)

The process of differentiation must increase the uncertainty in the result.

V.   LINEARIZED EQUATION FOR A DENSITY CORRECTION

A standard approach to the problem used by Twomey, Green, and Mateer is to assume that the ozone distribution is known approximately, and to obtain corrections by using the difference between the observed $F(D)$ and $\bar{F}(D)$ given by the approximate distributions (30-32). Suppose

$$w_2(y) = \bar{w}_2(y) + \delta w_2(y) \tag{5.1}$$

where $\bar{w}_2(y)$ is an approximate distribution, and $\delta w_2(y)$ is a correction term. If $\delta w_2(y)$ is small, it can be seen that

$$\delta F(D) = \beta D \int_0^\infty e^{-[\alpha w_1(y) + \beta \bar{w}_2(y)]D} \rho(y) \, \delta w_2(y) \, dy \tag{5.2}$$

Equation (5.2) is a linear integral equation for $\delta w_2(y)$ in terms of the deviation of the observed $F(D)$ from the $\bar{F}(D)$ given by $\bar{w}_2(y)$. This problem can by symmetrized and treated the same as the problem in the previous section. If Eq. (5.2) is divided by D, multiplied by $\exp[-D\{\alpha w_1(x) + \beta w_2(x)\}]\rho(x)$ and summed on the observations, the integral equation

$$\int_0^\infty H(x,y) \, w_2(y) = r_2(x) \tag{5.3}$$

is obtained, where

$$H(x,y) = \rho(x) \, \rho(y) \sum_i e^{-D_i[\alpha w_i(x) + \beta \bar{w}_2(x)]}$$

$$\cdot \, e^{-D_i[\alpha w_1(y) + \beta \bar{w}_2(y)]} \tag{5.4}$$

$$r_2(x) = - \sum_i (\beta D_i)^{-1} e^{-D_i[\alpha w_1(x) + \beta \bar{w}_2(x)]} \, \delta F(D_i) \tag{5.5}$$

Solution of Eq. (5.3) is equivalent to minimizing the sum

$$\sum_i \frac{1}{D_i} \left[\delta \bar{F}(D_i) - F(D_i) \right]^2 \tag{5.6}$$

(cf. Eq. (4.9)).

Equations (5.3) to (5.6) are not unique because various weights w_i can be assigned to the experimental observations and these would then appear in the sums of Eqs. (5.4) and (5.5). If it is assumed that the observations have a uniform percentage error, it would be reasonable to weight the terms in Eq. (5.6) by $F^{-2}(d_i)$ rather than by $1/d_i$; such a change of weighting could not be expected to affect the qualitative problems of solving the equation.

Eqs. (5.1) and (5.3) could, in principle, be solved iteratively by correcting $\bar{w}_2(y)$ by $\delta w_2(y)$ and solving for a new $\delta w_2(y)$. The instability problems discussed in the previous section will be present also for Eq. (5.3) but here the uncertainties in $r_2(x)$ are greater because $r_2(x)$ is determined by the differences $F(D_i) - \bar{F}(D_i)$. Therefore, only one or two terms in the eigenfunction expansion are likely to be significant.

It may be thought that by measuring at different wavelengths, this problem might be avoided because observations at shorter wavelengths presumably sample the higher regions of the atmosphere. It turns out that this is true to a limited extent. The authors have calculated the eigenvalues and first eigenfunction of $H(x,y)$ for the six shortest wavelengths used in the Nimbus 7 satellite. The "standard" ozone distribution was taken to be

$$\bar{w}_2(y) = C(1 + E^{a/h})/(e^{a/h} + e^{y/h}) \qquad (5.7)$$

where $h = 4.5$ km, $a = 22.5$ km, and $C = 0.32$ cm. The values of D_i were chosen to be 0.2, 0.4, ..., 4.0. The three largest eigenvalues of $H(x,y)$ at the six wavelengths are given in Table I. At the three shorter wavelengths the eigenvalues fall off more slowly and the first two terms in the eigenfunction expansion might be significant. At the two longer wavelengths $(\lambda_2/\lambda_1) < 0.02$ and probably only one term in the eigenfunction expansion can be significant.

The eigenfunctions of $H(x,y)$ corresponding to the largest eigenvalue for the five shortest wavelengths are shown in Fig. 10. The results confirm that the observations at short

TABLE I.   The Three Largest Eigenvalues of $H(x,y)$ Defined
in Eq. (5.4) at Six Wavelengths Used in the Nimbus 7
Satellite.

λ (Å)	λ_1	λ_2	λ_3
2556	1.10 (-5)	3.00 (-6)	3.34 (-7)
2736	2.75 (-5)	5.37 (-6)	9.14 (-7)
2831	1.05 (-4)	2.34 (-5)	4.26 (-6)
2877	9.59 (-4)	9.21 (-5)	1.29 (-6)
2923	1.02 (-2)	2.16 (-4)	3.56 (-5)
2976	6.48 (-2)	6.04 (-4)	1.08 (-4)

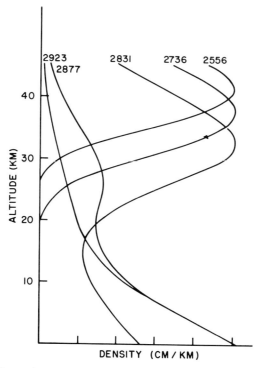

FIGURE 10.   Eigenfunctions of the kernel $H(x,y)$ of Eq. (5.4)
corresponding to the largest eigenvalue at the five shortest wavelengths used in the Nimbus 7 satellite.

wavelengths yield information about the density of ozone at high
altitude. The eigenfunctions are rather broad, however, and it
appears that it cannot be expected that data at a particular
wavelength can be used to infer the ozone density at a particular
altitude. It does seem, though, that corrections to $\bar{w}_2(y)$ at
high and low altitudes could be obtained from data at say
2735 $\overset{\circ}{A}$ and 2923 $\overset{\circ}{A}$.

The kernel $H(x,y)$ in Eq. (5.3) is expressible in the form

$$H(x,y) = \sum_i f_i(x) \, f_i(y) \tag{5.8}$$

Its eigenfunctions have been calculated by expanding

$$\phi_n(x) = \sum c_{n,i} f_i(x) \tag{5.9}$$

When this equation is substituted into the eigenvalue equation,
the result is

$$\sum_{i,j} c_{n,j} \, f_i(x) \int_0^\infty f_i(y) \, f_j(y) \, dy = \lambda_n \sum_i c_{n,k} \, f_i(x) \tag{5.10}$$

If the $f_i(x)$ are linearly independent, this relation reduces to

$$\sum_j h_{ij} c_{n,j} = \lambda_n c_{n,i} \tag{5.11}$$

where

$$h_{ij} = \int_0^\infty f_i(x) \, f_j(x) \, dx \tag{5.12}$$

Equation (5.11) is just the eigenvalue problem for the matrix
given by Eq. (5.12).

VI. A PARAMETRIZED MODEL APPROACH

The methods discussed in the previous two sections assume
that data are given for various D_i at fixed wavelength. In fact,
the data from the Nimbus 7 satellite is at a fixed D value, close

to 2, and at 12 different wavelengths.   A parametrized model has been used for the ozone distribution to investigate the extent to which the ozone distribution can be deduced from the observations of $F(D)$.

The model distribution was taken to consist of two terms of the type used in "The Analytic Method."   Thus, it is assumed that

$$w_2(y) = \sum_{i=1,2} c_i (a_i + 1) (a_i + e^{y/h_i})^{-1} \tag{6.1}$$

where c_i, a_i, and h_i constitute six parameters.   The ozone density corresponding to this relationship is

$$\rho_2(y) = \sum_{i=1,2} c_i (a_i + 1) e^{y/h_i} (a_i + e^{y/h_i})^{-2} / h_i \tag{6.2}$$

In view of the preceding discussions, it seems unlikely that the experimental observations can be used to determine more than three or four parameters if the model is fairly realistic.   It may be possible to augment the data with other information, that is, the ground level density and $w_2(0)$, the ground level value of $w_2(y)$.   It should be noted that Eq. (6.1) embodies an additional reasonable physical constraint, that ρ_2 and w_2 fall off exponentially at high altitude.

This approach was tested by generating simulated experimental observations from three assumed ozone distributions.   The three models were:

Model I:

$$w_2(y) = 50(150 + e^{y/0.7})^{-1} + 0.02 \ln(1 + 20e^{-y/0.5}) \tag{6.3}$$

$$\rho_2(y) = (50/0.7)e^{y/0.7}(150 + e^{y/0.7})^{-2}$$

$$+ 0.8(20 + e^{2y})^{-1} \tag{6.4}$$

Model II:

$$w_2(y) = 50(150 + e^{y/0})^{-1} + 0.02 \ln(1 + 20e^{-y/0.5})$$

$$+ 0.008e^{20}(e^{20} + e^{10y})^{-1}$$

$$- 0.016e^{20}(e^{20} + e^{5y})^{-1} \tag{6.5}$$

$$\rho_2(y) = (50/0.7)e^{y/0.7}(150 + e^{y/0.7})^{-2}$$

$$+ 0.8(20 + e^{2y})^{-1} + 0.08e^{20}e^{10y}(e^{20} + e^{10y})^{-2}$$

$$-0.08e^{20}e^{5y}(e^{20} + e^{5y})^{-2} \tag{6.6}$$

Model III:

$$w(y) = c\,\frac{a + b}{5^3}\,(25t^4 + 20t^2 + 8)\,e^{-5t^2/2} \tag{6.7}$$

$$\rho(y) = ct^5\,e^{-5t^2/2} \tag{6.8}$$

where $t = (y + b)/(a + b)$ and $a = 3.5$, $b = 1.0$, $c = 1.46$. The
units for y are taken as the atmosphere scale height, 6.42 km.
The ozone densities for these three models are shown in Figs.
11 to 13 by the solid curves.

The "experimental" data were generated by integrating
Eq. (4.1) numerically. This is not completely simple because
for large values of $\beta(\lambda)$ (short wavelengths), the factor
$e^{-D\beta w_2(y)}$ changes rapidly from 1 to 0 in a small interval in y.
Gauss-Laguerre integration on 32 points failed to give
satisfactorily consistent results. Therefore, a lower order
procedure, Simpson's rule, was used with interval subdivision
to test the accuracy of the results.

Two sets of experimental data were generated. One set was
to represent satellite data for the fixed $D = 2.0237$ value used
in the satellite and the eight shortest wavelengths observed in

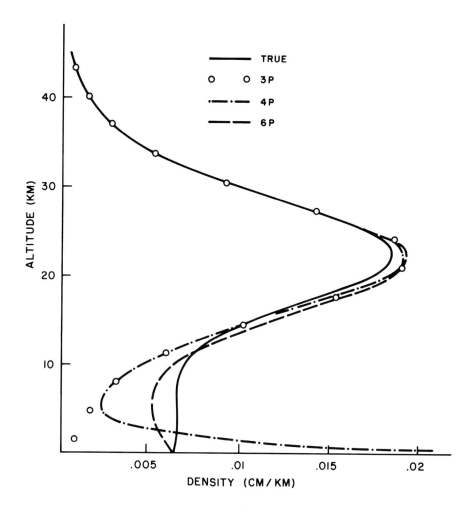

FIGURE 11. Various parametrized model fits to data
synthesized from the given ozone distribution of Model I. Solid
curve: actual distribution. Circles: three parameter fit to
data corresponding to satellite measurements. Dot-dash curves:
four parameter fit to data corresponding to satellite measurements and the total optical thickness. Dashed curves: six
parameter fit to data corresponding to satellite and ground
measurements, total optical thickness, and ground level ozone
density.

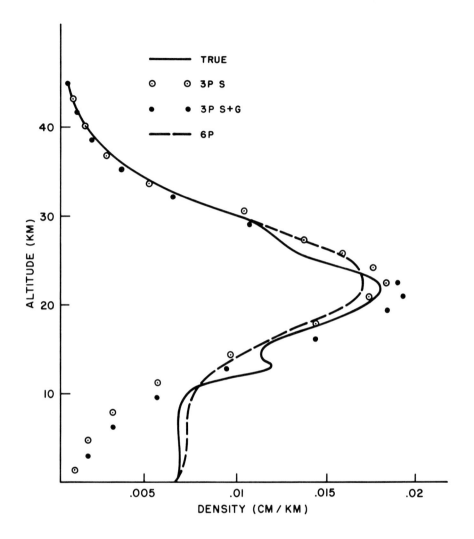

FIGURE 12. *Various parametrized model fits to data synthesized from the given ozone distribution of Model II. Solid circles: three parameter fit to data corresponding to satellite plus ground observations.*

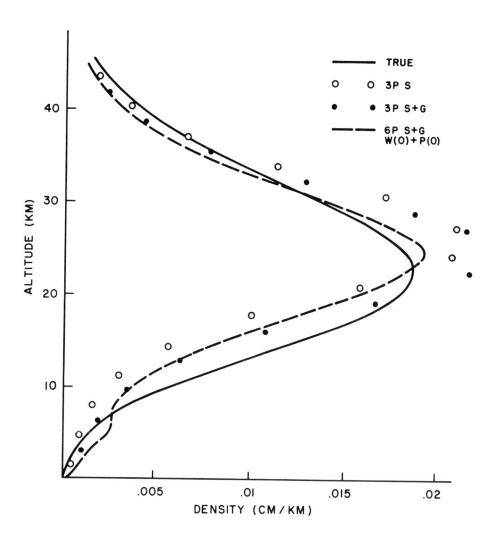

FIGURE 13.   *Various parametrized model fits to data*
synthesized from the given ozone distribution of Model III.

the satellite. The second set of data was to simulate possible
ground observations and was constructed for D = ± 0.5 and λ
values of 2975 Å, 3019 Å, 3058 Å, 3124 Å, and 3174 Å. Additional
ground observations at larger D values are also possible; this
choice was made to investigate whether the ground data can
profitably augment satellite data.

The parameters in Eq. (6.2) are adjusted to minimize the
quantity

$$S = \sum_i [\hat{F}(D_i) - F(D_i)]^2 / F(D_i)^2 \qquad (6.9)$$

where $F(D_i)$ are experimental data and $\hat{F}(D_i)$ are results constructed from the parametrized form. The choice of weighting
$F(D_i)^{-2}$ was made under the assumption of uniform percentage
error in the data.

A three parameter fit to satellite data alone, using only
one term in Eq. (6.2) or putting $c_2 = 0$ gave the results
indicated by circles in Fig. 11. The $\hat{F}(D_i)$ generated by the
fitted distribution agree with the $F(D_i)$ to better than 2 percent.
The results indicate that the satellite data cannot give
reliable information below about 12 km, but suggest that it may
be possible to determine the rate of exponential fall-off and
the position and magnitude of the maximum density.

The dot-dash curve gives a four parameter fit to the
satellite data together with $w_2(0)$. In this case the parameters
in Eq. (4.2) were restricted by $a_2 = 0$, $h_2 = 0.2$ with c_2
adjusted to fit $w_2(0)$. The fit to experimental data is again
better than 2 percent. This again indicates the failure of
satellite observations at low altitudes. The large peak at
ground level arises because of the small value of h_2. If h_2
had been chosen larger, the peak would be flattened out but
the agreement with experimental data shows that there is no
justification for any other choice of h_2.

The dashed curve shows a six parameter fit to the ground
and satellite data, together with assumed known values for
$w_2(0)$ and $\rho_2(0)$. There is some 20 percent disagreement at
about 6 km, but the overall agreement is quite good. In this
case the agreement between experiment and fitted results was
again better than 2 percent. It seems possible that this result
represents the best that can be achieved.

Results similar to the preceding are given in Fig. 12 for
Model II. The four parameter fit to the satellite data and
$w_2(0)$ are not included. The x's show the result of a three
parameter fit to satellite and ground data; in this case, however, the experimental data are reproduced only to about
4 percent accuracy. Again, it appears that the six parameter
fit to the experimental data, $w_2(0)$ and $\rho_2(0)$, is able to
recover the overall details of the ozone distribution.

These results may be open to question because the overall
functional nature of the (true) density is the same at medium
and high altitudes as that of the functional form with which
it is being fit. The most important feature of the assumed
density is perhaps the exponential behavior at large altitudes
and this is probably the nature of the true ozone distribution.
Other possible functional forms could probably be used in
place of Eq. (6.2), for example, a linear combination of
exponential functions. However, the form used is convenient
since, among other reasons, if only one term is used, $w_2(0)$ is
monotonic and $\rho_2(0)$ is constrained to be positive.

The importance of maintaining the correct functional form at
high altitude is illustrated in Model III (see Fig. 13). The
results of three parameter fits (one term from Eq. (6.2)) to the
satellite data and satellite data and ground data are shown,
together with the six parameter fit to satellite and ground
data together with $w(0)$ and $\rho(0)$. In none of these cases did
the computed $F(D_i)$ agree reasonably with the "experimental" data.

It is rather surprising that the six parameter fit was the worst at reconstructing the experimental data, and gave discrepancies up to 20 percent. The three parameter fit to the satellite and ground data gave 10 percent agreement, and the three parameter fit to the satellite data gave 7 percent agreement. In this case, the constraint to the total area, $w_2(0)$, was apparently difficult to meet and produced the large discrepancies in the fitting.

The results of a four paramenter fitting (with $a_2 = 0$, $h_2 = 0.2$) to three actual sets of satellite data and total ozone $w_2(0)$ are shown in Fig. 14. Only the six shortest wavelength results were used to avoid the multiple scattering problem. The density shown fits the experimental data to 1 percent to 1.5 percent. It is clear that the large low altitude peaks are spurious; they are present only to fit $w_2(0)$ but the information to improve this low level behavior is not available.

Finally, the detailed analyses given in this paper tend to confirm the earlier assessment made in 1965 (see "The Analytic Method") and at the previous remote sensing conference that satellite observations can only give a few parameters of a suitably chosen ozone distribution (31). The parametrized model approach provides a reasonably efficient procedure for extracting such parameters. Furthermore, it also provides an obvious physical scheme for organizing systematics of the data, in terms of the time and geographic variation of these parameters.

SYMBOLS

a	fit parameter (correlates with height parameter)
$A(y,z)$	symmetric kernel defined in Eq. (4.11)
$A_1(x,y)$	discrete analog of $A(x,y)$
b	$= \alpha w_i(0) + \beta w_2(0)$
B	total scattering power into detector

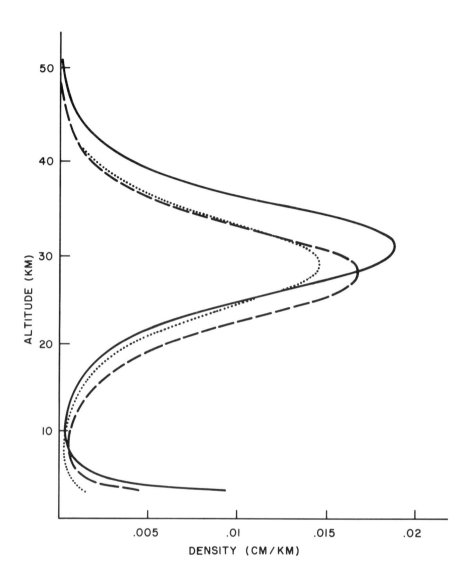

FIGURE 14.   Four parameter fits to three sets of actual
satellite.   The data fitted are the F(D) at the six shortest
wavelengths and the TOMS value for the total optical depth.

c   scaling parameter of "standard" ozone distribution

c_i, a_i, h_i ($i = 1,2$) six parameters for analytic representation of $w_2(y)$ (ozone thickness) (same character as c,a,h)

$c_{n,i}$ coefficients of expansion of nth eigenfunction $\phi_n(x)$ of H(x,y) in terms of $f_i(x)$

$[D_1, D_2]$ interval on which measurements of F(D) are made

$D_1, D_2, \ldots, D_N$ finite subset of $[D_1, D_2]$ for which F(D) is known

E   upper bound on $\int_0^b g^2(x)\, dz$

F(D)   integral of altitude dependent factor of transmitted direct flux

$F(D_i)$ results constructed from parametrized form

$\bar{F}(D)$ approximation to observed F(D)

$\bar{F}(D_i)$ experimental data

f   factor that represents departure of $\beta(\lambda)$ from smooth function

$f_i(x)\ f_i(y)$ ith term of expansion of symmetric kernel, H(x,y)

g(z) $= \dfrac{dw_1}{dz}(z)$; definition of g(z)

G(z) $= w_1(z)$; in effect, definition of g(z)

$G^{-1}(w_1)$ $= z$; inverse function of $w_1 = G(z)$

h   width parameter

h_a width parameter

$h_1(y)$ discrete analog of h(y)

h(y)   inhomogeneous term of integral equation, defined in Eq. (4.21)

H(x,y)   symmetric integral kernel

$H(\lambda)$ extraterrestrial solar flux

$k_a(\lambda)$ Rayleigh scattering coefficient

$k'(\lambda)$ ozone absorption coefficient

$k_2(\lambda)$ particulate (aerosol) scattering coefficient

$k_3(\lambda)$	ozone absorption coefficient
n_o	$= n(0)$
$n(y)$	"standard" ozone distribution
$N(0)$	total vertical air mass
$r_2(x)$	inhomogeneous term of integral equation
R	range to satellite
S	variational integral to be minimized
$T(\phi,y)$	transmission to satellite
$T(\alpha,v,\delta)$	explicit functional dependence of T given in Eq. (3.11)
v	ratio of ozone width parameter to Rayleigh width parameter, h/h_a
w	integral distribution
$w_a(0)$	total air mass from sea level upward
w_{ao}	adjusted constant, $w_{ap}\delta$
$w_3(y)$	equivalent ozone thickness
W	transmission weighted effective air thickness, $w_{ao}\tau_a$
$W(\theta,\phi,\lambda)$	W expressed as function of θ, ϕ, and λ
y_p	height parameter
$y(\tau)$	inverse function of $\tau(y)$
z	$= \alpha w_1(y) + \beta w_2(y)$; transformation of variables
α	total optical depth from ground of air ($\alpha = k_1(\lambda) \cdot w_1(0)$)
α	$= w_p k(\lambda)(\sec\theta + \sec\phi)$
$\{\alpha_n\}$	coefficients of eigenfunction expansion of $g(z)$
β	ozone absorption coefficient
$\beta(\lambda)$	proportional to $H(\lambda) k_a(\lambda)/k^v(\lambda)$ (refer to Eqs. (3.10) to (3.13))
$\beta(\theta,\phi,\lambda)$	contains angle dependence due to phase function and slant path (refer to Eqs. (3.10) to (3.13))

$\gamma(\alpha,x)$ incomplete gamma function

δ $= \exp y_p/h$; adjusted constant

$\delta F(D)$; $\delta w_2(y)$ correction terms

$\Delta\Omega$ solid angle accepted by detector (subtending effective scattering mass)

$\Delta\Omega'$ solid angle subtended by the effective scattering mass ΔM

Δy difference element of altitude

ϵ estimate of error in determination of image

ϵ upper bound on S

θ solar zenith angle

λ_n eigenvalue of $\phi_n(x)$

$\{\mu_n\}$ coefficients of eigenvalue expansion of $h(y)$

ϕ zenith angle of satellite

$\phi_n(x)$ eigenfunctions of $H(x,y)$

$\phi_n(x)$ orthonormal eigenfunctions of the symmetric kernel, $A(x,y)$

ψ scattering angle

$\rho_1(y)$ air density

$\tau(y,\theta,\phi,\lambda)$ $= \tau_a + \tau_2 + \tau_3$

$\tau_i(y)$ $= k_i(\lambda) w_i(y) (\sec\theta + \sec\phi)$; optical thickness for species i from sun to scattering element to detector

$\tau(0)$ total optical depth to and from sea level along slant path

ACKNOWLEDGMENT

We would like to thank Dr. F. Riewe and Dr. C. P. Luehr for helpful suggestions in relation to the approach in "The Linearized Problem" section and Mr. P. F. Schippnick for help with the manuscript.

REFERENCES

1. Singer, S. F., and Wentworth, R. C., *J. Geophys. Res. 62*, 299-308 (1957).

2. Kaplan, L. D., "Chemical Reactions in the Lower and Upper Atmosphere." Interscience, New York (1961).

3. Sekera, Z., and Dave, J. V., *Astrophys. J. 133*, 210-277 (1961).

4. Twomey, S., *J. Geophys. Res. 66*, 2153 (1961).

5. Green, A. E. S., Attenuation by Ozone and the Earth's Albedo in the Middle Ultraviolet. Report ERR-AN221, General Dynamics/Astronautics, San Diego, California (1962).

6. Green, A. E. S., *Appl. Opt. 3*, 203-208 (1964).

7. Mayfield, E. B., Rawcliffe, R. D., Friedman, R. M., and Meloy, G. E., Experiment and Preliminary Results of Satellite Determinations of the Vertical Distribution of Ozone and of the Albedo in the Near Ultraviolet. Aerospace Corporation, TDR-169(3260-50)TN-2, El Segundo, Los Angeles, California (1963).

8. Friedman, R. M., Rawcliffe, R. D., and Meloy, G. E., *J. Geophys. Res. 68*, 6419-6423 (1963).

9. Twomey, S., and Howell, H. B., *Mon. Wea. Rev. 91*, 659-664 (1963).

10. Barth, C. A., *Trans. Am. Geophys. Union, 43*, 436 (1962).

11. Hennis, J. P., Fowler, W. B., and Dunkelman, L., *Trans. Am. Geophys. Union, 43*, 436 (1962).

12. Yakoleva, A. V., Kurdryavtseva, L. A., Britaev, A. S., Gerasev, V. F., Kachalov, V. P., Kuznetsov, A. P., Pavlenko, N. A., and Iozenas, V. A., *Planet. Space Sc. 11*, 709-721 (1963).

13. Green, A. E. S. (ed.), "The Middle Ultraviolet: Its
     Science and Technology." Wiley, New York (1966).

14. Green, A. E. S., in "The Middle Ultraviolet: Its Science
     and Technology" (A. E. S. Green, ed.), pp. 118-129. Wiley,
     New York (1966).

15. Green, A. E. S., and Martin, J. D., in "The Middle Ultraviolet:
     Its Science and Technology" (A. E. S. Green, ed.), pp. 140-
     157. Wiley, New York (1966).

16. Dowling, J., and Green, A. E. S., in "The Middle Ultraviolet:
     Its Science and Technology" (A. E. S. Green, ed.), pp. 130-
     139. Wiley, New York (1966).

17. Barth, C. A., in "The Middle Ultraviolet: Its Science and
     Technology" (A. E. S. Green, ed.), pp. 177-218. Wiley,
     New York (1966).

18. Barth, C. A., and Mackey, E. F., *IEEE Trans. Geosci.
     Elec. GE-7*, 114 (1969).

19. Anderson, G. P., Barth, C. A., Cayla, F., and London, J.,
     Ann. Geophys. 25, 341 (1969).

20. London, J., and Frederick, J. E., *J. Geophys. Res. 82*,
     2543-2556 (1977).

21. Iozenas, V. A., *Geomagnetizm i Aeromiya, 8*, 508-513
     (1968).

22. Heath, D. F., Mateer, C. L., and Krueger, A. J., *Pure
     Appl. Geophys. 106-108*, 1238 (1973).

23. Heath, D. F., and Krueger, A. J., *Optical Engineering, 14*,
     323 (1975).

24. Chapman, R., Geophysics Corporation Report, 61-35-A Af 19
     (604)-7412, Lexinglon, Massachussetts (July 1, 1961).

25. Ban, G., LAS-TR-199-37, Laboratories for Applied Science,
     University of Chicago, (Nov. 1962).

26. Brinkman, R. T., Green, A. E. S., and Barth, C. A., *Trans.
     Am. Geophys. Union, 47*, 126 (1966).

27. Hille, E., and Tamarkin, J. D., *Acta Math.* *57,* 1 (1931).
28. Miller, K., *SIAM J. Math, Anal.* *1,* 52 (1970).
29. Bertero, M., De Mol, C., and Viano, G. A., *J. Math. Phys.* *20,* 509 (1979).
30. Twomey, S. S., *in* "Inversion Methods in Atmospheric Remote Sounding" (A. Deepak, ed.), p. 41. Academic Press, New York (1977).
31. Green, A. E. S., and Klenk, K. F., *in* "Inversion Methods in Atmospheric Remote Sounding" (A. Deepak, ed.), p. 297. Academic Press, New York (1977).
32. Mateer, C. L., *in* "Inversion Methods in Atmospheric Remote Sounding" (A. Deepak, ed.), p. 577. Academic Press, New York (1977).

DISCUSSION

Planet: You sound optimistic that something like this SBUV can obtain ozone profiles down to the ground. Is that correct?

Green: Well, with a little more flexibility at the ground, I should say that the TOMS experiment which is going along with the SBUV gives you total ozone. This is a constraint on your profile in that the density profile has to integrate to the total. Thus TOMS gives you some information below the peak but not very great detail, at least for the analytic model that we have assumed. Below the ozone peak there is quite a bit of opportunity to get a variety of profiles, yet your fit to the Q function data will be very accurate. You might hit it within 2% and still have quite a bit of deviation on the determination below the peak. Now I should say these inversions were based upon six wavelengths, backscattered data, and the next speaker will talk about measurements a little longward of those six wavelengths. They will have to worry about the multiple scattering--they will go into the 300 nm region; whereas, we use six wavelengths shortward of 300 nm.

Planet: You mentioned using the TOMS data. Could you not also use total ozone data from any other type of satellite measurement?

Green: Yes.

Bhartia: The SBUV does have total ozone measuring channels. It has twelve wavelengths, four of which measure total ozone, and eight are used for profiling. So you don't really need TOMS to know the total ozone in the SBUV field of view.

Green: These longer wavelengths which are in the multiple scattering regime can be used with a different technique similar to Dobson to get a total ozone out of the SBUV. I thought, however, the more detailed total ozone analysis has come from TOMS. But you are right. If I understand the substance of Dr. Toldalagi's presentation in the previous session, I think that the essence is that if you put geophysical information into the inversion process, like your climatological model, the more you put in, the more likely it is that your profiling will be meaningful and close to the truth. So this is a similar picture. We are using relatively arbitrary, but fairly convenient flexible, splines that from a statistical point of view characterize many reports of the ozone distribution as measured by balloons and rockets. It would be better if one had a good model of ozone concentrations versus altitudes and fed that model into the inversion scheme. We should get closer to the truth.

ROLE OF MULTIPLE SCATTERING IN OZONE PROFILE
RETRIEVAL FROM SATELLITE MEASUREMENTS
IN THE ULTRAVIOLET

S. L. Taylor, P. K. Bhartia, V. G. Kaveeshwar
and K. F. Klenk

Systems and Applied Sciences Corporation
Riverdale, Maryland

Albert J. Fleig

Goddard Space Flight Center
Greenbelt, Maryland

C. L. Mateer

Atmospheric Environment Service
Downsview, Ontario

The retrieval of the ozone profile from satellite ultraviolet
measurements can be extended to greater depths when multiple
scattering is taken into account. The sensitivity of the
multiple-scattered wavelength radiances to geophysical variables
are discussed and results of profile inversions of Nimbus 4
backscatter ultraviolet data for coincident ground-truth measurements with and without multiple scattering are presented.

I. INTRODUCTION

The Backscatter Ultraviolet (BUV) experiment uses a nadir-

viewing double monochromator to measure the solar ultraviolet

(UV) backscattered by the earth's atmosphere and surface in

219

twelve wavelength bands of 10 Å half-width between 2550 Å and
3400 Å (1). The longer wavelengths (above 3100 Å) penetrate
to the surface and are used for total ozone determination and
the shorter than 3100 Å bands are used for ozone profile
inversion.

For the shortest BUV wavelengths, the ozone optical thickness
is much greater than the air optical thickness; thus, the backscattered radiance to a very good approximation can be computed
by using a single-scattering radiative transfer model. Krueger,
Mateer and Heath used the single-scattering approximation in
the Yarger pressure-increment formulation in the initial profile
inversions of the BUV-4 data (2-5). At the longer wavelengths,
the multiple scattered and surface reflected radiance is a
significant fraction of the total radiance and cannot be
neglected. Consequently, in a profile inversion scheme limited
to single-scattering only, the shortest five- or six-wavelength
channels (below 2800 Å) can be used. The corresponding depth
to which the ozone profile can be retrieved is roughly 3 mbar
to 5 mbar (35 km to 40 km). It is necessary to account for
multiple scattering in order to use the information at the
longer wavelengths and increase the depth to which the ozone
profile can be determined.

The multiple scattered and reflected contributions to the
backscattered radiance at the longer profile wavelengths--2922 Å,
2975 Å, 3019 Å and 3058 Å--were computed and were studied to
determine how these contributions depend on total columnar
ozone, the ozone profile shape, and the surface reflectivity.
In this paper some results of these sensitivity studies are
described and a scheme for accounting for multiple scattering
in profile inversion is presented. Finally, how the inversion
is improved at greater depths when multiple scattering is taken
into account is shown.

II. THE MULTIPLE SCATTERED AND REFLECTED RADIANCE

Following Dave, the total radiance I backscattered from an atmosphere with an underlying surface of reflectivity R can be conveniently expressed as the sum of the purely atmospheric backscattered light I_o and a second term which accounts for the light which has been reflected by the surface as in the expression

$$I = I_o + \frac{RT}{1 - RS^b} \tag{1}$$

where T is called the atmospheric transmission and S^b is the fraction of the reflected light that the atmosphere scatters back to the surface (6). The second term accounts for directly and diffusely transmitted reflected light. The atmosphere is considered to consist of only Rayleigh scatters and ozone absorbers. The quantities I_o and T depend on the solar zenith angle θ_o, the total columnar ozone Ω, the vertical distribution of the ozone, and the pressure of the reflecting surface P_o. The quantity S^b depends on P_o, Ω, and ozone distribution but not on solar zenith angle. The term I_o can be expressed as the sum of the single-scattered I_{SS} and multiple-scattered radiance I_{MS}. The single-scattered intensity is easy to compute but computation of the multiple scattered radiance is time consuming and expensive since multi-dimensional integrations are involved.

The multiple scattered plus reflected (MSR) contribution to the total radiance can be studied as an absolute value I_{MSR} as in

$$I = I_{SS} + I_{MSR} \tag{2}$$

An iterative solution to the auxiliary equations to compute the MSR radiances for the work presented in this paper is used (5).

III. SENSITIVITY OF THE MULTIPLE SCATTERED AND REFLECTED
      RADIANCES TO OZONE AMOUNT AND DISTRIBUTION

The amount of multiple scattered and reflected (MSR)
radiation becomes significant (greater than 1% of the total
radiance) only at wavelengths greater than 2950 $\overset{o}{A}$. The absolute
intensity of the MSR radiation and the intensity relative to
the single-scattered intensity depend strongly on the stratospheric ozone amounts and the solar zenith angle. Increasing
the solar zenith angle or ozone amount increases the slant
path optical thickness and thus prohibits significant amounts
of radiation from reaching the surface and the denser air of
the troposphere where reflection and multiple scattering occur.
Thus, the largest MSR contributions will occur for overhead
sun and small total ozone amounts.

Although the multiple scattered component of the radiation
exhibits a strong dependence on total ozone, it is largely
insensitive to the distribution of ozone when the multiple
scattering is significant.

To illustrate this point, consider the two ozone distributions having the same integrated ozone amounts shown in Fig. 1
and the percentages listed in Table I. The profiles have
significantly different shapes, yet the multiple scattered
intensities are nearly the same (column 3) although the percentage multiple scattering is large (column 2). The single
scattered intensities (column 4) on the other hand are significantly different. The pattern emerging from Table I is valid
generally. As the solar zenith angle is increased, the multiple
scattering takes place in layers higher in the atmosphere and,
as a result, becomes more sensitive to the shape of the ozone
profile. Fortunately, the magnitude of the MSR intensity
relative to the SS intensity, also drops making the MSR less
important in the inversion.

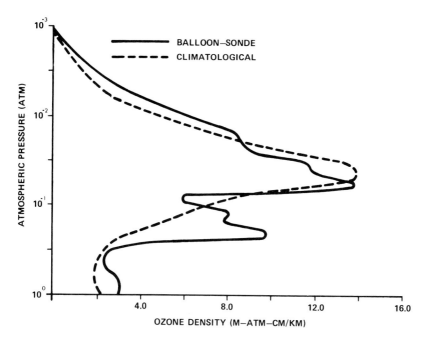

FIGURE 1. *Ozone density profiles for a climatological pro-file (dash curve) and for a balloon-sonde profile (solid curve) both having integrated ozone amounts of 0.346 atm-cm.*

TABLE 1. *Percentages of Multiple Scattering and Percentage Differences in Single and Multiple Scattering for 3058 Å and the Profiles in Fig. 1.*

Solar zenith angle	$100 \times \dfrac{I_{MS}^{(1)}}{I^{(1)}}$	$100 \times \dfrac{I_{MS}^{(2)} - I_{MS}^{(1)}}{I^{(1)}}$	$100 \times \dfrac{I_{SS}^{(2)} - I_{SS}^{(1)}}{I^{(1)}}$
0.0	45.0	−0.4	11
60.3	25.0	1.4	15
77.1	6.4	1.7	11

IV.   PROFILE INVERSION USING THE LONGER WAVELENGTHS

The backscattered radiance at longer wavelengths depends on
many variables including solar zenith angle, total ozone, surface
reflectivity, surface pressure, and profile shape.   Errors in any
of these will lead to errors in the inversion.   The solar zenith
angle for a BUV sample is determined to within a few minutes of
a degree by image location and solar ephemeris data.   The total
ozone coincident with the profile scan is determined to within
2% to 3% by using the total ozone wavelength radiances.   Also,
the coincident effective surface reflectivity is determined from
radiance measurements just outside the ozone absorption band to
within a few percent.   The surface pressure depends on whether
terrain or clouds occupy the field of view (FOV).
(Inhomogeneous FOVs present an additional problem which is not
addressed in this paper.)   For low reflectivities, terrain may
be assumed and the average terrain pressure can be determined
from global terrain tables to within approximately 50 mbar.   For
high reflectivities, cloud top heights may vary and snow and
clouds at high latitudes cannot be distinguished effectively,
but the backscattered radiance is fairly insensitive to the
pressure at large reflectivities.   The impact of the uncertainty
in the pressure on the retrievals is being assessed and will be
presented at a later time.

The inversion consists of selecting a first guess profile
from a set of climatological profiles based on latitude and
total ozone.   The MSR contribution to the total measured radiance
is estimated by using precomputed tables based on the first
guess.   The measured radiances are corrected by subtracting the
MSR component and the resulting radiances are used in an
inversion scheme based on the single scattering approximation
(2).

For most of the BUV samples, the largest uncertainty in the MSR radiance relative to the total radiance is due to the 2% to 3% uncertainty in the total ozone. How errors in the MSR contribution propagate through the inversion has been studied and a wavelength cutoff criterion has been devised based on the sensitivity of the retrieved profile to errors in the MSR correction due to this uncertainty in total ozone. A wavelength measurement is not used in the inversion where these errors propagate into unacceptable errors in the retrieved profile. A more complex wavelength cutoff scheme may be necessary, however, to span the whole range of geophysical possibilities for BUV-FOVs. By using this cutoff scheme, the MSR fraction of the total backscatter radiance can be estimated to within 1% to 2%.

V.   RESULTS AND CONCLUDING REMARKS

When the MSR contribution is significant, the MSR radiance is fairly independent of the profile shape. Because of this, there is little advantage in solving the radiative transfer equation for each BUV sample. The precomputed table of radiances which has been constructed by the authors can be used to obtain the MSR contribution corresponding to the first guess. Computationally, this is a very efficient process and allows the multiple-scattering inversion algorithm to be operationally feasible.

The scheme for accounting for MSR using precomputed tables has been implemented into the Nimbus 4 profiling algorithm. In Figs. 2, 3, and 4, profiles retrieved from BUV radiances are compared with coincident rocket sonde profiles (obtained from A. Krueger by personal communication). Profiles using single-scattering only (SS) and the multiple scattering tables (MS) are shown. For the Point Mugu (34.1° N, 119.1° W) coincidence in Fig. 2 and the Panama (9.3° N, 80° N) coincidence in Fig. 3,

the SS inversion used only 5 wavelengths, whereas the MS
inversion used 6 wavelengths. The 7th (3019 Å) and the 8th
(3058 Å) wavelengths in this case were rejected by the cutoff
scheme mentioned previously. For the Primrose Lake coincidence
(54.8 N, 110.1 W) in Fig. 4, 6 wavelengths were used in the SS
inversion whereas all 8 wavelengths were used in the MS
inversion. The arrows in the figures indicate the peak of the
single-scattering contribution function. Noticeable improvement
is observed at the lower levels, 7 mbar to 20 mbar for the
Panama and Point Mugu coincidences and down to 40 mbar for the
Primrose Lake coincidence.

REFERENCES

1. Nimbus Project, The Nimbus IV User's Guide. National
    Space Science Data Center, Goddard Space Flight Center,
    Greenbelt, Maryland (1970).

2. Krueger, A., Heath, D., and Mateer, C., *Pure Appl. Geophys.*
    106-108, 1254 (1973).

3. Mateer, C. L., *in* "Mathematics of Profile Inversion" (L. Colin,
    ed.). NASA TM X-62, 150 (1972), p. 1-2, available NTIS,
    Springfield, Va.

4. Mateer, C. L., *in* "Inversion Methods in Atmospheric Remote
    Sounding" (A. Deepak, ed.), p. 577. Academic Press, New York
    (1977).

5. Yarger, D. N., *J. Appl. Meteor. 9*, 921-928 (1970).

6. Dave, J. V., *Astrophys. J. 140*, 1292-1303 (1964).

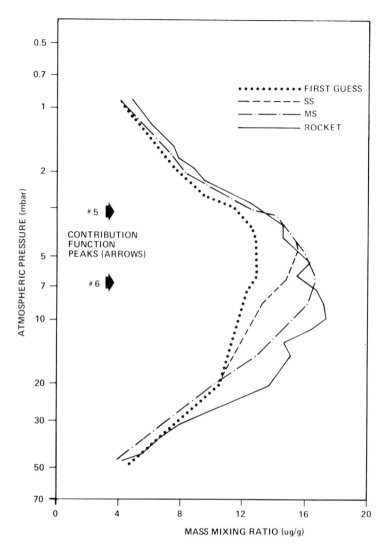

FIGURE 2. Ozone mass mixing ratios profiles for Nimbus 4 BUV data for the single-scattering inversion (dashed curve) using up to and including the 5th wavelength channel (i.e., 2922 Å), the inversion with multiple scattering (dot-dashed curve) using up to and including the 6th wavelength (i.e., 3975 Å) and the coincident rocket profile (solid curve) at Point Mugu (6/18/70). The initial guess is shown as the dotted curve. (1 bar = 100 kPa)

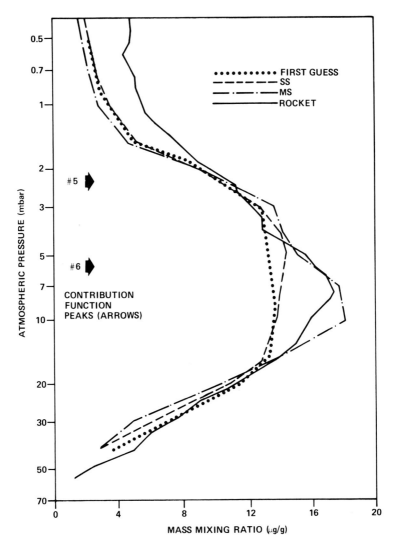

FIGURE 3.   Ozone mixing ratios profiles for Nimbus 4 BUV data for the single-scattering inversion (dashed curve) using up to and including the 5th wavelength channel (i.e., 2922 Å), the inversion with multiple scattering (dot-dashed curve) using up to and including the 6th wavelength (i.e., 2975 Å) and the coincident rocket profile (solid curve) at Panama (11/16/70). The initial guess is shown as the dotted curve.   (1 bar = 100 kPa)

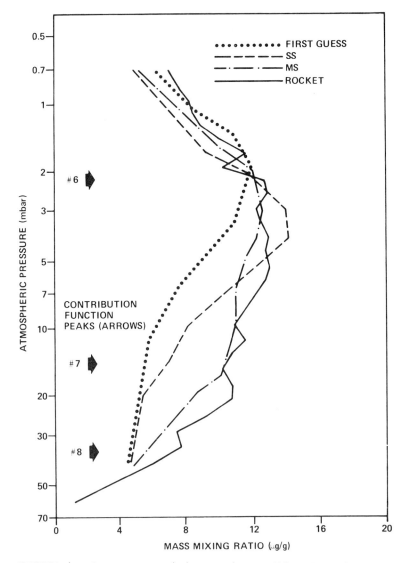

FIGURE 4. *Ozone mass mixing ratio profiles for Nimbus 4 BUV data for the single-scattering inversion (dashed curve) using up to and including the 6th wavelength (i.e., 2975 Å), the inversion with multiple scattering (dot-dashed curve) using all eight profiling wavelengths and the coincident rocket profile (solid curve) at Primrose Lake (10/17/70). The initial guess is shown as the dotted curve. (1 bar = 100 kPa)*

DISCUSSION

Deepak: Could you please explain what you mean by I_R and I_{MS}.
My question is: doesn't the I_{MS} include also to some extent the
reflectivity in your computations?

Taylor: No. I_{MS} is computed by assuming a surface of zero
reflectivity.

Deepak: But isn't that somewhat unrealistic, since the multiple
scattering contribution is sensitive to reflectivity. That is
an approximation you make in the theory, is it not?

Bhartia: The I_R component has full multiple scattering in it
but it is the light that has gone through at least one reflection
by the surface; whereas, I_{MS} is the component which has not gone
through any reflection from the surface. That is the only
difference. But both of them have multiple scattering in them.
For our purpose, it is convenient to separate them this way
because we get a reflectivity measured by the satellite that
can be used to compute I_R.

Green: Do you not use the four long wavelengths in your production
routines?

Taylor: Yes, ·ᐨe do. The selection of the first guess is based
on the total ozone. Also, the selection of the multiple scattering correction factors is based on the total ozone and
reflectivity which are derived using the four longest channels.
You have to have a good total ozone value to make a good
correction.

Planet: On one slide, you used five channels for the single
scattering, and on the next slide, you used six channels for the
single scattering. Do you expect any gain in accuracy or better
fit to the data in going from five to six channels in a single
scattering?

Taylor: The difference between the two scans is that there was
considerable difference in the solar zenith angle. The last
slide that used six was at 60°; the first slide was at, I think,
about 5° solar zenith angle. And so that when you go to the
longer solar zenith angles, you can pick up longer wavelengths
where they are still just being singly scattered. As soon as
you get too far into the atmosphere, you are going to have to
account for the multiple scattering.

Zardecki: How many scattering events are accounted for in your
correction?

Bhartia: We use a code developed by Dr. J. V. Dave of IBM. It computes scattering up to 6 orders and then it uses a geometric series approximation to compute the contributions from higher orders of scattering.

INTERPRETATION OF NO_2 SPIRE SPECTRAL DATA USING
THE AFGL FASCODE COMPUTER MODEL

H. J. P. Smith

Visidyne, Inc.
Burlington, Massachusetts

R. M. Nadile and A. T. Stair

Air Force Geophysics Laboratory
Hanscom Air Force Base, Massachusetts

D. G. Frodsham and D. J. Baker

Utah State University
Logan, Utah

*A rocket experiment named SPIRE was launched from Poker Flat,
Alaska at local dawn on 28 September 1977. The payload included
two CVF spectrometers covering the 1.4 to 6.7 μm and 8.7 to 16.5 μm
wavelength regions. The instrument was programmed to perform a
number of spatial scans through the limb on both sides of the dawn
terminator. The data obtained have been reduced and are currently
being analyzed and interpreted. In this paper, representative
samples of the reduced data are presented, together with some of
the results obtained to date. Emphasis is placed on the use of
the AFGL FASCODE computer program to interpret the observed spectral feature near 6.2 μm in the night side data which is not
observed on the day side. This feature has been identified as
being due to the NO_2 (ν_3). The FASCODE program has been used to
estimate the vertical profile of NO_2. These results are presented and discussed.*

I. INTRODUCTION

The infrared spectral results presented here were obtained
with an exoatmospheric rocket payload which measured infrared
H_2O emission in an earth limb viewing mode. This *SP*ectral
*I*nfrared *R*ocket *E*xperiment (SPIRE) was conducted by the U.S.
Air Force Geophysics Laboratory (AFGL) under the sponsorship of
the Defense Nuclear Agency to test models of sunlit backgrounds
in the auroral region. In addition to the primary mission, SPIRE
performed a wide ranging survey of both sunlit and nighttime backgrounds including emission spectra of CO_2, O_3, OH, NO_2, HNO_3, O_2,
H_2O, and aerosol scattering.

II. MEASUREMENT SYSTEM

The SPIRE payload was launched on a Talos Castor rocket just
before dawn on September 28, 1977 from the Poker Flat, Alaska,
rocket range (1,2). Two cryogenically cooled circular variable
filter (CVF) spectrometers, coaligned and equipped with high off-
axis rejection telescopes, provided wavelength coverage from 1.4
to 6.7 and 8.7 to 16.5 µm. In order to provide a visible benchmark for the infrared aerosol measurements, a dual channel
photometer completed the sensor complement.

A programmable attitude control system maneuvered the payload
through 12 vertical spatial scans of the earth limb, as shown in
Fig. 1. The measurements are divided into three phases, with the
first seven limb scans forming a tight pattern about the terminator.
As SPIRE approached apogee, the eighth limb scan was made in the
earth's shadow to provide a quiescent benchmark for the sunlit
measurements. After a long azimuth traverse back across the
terminator, scans 9 to 12 examined the effect of several solar
incidence angles on both high altitude emissions and low altitude
infrared aerosol scattering.

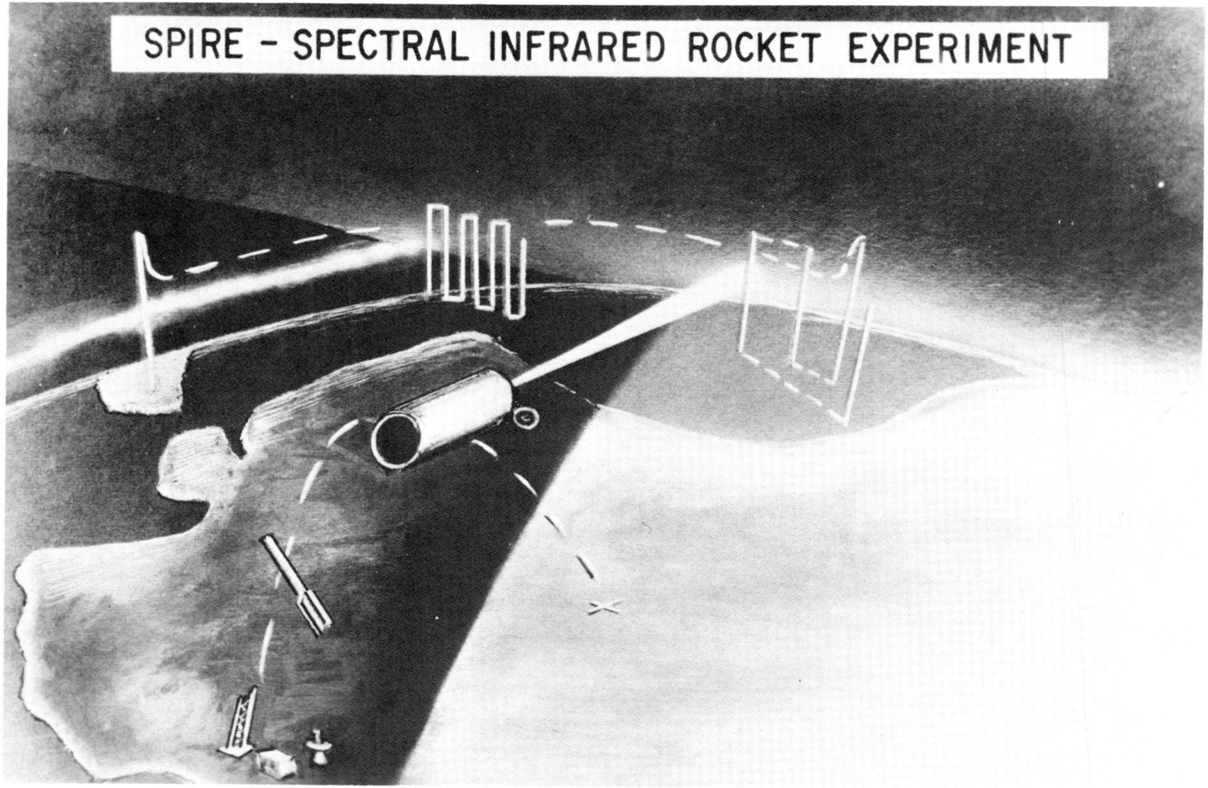

SPIRE – SPECTRAL INFRARED ROCKET EXPERIMENT

FIGURE 1. Sketch of SPIRE Experiment.

Construction details and method of operation of the rocketborne CVF spectrometer have been reported by Rogers et al. (3)
and Wyatt (4). Figure 2 shows a cutaway view of the liquid He
cooled spectrometer (HS-2).

All three instruments contained matched telescopes with nominal
1/4-degree full angle fields of view, coaligned to within 0.04°.
The constant angular field of view resulted in a variable footprint during each spatial scan varying from 3.5 km at an 80 km
tangent altitude expanding to 6 km just above the hard earth limb.
An overview of the twelve spatial scans in geographic perspective
is shown in Fig. 3 with the limits of the auroral oval superimposed.
Figure 4 shows the same geometry projected on a near-simultaneous
Defense Meteorological Satellite Program (DMSP) infrared weather
photo (10 to 14 μm).

Two star sensors and a television camera were to provide
redundant aspect information to accurately monitor payload
orientation and thus determine the tangent height intercept of the
sensor line of sight. Failure of the television camera shortly
after lift-off combined with insufficient baffling of the celestial
aspect sensors resulted in limited aspect data on Jupiter and Mars
for two spatial scans. This was sufficient to fix the azimuth
orientation and confirmed a faster than planned pitch scan rate
of 0.50° per sec. Determination of the tangent altitude critical
to the analysis was based on the infrared data itself utilizing
known altitude profiles of OH as measured by numerous rocket
probes (5,6), and balloon-borne measurements of HNO_3.[1] The
method used to apply appropriate Van Rijn factors has been described by Grieder et al. (7), and results in an absolute tangent
height uncertainty of $\pm$ 4 km. However, the accuracy of the relative shape of the tangent height profiles is much better and is

[1]*Murcray, D. G., private communication.*

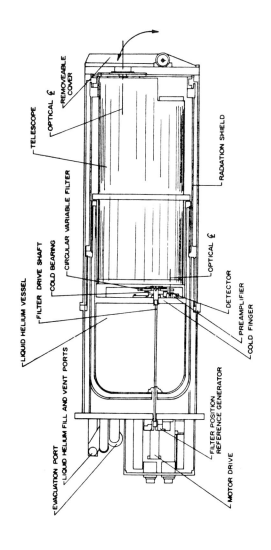

FIGURE 2. Cutaway view of the liquid He cooled spectrometer (HS-2).

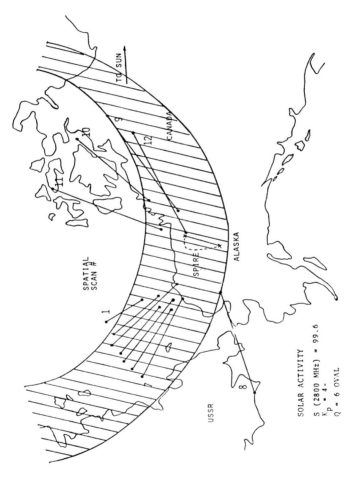

FIGURE 3.   *Geographic Perspective of SPIRE scans with auroral oval.*

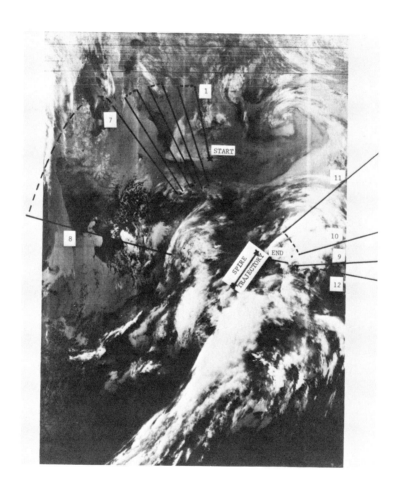

FIGURE 4. Projection of SPIRE spatial scans on DMSP IR weather photo.

illustrated by the small spread in the data over all twelve
spatial scans for species with little or no diurnal variation.

Specifications and test results for the telescopes have been
reported by Williamson (8) and Nadile et al. (9) and were shown
to have better than 10^{-8} rejection of point sources at 4° off-axis.
Despite the excellent point source rejection properties of the
telescopes, some near-field leakage of the earth limb occurred
for tangent heights below 10 km. At tangent altitudes above
160 km, the integrated earth radiance through the telescope far
field established a background limit for some of the LWIR measurements. Rejection of solar radiation was sufficient for all spatial
scans save the top of the twelfth which was intentionally chosen
to approach the sun at the end of the flight.

During rocket ballistic flight and prior to commencing the
measurement phase of SPIRE, the CVF spectrometer viewed a cold
cover which provided an excellent means to determine the in-flight
noise equivalent spectral radiance (NESR) of each sensor. Sensor
output voltage data obtained for fifty consecutive spectral scans
during this period were used to compute a no radiation voltage
mean and standard deviation value as a function of wavelength.
The mean value was used to correct the data for electronic voltage
offset and the standard deviation was used to compute the NESR of
the total system which included the contributions of noise from
the sensor, telemetry, recording and digitization systems. A plot
of the NESR as a function of wavelength for each CVF sensor is
shown in Fig. 5. Spectral resolution based on a preflight
laboratory calibration is also shown and is equivalent to ~ 35 cm^{-1}
in the NO_2 (ν_3) emission region.

III. MEASUREMENT AND COMPARISON WITH MODELS

The SPIRE CVF spectrometers obtained spectra corresponding to
many tangent heights on both day and night spatial scans. In this

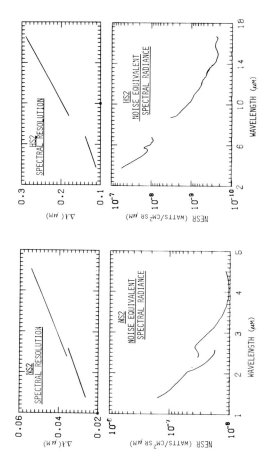

FIGURE 5.   Plot of NESR for each CVF sensor.

paper, data are analyzed for two tangent heights, H_T = 40 km and
H_T = 24 km. For completeness, however, in Fig. 6, one spectral
scan is shown at a higher tangent height, namely, H_T = 130 km.
The NO (1-0) band at $\approx$ 5.4 μm is apparent as is the CO_2 15-μm band.
Complete discussion of the SPIRE data will appear in a forth-
coming paper.[2]

Figures 7 and 8 show typical spectra for $H_T \approx$ 40 km for night
and day scans, respectively. The corresponding spectra for
$H_T \approx$ 24 km are shown in Figs. 9 and 10. Starting at the
higher wavelength end of the spectrum, one clearly identifies
the 15-μm bands of CO_2 which saturate the SPIRE detectors. The
O_3 (ν_3) bands are seen at $\approx$ 9.6 μm at both altitudes and for
H_T = 24 km: the HNO_3 (ν_5) bands are seen at $\approx$ 11.3 μm. In the
lower portion of the CVF spectra, the P and R branches of H_2O are
seen at $\approx$ 6.5 and $\approx$ 5.85 μm at both tangent heights. For the
night scans, a spectral feature appears between the two water
vapor peaks which has been identified as the NO_2 (ν_3) band. This
band nearly disappears on the day scans. There is a feature on
the day scans near $\approx$ 5.3 μm which may be due to the NO fundamental.
A combination of O_3 and CO_2 bands is seen at $\approx$ 4.8 μm. Finally,
the 4.3-μm peak of CO_2 is clearly seen. The day scan at 24 km
shows a considerably larger peak than the night side. This
result seems to be attributable to solar fluorescence.

The water vapor ν_2 band is observed at lower tangent heights
and is shown in Figs. 7 and 8 ($H_T \approx$ 40 km) for both day and night.
One can clearly see a feature at $\approx$ 6.1 μm which is stronger at
night. This has been identified as being due to the ν_3 band of
NO_2. Some calculations confirming this result are presented
later in this paper. The NO_2 is photodissociated by solar

---

[2]*Stair, A. T. et al. (1980) to be published.*

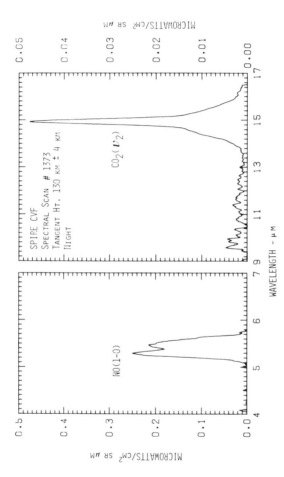

FIGURE 6.   Spectral scan for $H_T \approx 130$ km (Night).

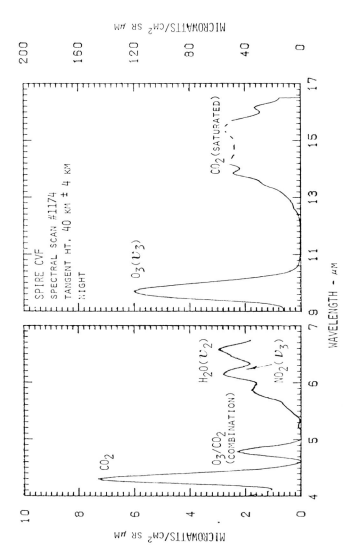

FIGURE 7. Spectral Scan for $H_T \approx 40$ km (Night).

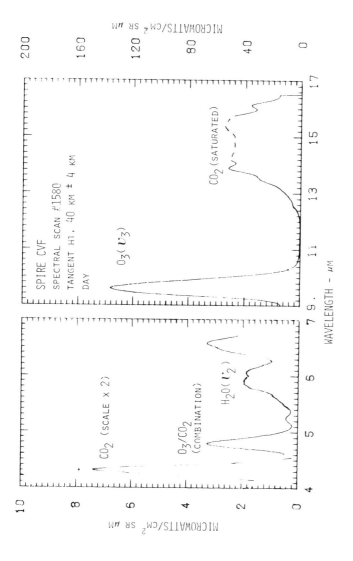

FIGURE 8. Spectral Scan for $H_T \approx 40$ km (Day).

245

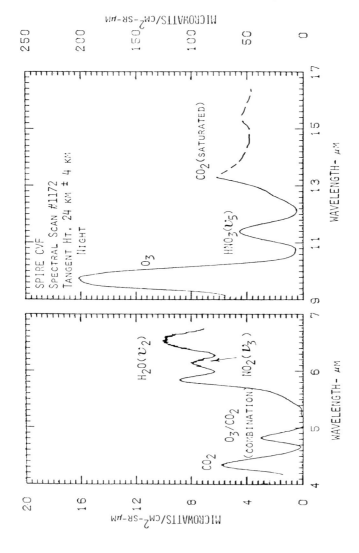

FIGURE 9. Spectral Scan for $H_T \approx 24$ km (Night).

246

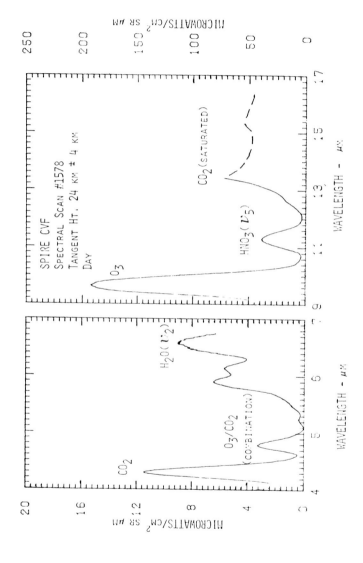

FIGURE 10. *Spectral Scan for $H_T \approx 24$ km (Day).*

247

radiation during the day into NO and O. After local sunset, the
NO product of the photodissociation can recombine in sufficient
amount to provide the thermal radiation observed. Figures 9 and
10 show the day side and night side spectral scans for $H_T \approx 24$ km.
The nighttime NO_2 is apparent.

At this point, several vertical profiles of the radiation
observed are presented in order to illustrate further the type of
data obtained. It is restated again that the footprint of the
optics provided a geometric tangent height uncertainty of
$\approx \pm 4$ km. This fact should be borne in mind in interpreting the
figures presented. The H_2O (ν_2) (6.5-μm peak) data for four SPIRE
spatial scans are shown in Fig. 11. On the plot, the results of
two AFGL computer models are also shown, the Degges Limb model at
high altitude (H $\gtrsim$ 60 km) (10,11), and the LOWTRAN 4 model at
lower altidues (H_T < 50 km) (12). The LOWTRAN 4 agreement with
the SPIRE data is good for 40 $\gtrsim$ H_T $\gtrsim$ 10 km even though it is
expected that the LOWTRAN assumption of strong line absorption
is not very good at higher altitude (H_T $\gtrsim$ 30 km). This is the
reason that the LOWTRAN results fall off the data above $H_T \approx 40$ km.
Further discussion of these results are forthcoming in a later
paper.

The primary objective of this paper is to discuss the NO_2 (ν_3)
band measurements at $\approx$ 6.15 μm. Figures 9 and 10 show the day
and night spectra for tangent height $H_T \approx 24$ km and Figs. 7 and 8
the corresponding results for $H_T \approx 40$ km. The night spectra show
clearly a peak near 6.15 μm between the P and R branches of the
H_2O (ν_2) band, which was identified as being due to the ν_3 band
of NO_2.

The LOWTRAN 4 model does not contain absorption or emission
due to NO_2. Thus, in order to identify this feature more posi-
tively, it was decided to exercise the AFGL FASCODE line-by-line
computer code (13) by using the trace-gas line data of Rothman
et al. (14). This code performs a true line-by-line absorption

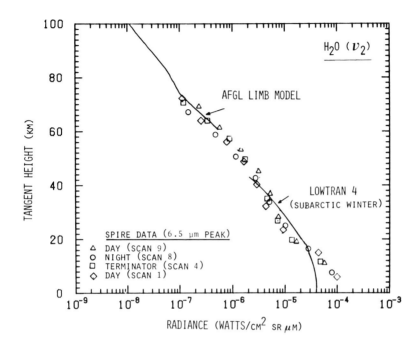

FIGURE 11. *SPIRE data altitude profile for* H_2O *peak at* ≈ 6.5 μm.

calculation by using the data of the AFGL line compilation (15).
A Voight line profile is used. By assuming local thermodynamic
equilibrium, the absorption data may be transformed into an
equilibrium radiance. This transformation is performed for an
arbitrary number of layers in a spherically layered atmosphere
for any geometry. The bookkeeping of changing line widths as
the pressure and temperature vary is handled automatically in
FASCODE.

Although FASCODE is a rather fast line-by-line code, there is
still a relatively large amount of computer time required to make
a run. Our choice was to compare day-night results at H_T = 24
and H_T = 40 km. For simplicity in setting up the calculations,
the choice was made to make the runs with a constant mixing ratio η
at all altitudes for each run. The code was then executed to
determine the predicted equilibrium spectral radiance in the limb.
The mixing ratio was then adjusted until a satisfactory agreement
with the observed data was obtained. For daytime, it was found
that a volume mixing ratio $\eta = 4 \times 10^{-10}$ gave good results for
both altitudes. At night, the final values used were $\eta = 4 \times 10^{-9}$
at H_T = 24 km and $\eta = 2 \times 10^{-9}$ at H_T = 40 km. It seems clear that
further refinement of these values for η could improve the agreement but it was decided that it would not be worth the computational effort at this time. Note that the dominant contribution
for a limb viewing geometry always comes from the region close to
the tangent height simply because of the very long path through
the layer bounded below by the tangent height. Thus, the results
are treated as if a variable mixing-ratio profile had been
used.

Figures 12 and 13 show the results for H_T = 24 km and H_T =
40 km, respectively. (Note the shift to wavenumbers rather than
wavelength.) The agreement is very good. Notice, however, that
the night side feature near 1710 cm^{-1} does not seem to be predicted by FASCODE. Tentatively, this feature is attributed to an

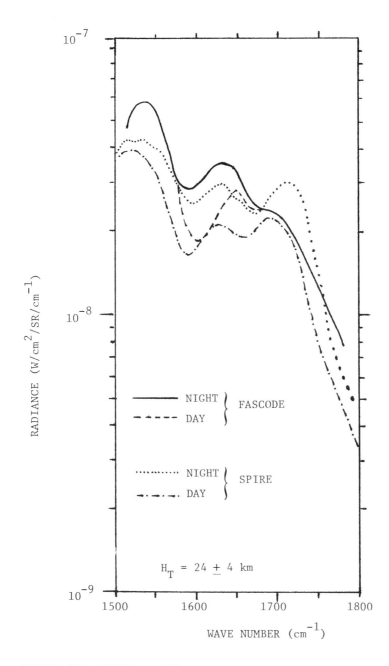

FIGURE 12.   FASCODE Radiance Prediction Compared with SPIRE
Data at H_T = 24 km.

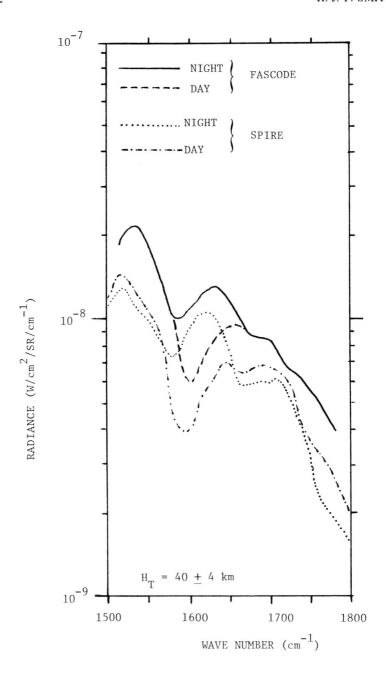

FIGURE 13.   FASCODE Radiance Prediction Compared with SPIRE Data at H_T = 40 km.

O_3 $(\nu_2 + \nu_3)$ intercombination band which is not as yet included in the AFGL line atlas.[3] This assumption is plausible since the feature is missing during the day and weaker at 40 km by which altitude O_3 is falling off from its peak near 25 km.

The two values of the NO_2 mixing ratio used at 24 and 40 km are in reasonable agreement with the results of earlier measurements as shown in Fig. 14 as adapted from Fontanella et al. (16). The reader is referred to the latter reference for details of the earlier measurements. The curve marked 1 is the result of Fontanella et al. (16), that marked 2 is a theoretical calculation due to Lazrus et al. (17), and 3 is a measured profile from Brewer et al. (18). Other experimental points are deduced from Murcray spectra (19) by Ackerman and Muller (20) in the ν_3 and $\nu_1 + \nu_3$ band (o); measured by Farmer et al. (21) in the $\nu_1 + \nu_3$ band ◆ ; and measured by Harries and Stone (22) in the far infrared □ . The shaded regions indicate the reported uncertainties of the measurements. The straight lines give approximate constant mixing ratios as indicated.

IV. CONCLUDING REMARKS

The SPIRE experiment has provided a wealth of spectral limb measurements for both day and night geometries. The interpretation and analysis of the observations are currently under intensive study. The results of the NO_2 FASCODE computations show that earth limb infrared spectral measurements can be used to determine species concentrations as a function of altitude. This NO_2 infrared spectral data showed a strong day/night variation with a nighttime value $\eta = 4 \times 10^{-9}$ at 24 km, in good agreement with previous measurements. The SPIRE limb viewing measurement technique extends these data to higher altitudes with a nighttime mixing ratio of $\eta = 2 \times 10^{-9}$ at 40 km.

[3]Private communication, L. Rothman, AFGL.

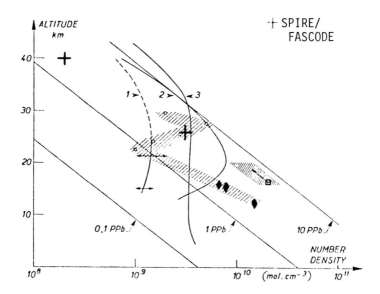

FIGURE 14. Comparison of SPIRE NO_2 concentrations with
other results (16). (See text for identification of curves.)

ACKNOWLEDGMENT

We want to acknowledge the help of W. Grieder, C. Foley, and R. Hegblom of Boston College in data processing.

REFERENCES

1. Nadile, R. M., Stair, A. T., Jr., Grieder, W. F., Wheeler, N. B., Frodsham, D. G., and Baker, D. J., "SPIRE-Mission Performance Evaluation," AFGL-TM-12, U.S. Air Force, 1978.
2. Stair, A. T., Jr., Nadile, R. M., Frodsham, D. G., Baker, D. J., and Grieder, W. F. Paper presented at the Topical Meeting on Atmospheric Spectroscopy, Optical Society of America, Keystone, Colorado, 1978.
3. Rogers, J. W., Stair, A. T., Jr., Wheeler, N. B., Wyatt, C. L., and Baker, D. J., "LWIR Measurements from the Launch of a Rocketborne Spectrometer into an Aurora," AFGL-TR-76-0274, U.S. Air Force, 1976.
4. Wyatt, C. L., *Appl. Opt. 14*, 3086-3091 (1975).
5. Baker, D. J., Conley, T. D., and Stair, A. T., Jr., *EOS, 58*, 460 (1977).
6. Rogers, J. W., Murphy, R. E., Stair, A. T., Jr., Ulwick, J. C., Baker, K. D., and Jensen, L. L., *J. Geophys. Res. 78*, 7023-7031 (1973).
7. Grieder, W. F., Baker, K. D., Stair, A. T., Jr., and Ulwick, J. C., "Rocket Measurements of OH Emission Profiles in the 1.56 and 1.99 μm Bands," AFCRL-TR-76-0057, ERP No. 550, HAES Report No. 38, U.S. Air Force, 1976.
8. Williamson, W. R., "High Rejection Telescopes," Honeywell Radiation Center Report No. 77-1-12, 1977.

9. Nadile, R. M., Wheeler, N. B., Stair, A. T., Jr., Frodsham,
    D. G., and Wyatt, C. L., "SPIRE--Spectral Infrared Experiment,"
    SPIE Proceedings, Vol. 124, Modern Utilization of Infrared
    Technology III, 1977.

10. Degges, T. C., "A High Altitude Infrared Radiance Model,"
    AFCRL-TR-74-0606, U.S. Air Force, 1974.

11. Degges, T. C., and Smith, H. J. P., "A High Altitude Infrared
    Radiance Model," AFGL-TR-77-0271, U.S. Air Force, 1977.

12. Selby, J. E. A., Kneizys, F. X., Chetwynd, J. H., Jr., and
    McClatchey, R. A., "Atmospheric Transmittance/Radiance:
    Computer Code LOWTRAN 4," AGFL-TR-78-0053, ERP No. 626,
    U.S. Air Force, 1978.

13. Smith, H. J. P., Dube, D. J., Gardner, M. E., Clough, S. A.,
    Kneizys, F. X., and Rothman, L. S., "FASCODE--Fast
    Atmospheric Signature Code (Spectral Transmittance and
    Radiance)," AFGL-TR-78-0081, U.S. Air Force, 1978.

14. Rothman, L. S., Clough, S. A., McClatchey, R. A., Young,
    L. G., Snider, D. E., and Goldman, A., *Appl. Opt. 17*, 507
    (1978).

15. McClatchey, R. A., Benedict, W. S., Clough, S. A., Burch,
    D. E., Calfee, R. F., Fox, K., Rothman, L. S., and Garing,
    J. S., "AFCRL Atmospheric Absorption Line Parameters
    Compilation," AFCRL-TR-73-0096, ERP No. 434, U.S. Air Force,
    1973.

16. Fontanella, J.-C., Girard, A., Gramont, L., and Louisnard, N.,
    Appl. Opt. 14, 825-839 (1975).

17. Lazrus, A. L., Gandrud, B. W., and Candle, R. O., *Appl.
    Meteorol. 11*, 389 (1972).

18. Brewer, A. W., McElroy, C. T., and Kerr, J. B., *Nature, 246*,
    129 (1973).

19. Murcray, D. G., Murcray, F. H., Williams, W. J., Kyle, T. G.,
    and Goldman, A., *Appl. Opt. 8*, 2519 (1969).

20. Ackerman, M., and Muller, C., *Nature, 240,* 300 (1972).

21. Farmer, C. B., Invited Review, IAGA, Kyato, 1973.

22. Harries, J. E., and Stone, W. B., "Proceedings of the Second CIAP Conference," U.S. Department of Transportation, TSC-OST73-4, p. 78, 1972.

DISCUSSION

Rodgers: You said you reduced the water vapor by about 20 percent.
Can you tell us what you reduced it to?

Smith: Unfortunately, I don't remember the number offhand, but
perhaps you are familiar with the LOWTRAN atmospheres.

Keafer: In one of your charts, you mentioned about a CO_2 shoulder
and the theories on that. Can you tell us what the essence of
those theories are?

Smith: I am afraid I really cannot. But, Tom Degges, a colleague
of mine at Visidyne, and Jack Kumer, at Lockheed, Palo Alto, are
working on those theories.

PROGRESS IN PASSIVE MICROWAVE REMOTE SENSING:
NONLINEAR RETRIEVAL TECHNIQUES[1]

David H. Staelin

Research Laboratory of Electronics
Massachusetts Institute of Technology
Cambridge, Massachusetts

A variety of nonlinear retrieval methods have been applied
to passive microwave remote sensing problems.   These problems
can be characterized in part by the degree to which their underlying physics and statistics can be understood and characterized
in a simple way. Four examples of varying complexity are considered here; the simplest problem requires only analytic
expressions for retrievals, whereas the most complex problem has
been handled only with pattern classification techniques. The
four examples are: (1) Doppler measurements of winds at 70 to
100 km, (2) retrieval of atmospheric water vapor profiles using
the opaque 183-GHz water vapor resonance, (3) retrieval of snow
accumulation rate by means of combined theoretical and empirical
procedures, and (4) classification of diverse polar terrain by
means of pattern recognition techniques.

[1]This work was performed in part under NASA Contract
NAS5-22929.

I. INTRODUCTION

Passive microwave remote sensing has advanced in several
directions in the 3 years since an earlier review of microwave
inversion methods (1). The first quantitative retrieval methods
employed were described in that review; they were generally
linear techniques, often based on regression analyses of computer-
simulated physical models and their theoretical emission spectra.
Such microwave sensing of atmospheric temperature profiles and
water vapor abundances can be handled well with linear estimators
because there is an approximately linear relation between the
physical parameters of interest and the observable brightness
temperatures at 15 to 70 GHz, and because the statistics of
atmospheric temperature and humidity are approximately jointly
Gaussian.

More sophisticated nonlinear techniques are now being
developed to handle these basic temperature and humidity retrieval
problems; some are discussed in this workshop by Rosenkranz and
Baumann, Toldalagi, and Hogg *et al*. (2-4). This paper reviews
several other applications of passive microwave techniques, each
characterized by an equation of radiative transfer which is
nonlinear or even unknown; the associated retrieval techniques
are also nonlinear.

Four such examples are discussed here: (1) direct sensing
of wind velocities by means of microwave Doppler-shift measurements, (2) retrieval of atmospheric water vapor profiles using
the opaque band in the 183-GHz water vapor resonance,
(3) retrieval of snow accumulation rate by means of a quasi-
theoretical, quasi-empirical technique, and (4) classification
of polar terrain by means of pattern recognition techniques.

II.   ANALYTIC PROCEDURES: WIND VELOCITY MEASUREMENTS

Wind velocity is a very important meteorological parameter, but it is difficult to sense remotely. One approach is to measure the Doppler shift of an atmospheric emission line; the accuracy of the measurement depends largely on the ratio of the line width to the expected frequency shift.

Waters et al. (unpublished proposal, JPL , 1978) have proposed a limb-scanning satellite experiment which would determine the vector velocity of winds at altitudes of 70 to 100 km by means of the 118-GHz resonance of oxygen. A simple analysis suggests the accuracy. Imagine a two-channel microwave radiometer, one channel centered on each wing (ν_i) of the resonance near the half-intensity point. The two observed brightness temperatures $T_B(\nu_i)$ are unequal in the presence of a Doppler shift $\Delta\nu$ by an amount:

$$\Delta T_B = 2\Delta\nu \left(\frac{\partial T_B(\nu)}{\partial\nu}\right)_{\nu_i} \tag{1}$$

Since the wind velocity v is proportional to $\Delta\nu$, and therefore to ΔT_B, the root-mean-square (rms) accuracy of the wind retrieval is proportional to the receiver sensitivity ΔT_{rms} as well as to the line shape

$$\left(\frac{\partial T_B(\nu_i)}{\partial\nu}\right)_{\nu_i}^{-1}$$

This accuracy is approximately 3 ms^{-1} for a 100 K line 300 kHz wide at 118 GHz, if the receiver sensitivity is $\approx$ 1 K.

Other practical considerations also limit the accuracy of wind determination. Because of the very high velocity of the spacecraft, it is essential to know the antenna pointing angle to $\approx$ 0.01° if Doppler errors less than 0.3 ms^{-1} are to be incurred.

An error in local oscillator frequency of only 100 Hz (one part in 10^9) would also yield a 0.3 ms^{-1} error.

The retrieval procedure is quite simple. First, the frequency of the Doppler-shifted line is determined, and then a sequential pair of such frequency shifts measured for a single air parcel is interpreted analytically as a velocity vector which lies in the plane determined by the air parcel and the two space-craft locations where the Doppler measurements were made. The two lines of sight for these observations would normally be at right angles. More than two frequency bands must normally be observed to ensure success because the Doppler shifts due to wind will generally be on the order of 10 kHz, much less than the spectral line width, and will be several line widths because of the spacecraft velocity. Doppler shifts would be determined by comparison of the observed spectrum with that expected on the basis of spacecraft velocity and geometry alone; the wind effect would be manifest as small linear offsets from the zero-wind spectrum. Thus, simple analytic calculations comprise the bulk of the retrieval procedure.

The reason the 118-GHz oxygen line is attractive for this purpose is (1) the uniformity of the oxygen mixing ratio and excitation level, (2) its high frequency, which increases the Doppler offset, (3) its narrow line width above 70 km and simple Zeeman splitting, and (4) the availability of sensitive receivers at millimeter wavelengths.

III. ITERATIVE PROCEDURES: WATER VAPOR NEAR 183 GHz

The atmospheric water vapor line at 183.310 GHz is opaque at the center, and this results in a highly nonlinear relation between water vapor density and brightness temperature, especially over the ocean. Over the ocean the low brightness temperatures which result from surface reflectivity increase with increasing

water vapor, until the line becomes opaque at an altitude sufficiently high that the brightness starts to diminish. With still
higher humidities, particularly near the very opaque center of
the line, the central line core contributed by the stratosphere
can be seen in emission against the colder temperatures of the
upper troposphere. In such a case the derivative of brightness
temperature with respect to humidity is approximately zero for
two different water vapor abundances; this condition complicates
accurate retrievals.

Linear estimators for water vapor ρ can be written in the
form

$$\hat{\rho} = \sum_i k_i T_B(\nu_i) \tag{2}$$

where T_{B_i} are the observed brightness temperatures at frequencies
ν_i, and k_i is a constant. In general, the k_i are, in part,
proportional to $\partial \rho / \partial T_B(\nu_i)$, which, in the foregoing example, has
two poles as the humidity increases. In such nonlinear situa-
tions linear retrieval techniques are clearly inferior to
nonlinear estimators that conform to the physics involved.

Iterative techniques have long been used for remote sensing
retrievals, as discussed by Fleming (5) in the previous workshop
on inversion methods. Two efforts to use iterative procedures
for 183-GHz water vapor retrievals are those of Schaerer and
Wilheit (6), and Baumann, unpublished.

Schaerer and Wilheit retrieve the water vapor profile by
(1) guessing one, (2) deriving differential humidity weighting
functions G(h) for that profile, such that the change in brightness temperature ΔT_B is

$$\Delta T_B \simeq \int_0^\infty G(h) \ \Delta R(h) \ dh \tag{3}$$

where $\Delta R(h)$ is an incremental change in humidity profile and h
is height, (3) calculating the emission spectrum based on the
guessed profile and comparing this value with the observed data,
(4) correcting the guessed profile on the basis of the differences
between the two brightness spectra at the specific altitudes of
the weighting-function maxima, and (5) iterating until the two
brightness spectra agree within $\simeq 1$ K; four iterations are
generally sufficient. This procedure assumes that the temperature profile is known. The water vapor weighting functions are
generally narrower (~ 6 km) than those for temperature (~ 10 km)
because the width is related to the scale height of water vapor
rather than to that of pressure.

The approach studied by Bauman was a Newtonian iteration
procedure where the $n + 1$ estimate for the parameters $\hat{\underline{p}}_{n+1}$ is

$$\hat{\underline{p}}_{n+1} = \hat{\underline{p}}_n + \left(\underline{\underline{\nabla\hat{T}}}_B^T \; \hat{\underline{T}}_B + \gamma \underline{\underline{H}} \right)^{-1} \underline{\underline{\nabla\hat{T}}}_B^T \left(\underline{T}_B - \hat{\underline{T}}_B (\hat{\underline{p}}_n) \right) \qquad (4)$$

and where $\underline{T}_B$ is the observation vector, $\underline{\underline{\nabla\hat{T}}}_B$ is a Jacobian matrix,
the distance $\nabla\underline{p} = \hat{\underline{p}}_{n+1} - \hat{\underline{p}}_n$ between guesses is determined by
minimizing $\Delta\underline{p}^T \underline{\underline{H}} \Delta p$, the error in the estimate $(\underline{\underline{\nabla\hat{T}}}_B \Delta\underline{p} - \Delta\underline{T}_B)^T$.
$(\underline{\underline{\nabla\hat{T}}}_B \Delta\underline{p} - \Delta\underline{T}_B)$ is held constant, and $\Delta\underline{T}_B$ is defined as
$\underline{T}_B - \hat{\underline{T}}_B (\hat{\underline{p}}_n)$.

To simplify these computations, Baumann has also simplified
the transmittance function as

$$e^{-\left(\int_h^\infty \alpha_{H_2O}(h)\,dh \right)} \simeq e^{-Be^{Ah}} \qquad (5)$$

where A is a function of water-vapor scale height and B is a
function of all the parameters. This is a good approximation if
the water vapor is exponentially distributed in altitude. By
using this procedure he has retrieved exponential water vapor

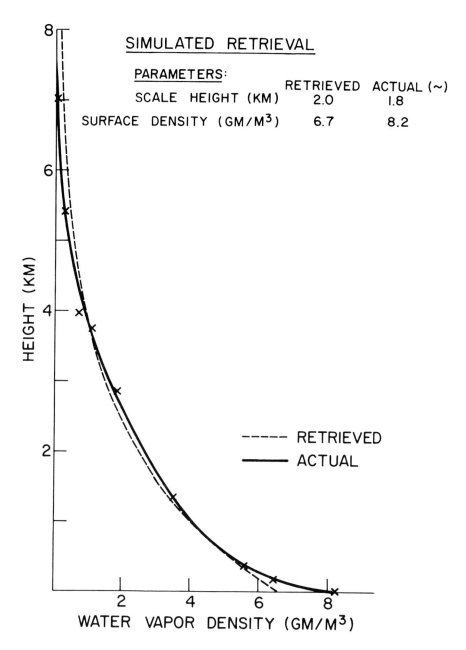

FIGURE 1. *Sample retrieval of the scale height and surface density of an exponential water vapor profile; noiseless observations from space at 140 and 160 GHz were assumed.*

distributions based only on observations at 140 and 160 GHz.
A typical simulated retrieval is shown in Fig. 1 for zero
receiver noise. Receiver sensitivities of ~ 1 K would produce
rms errors in the retrieved water vapor scale heights and
surface densities of ~ 0.3 km and ~ 1 gm/m^3, respectively,
depending upon the particular sounding.

By applying such a procedure to different altitude regions
in the atmosphere, a complete profile extending to the stratosphere could be pieced together. The disadvantages of this
procedure are its assumptions of exponential humidity distributions and transmittance functions, and the advantage is computational simplicity. Conversely the use of more precise piecewise
approximations to complex water vapor profiles would significantly
increase the computational burden.

IV. UNCERTAIN PHYSICS: REGRESSION FOR SNOW ACCUMULATION RATE

Although microwave retrievals of atmospheric parameters are
generally based upon well understood physics, the same is
generally not true of snow, ice, or other terrain. Unfortunately,
to ignore all physics and use only empirical regression techniques is undesirable because there are too many degrees of freedom in the microwave data, in the atmospheric conditions, and in
the range of parameters characterizing terrain. If there are too
many degrees of freedom, then the size of the data set used for
regression must be impractically large. Therefore, it is sometimes best to combine physics and empirical procedures in one
inversion procedure.

One way to reduce the number of degrees of freedom is to
invoke all reasonably well-known elements of the physical models,
and to perform empirical regressions only to determine a few
uncertain constants in those equations. In this way very complex
problems can be simplified to the point where only a small amount

of empirical data is required to produce an acceptable relation between the observations and the associated physical reality.

An example of a procedure which combines both physics and empirical relations is the method developed by Rotman, Fisher, and Staelin (paper A) submitted to the Journal of Glaciology (1979), for the retrieval of snow accumulation rates over Antarctica and Greenland. In this case, purely empirical relations would not be best because the ground truth data are too limited, and pure physics would require arbitrary assumptions about internal wave propagation and scattering in the snow.

The important physical variables are (1) the snow temperature profile $T(z)$, which is assumed to equal a constant T_{10} plus an exponential with depth; (2) the snow accumulation rate A; (3) the absorption coefficient K_a which is assumed to be a classical known function of frequency ν and T_{10} alone; and (4) the scattering cross section, which is assumed to increase linearly with depth and to depend upon A, T_{10}, z, and snow density ρ_o as

$$K_s = C_1 \frac{z\rho_o}{A} \exp(- C_2/T_{10})$$

where C_1 and C_2 are the two unknown constants to be determined empirically. Thus, simple physical assumptions reduce this problem to estimation of two constants.

The constants C_1 and C_2 were determined by adjusting them to produce the minimum mean square difference between (1) the brightness temperatures observed by the Nimbus-6 microwave spectrometer experiment (7; Rotman et al., op cit.); and (2) those temperatures predicted on the basis of the ground-truth in situ measurements of T_{10} and A for most of Antarctica.

Once the constants are determined, there is a simple relation between the desired A, the assumed value of T_{10}, and the brightness temperature observed at the chosen frequency.

The observed Nimbus-6 brightness temperatures at 31 GHz were used
to obtain the retrieved snow accumulation rate,map for Antarctica
shown in Fig. 2. The agreement between this map and the associated ground truth map is very good except near 72°S, 150°E, where
the ground truth contours tended to be parallel to the coast
rather than to lines of constant latitude. The same retrieval
algorithm worked as well or better in Greenland, although no such
ground truth data was employed in establishing the retrieval
algorithm.

V.   UNCERTAIN PHYSICS: CLUSTERING TECHNIQUES FOR POLAR TERRAIN

     In certain situations, the radiative physics is sufficiently
unknown that not even the quasi-physical techniques of the
previous section are warranted. This is probably the case for
sea ice when various ice ages, temperatures, snow depths, brine
contents, and remelting histories are involved; there are too
many unknowns to be retrieved from such a small set of measurements, and appropriate models for radiative transfer in
partially remelted sea ice, for example, have not been well
tested.

     In such complex situations it is often useful merely to
classify the received data into categories, the physical
significance of which may then become more apparent. One
method is cluster analysis, which separates the observed data
vectors into groups; the Euclidean difference between the
vectors in a single group "cluster" is small.

     Such a technique was applied by Rotman, Fisher, and Staelin
(Paper B, submitted to the Journal of Glaciology, 1979) to the
analysis of polar snow and ice. The data consisted of the 22.2
GHz and 31.6 GHz observations from the SCAMS instrument on
the Nimbus-6 satellite. By accumulating data over 1-week
intervals, each spot in the 70° to 80° latitude band could be
observed at each of seven view angles. In this fashion, 900

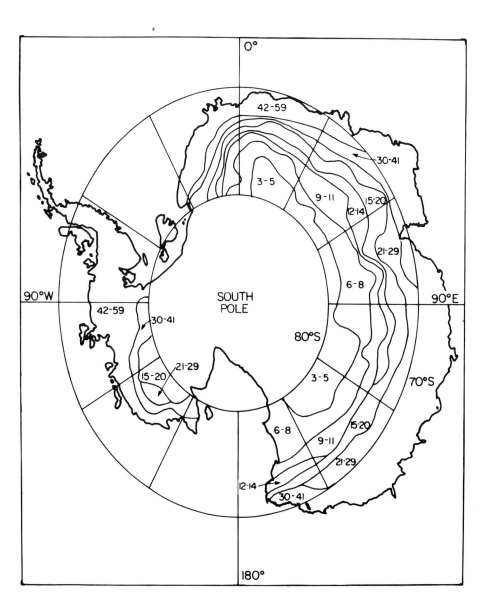

FIGURE 2. Retrievals of snow accumulation rate $(g\ cm^{-2}\ yr^{-1})$ in Antarctica based on 31.6 GHz observations from SCAMS in 1975–1976.

locations in the Arctic and 900 in the Antarctic were observed
in September, November, and January during the period 1975 to
1976.

Each point was characterized by a 14-element data vector
representing observed brightness temperatures at two frequencies
and seven angles. The seven zenith angles at the terrestrial
surface were: $0°$, $8.4°$, $16.9°$, $25.6°$, $34.4°$, $43.5°$, and $53.3°$.

A discussion of clustering techniques has been given by
Nagy (8). Rotman et al. used an agglomerative hierarchical
approach in which the clustering process was initiated by
treating each vector as a one-member cluster. The Euclidean
distance between each possible pair of vectors in the data set
was computed and stored as a similarity measure. Then the
similarity threshold for which two clusters were combined was
increased monotonically to form new clusters, until all the
vectors in the data set were combined into one cluster. All
the clusters at a given intermediate threshold could be
reproduced later. Many options exist for computing the
distances between newly formed clusters. The technique used
was complete-link clustering, in which the similarity between
any two clusters was defined as the greatest Euclidean distance
between any of the 14-dimensional vectors in each cluster.
In single-link clustering the least distances would be used
(9).

Such unsupervised clustering techniques have the advantage
of providing "objective" indications of the information imbedded
in multi-dimensional data sets, which otherwise are difficult
to study. The disadvantage of such techniques is that there is
no guarantee that the resulting clusters will have useful
physical significance. In the study of Rotman et al. (paper B)
the physical significance of the clusters is often apparent,
with classifications of open ocean, new sea ice, old sea ice,

firn, etc., generally being clear. Gradations of multi-
year ice are also discernible.

In Fig. 3 are plotted the results of classifying the Arctic
into 14 clusters for November 1975. The clusters are generally
associated with the age of old sea ice, the accumulation rate
of firn over Greenland, the character of new sea ice, etc.

Another important use for clustering is the identification
of significant geophysical signatures. Such analysis of the
Arctic data makes it clear, for example, that new sea ice mixed
with open water can readily be distinguished from old sea ice
on the basis of the angular dependence alone, as well as upon
the two-frequency signatures at nadir. Thus, clustering
techniques can be used in their own right, or merely as a
prelude to more complete understanding of a data set and
development of appropriate physical models.

SYMBOLS

A	snow accumulation rate (g cm^{-2} yr^{-1})
A	adjustable parameter
B	adjustable parameter
$G(h)$	water vapor weighting function
h	height
K_a	attenuation coefficient in snow
k_i	constant
$\hat{\underline{p}}_j$	jth iterated estimate of the parameter vector $\underline{p}$
$T_B(\nu)$	brightness temperature (K)
$\underline{T}_B$	brightness temperature observation vector
$\hat{\underline{T}}_B$	estimated brightness temperature vector
T_{10}	snow temperature at 10-meter depth
z	depth in snow
α_{H_2O}	water vapor absorption coefficient (nepers m^{-1})

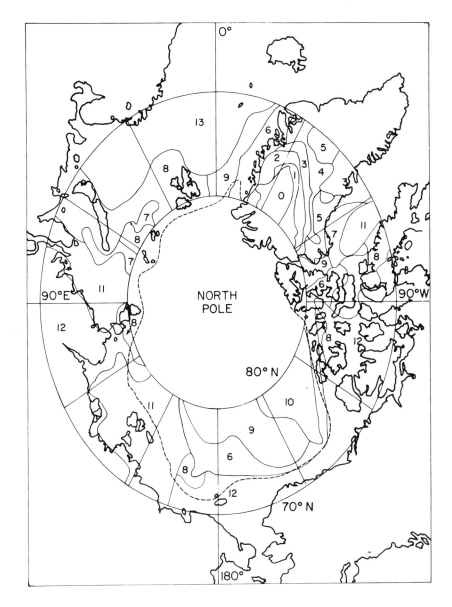

FIGURE 3.   Cluster analysis of 14-element microwave data
vectors obtained by SCAMS in November 1975.   The cluster numbers
are sequenced arbitrarily and the dashed line represents the
nominal multi-year sea ice limit.

γ adjustable parameter

$\Delta\nu$ Doppler shift (Hz)

$\Delta\underline{p}$ change in parameter vector $\underline{p}$

$\Delta R(h)$ incremental change in water vapor profile

ΔT_B change in brightness temperature

$\underline{\underline{\Delta\hat{T}}}_B$ Jacobian matrix for estimated brightness temperature

ρ_o snow density

ACKNOWLEDGMENT

The author is indebted to W. T. Baumann, S. R. Rotman, J. W. Waters, and T. T. Wilheit for discussions of recent progress prior to publication.

REFERENCES

1. Staelin, D. H., *in* "Inversion Methods in Atmospheric Remote Sounding" (A. Deepak, ed.), p. 361. Academic Press, New York (1977).

2. Rosenkranz, P. W., and Baumann, W. T., Inversion of Multiwavelength Radiometer Measurements by Three-Dimensional Filtering *in* "Remote Sensing of Oceans and Atmospheres" (A. Deepak, ed.). Academic Press, New York (1980).

3. Toldalagi, P. M., Some Adaptive Filtering Techniques Applied to the Passive Remote Sensing Problem *in* "Remote Sensing of Oceans and Atmospheres" (A. Deepak, ed.). Academic Press, New York (1980).

4. Hogg, D. C., Little, C. G., Decker, M. T., Guiraud, F. O.,
    Strauch, R. G., and Westwater, E. R., Design of a Ground-
    Based Remote Sensing System Using Radio Wavelengths to
    Profile Lower Atmospheric Winds, Temperature, and Humidity
    in "Remote Sensing of Oceans and Atmospheres" (A. Deepak,
    ed.). Academic Press, New York (1980).

5. Fleming, H. E., *in* "Inversion Methods in Atmospheric
    Remote Sounding" (A. Deepak, ed.), p. 325. Academic Press,
    New York (1977).

6. Schaerer, G., and Wilheit, T. T., *Radio Science, 14,* 371
    (1979).

7. Staelin, D. H., Rosenkranz, P. W., Barath, F. T., Johnston,
    E. J., and Waters, J. W., *Science, 197,* 991 (1977).

8. Nagy, G., *Proc. I.E.E.E. 56,* 836 (1968).

9. Anderberg, M. R., "Cluster Analysis for Applications,"
    Academic Press, New York (1973).

DISCUSSION

Deepak: Could you tell us how you could distinguish the old ice from the new ice? Is there a change in the refractive index?

Staelin: Let's talk about the evolution of sea ice. New ice is typically frozen with a lot of brine pockets in it, so that its salt content is fairly high. When snow falls on that new ice, the brine is immediately concentrated by capillary action in that snow layer just above the ice. This can produce a very good black body for microwaves. Now as the ice ages through tempera- ture cycling, that salt is concentrated, drainage channels form, and the highly saline solution drains out. So over a period of time the salt concentration in the ice diminishes leaving behind a lacey network of voids and snow. That low-loss medium scatters very well and is strongly frequency dependent. Thus old ice has a cooler appearance because it scatters well the cold microwave radiation from space; the brightness is strongly frequency dependent, typically being coldest at wavelengths near 1 cm.

Deepak: How do you use your multiangle scattering measurements to distinguish the two?

Staelin: The angular dependence effect is primarily used to distinguish the fraction that is open water.

Westwater: One of the ways to handle a nonlinear problem in a linear fashion is to use higher order moments of the brightness temperatures as predictors. Did you compare this method in the water vapor retrievals versus the iterative method that you were discussing?

Staelin: No. No quantitative comparisons of that type have been made for the opaque water line to my knowledge. I might point out, by the way, that the opaque water line problem we have in the 183 GHz line is very similar to the problem faced, for example, with the 6.7 μm water band in the infrared, which is also an opaque water problem.

Green: A technical question. You showed a lot of substances which needed different frequencies. Do you have the technology that could scan the frequencies?

Staelin: The instrument that is being proposed by Dr. J. Waters and his colleagues envisions a multichannel system, as I recall, with several local oscillators and then several filter banks-- one following each I.F. strip; so it is actually several spectrom- eters--one devoted to each line. Recent advances in the state- of-the-art in microwave technology is making that ambitious system possible.

Chahine: What is the accuracy of your retrieved water vapor--
not the total amount but the rms distribution?

Staelin: Neither of the groups looking at the 183 GHz line
retrievals have yet come up with numerical estimates of the
accuracy. One might guess very roughly that the accuracy would
be perhaps 10 or 20 percent.

Chahine: You refer to a Wilheit nonlinear inversion method.
Has it been published?

Staelin: He has a preprint of his method. I have not seen it
in print yet. [1]I understand that it is to be published in
Radio Science.

Keafer: You referred to limb scanning and the line-of-sight
problem. I wonder whether you could tell us what the vertical
resolution or field of view was?

Staelin: In the Waters' proposal, the resolution depended on
which frequency was used because it was a diffraction-limited
antenna. I think the vertical resolution of that beam width for
the configuration he proposed ranged from perhaps 2 to 8 km
depending on the frequency. I have a report here which provides
a more exact answer to that question.

Abbas: A question about microwave limb scanning measurements of
Joe Waters. Could you tell me what is the range over which the
ClO measurements are proposed and also the range for the wind
velocity measurements?

Staelin: The wind measurements covered the altitude range 70 to
110 km and the ClO measurements were 20 to 40 km. One can
change the instrument specifications--use different resonances,
different antennas, and different techniques, and expand the
range of altitudes which could be monitored--but this is what he
and his colleagues have proposed for this instrument.

[1] Schaerer, G., and Wilheit, T. T., *Radio Science, 14,* 371 (1979).

INVERSION OF MULTIWAVELENGTH RADIOMETER
MEASUREMENTS BY THREE-DIMENSIONAL
FILTERING[1]

P. W. Rosenkranz and W. T. Baumann

Research Laboratory of Electronics
Massachusetts Institute of Technology
Cambridge, Massachusetts

Remote sensing data from satellites typically have three
dimensions: scan position, spacecraft position, and wavelength.
Inversion of the radiometric data to infer geophysical parameters
is a filtering problem in which the dimension of wavelength (or
channel number) is transformed into a dimension of geophysical
parameters, and the most general solution is a three-dimensional
filter. Linear filters have the advantages of computational
speed and easily described transfer functions; but often the
measurements are nonlinear functions of the parameters to be
inferred. To the extent that the nonlinear inversion problem is
overdetermined, it can be modeled by a critically determined
linear problem. As an example, inversion of Scanning
Multichannel Microwave Radiometer (SMMR) data by means of a
three-dimensional Wiener Filter is described. Atmospheric water
vapor content, rain liquid water content, surface wind speed and
surface temperature are the parameters inferred from the measurements. Nonprecipitating liquid water and water vapor scale
height are also modeled but not retrieved. The a priori
statistics on which the filter is trained have the effect of
governing the selection of a trade-off point of noise as a
function of resolution (in all three retrieval dimensions).

[1]This research was supported by NASA contract NAS5-22929

I.   INTRODUCTION

It has long been recognized (e.g., Ref. 1) that atmospheric remote sounding belongs to the class of inverse filtering
problems. Attention has been given to the retrieval of geophysical parameters from radiometric measurements at several
wavelengths; in the case of a temperature profile, this
corresponds to inverse filtering in the vertical direction.
Real measurements also are averaged by a kernel in the two
horizontal directions. Because of corruption of the measurements by noise, a general approach to geophysical parameter
estimation suggests use of a three-dimensional inverse filter.
Some steps in this direction have been made (2,3).

Numerical filters can be classified by structure, e.g.,
linear or nonlinear, shift-invariant, recursive, etc., and also
by the design criterion, e.g., maximum *a posteriori* probability,
minimum expected square error, etc. Linear filters almost
always impose less of a computational burden than nonlinear
filters, and their characteristics are more susceptible to
analysis. In this paper linear shift-invariant minimum-mean-
square error filtering, which derives from the classical work
by Wiener will be discussed (4).

What happens when a linear filter is applied to data from
a nonlinear process? Consider a one-dimensional, one-parameter
problem. If there is only one observable, then the full impact
of the nonlinearity is transmitted into the estimate of the
parameter. If there are two observables, however, b_1 and b_2, a
linear solution $\hat{p}(b_1, b_2)$ is represented by a plane in (b_1, b_2, p)
space. (See Fig. 1.) The true relation (in the absence of
noise) between the parameter and the two observables is
represented by a curved line in (b_1, b_2, p) space; the addition
of noise to the observables is represented in Fig. 1 by a cloud of

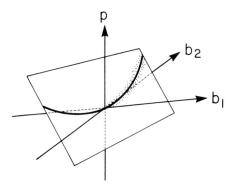

*FIGURE 1. Inferring one parameter from two observables,
which are related to the parameter nonlinearly.*

dots around the curve. Let us assume that b_1 and b_2 can be
approximated by distinct quadratic functions of p over the range
of interest; then this curve is a parabola. A linear least-
square error solution plane will be oriented so as to minimize
the root mean square distance in the p direction of the dots
from the plane. Since a plane can contain a parabola, the
solution curve in the limit of high signal-to-noise ratio will
approach the exact inverse solution. In the presence of noise,
the solution plane is reoriented to compromise between errors
due to nonlinearity and errors due to noise.

The approach for implementation of a spatial filter on
satellite data is to take the two-dimensional Fourier transform
of radiometric images from each channel, perform a matrix
multiplication in transform space to obtain estimates of
parameters, and then inverse Fourier transform. This approach
permits utilization of the computational advantages of a Fast
Fourier Transform algorithm and also simplifies the job of
working with images of unequal resolution.

II. DESCRIPTION OF THE INSTRUMENT AND DATA FORMAT

As a specific example, inversion of Scanning Multichannel
Microwave Radiometer (SMMR) measurements will be considered.
This instrument measures thermal emission from the earth in
vertical and horizontal polarization at frequencies of 6.6,
10.7, 18, 21 and 37 GHz (5). The antenna is an offset-fed
paraboloid with projected aperture diameter of 79 cm. The
reflector rotates sinusoidally about an axis perpendicular to
the earth's surface, with a peak-to-peak travel of 50°. The
antenna beam points at 42° from nadir (48.8° earth incidence)
so an image of the earth's surface is constructed from conical
scans. The data that have been used were measured by the
instrument on SeaSat and were calibrated and averaged by the
Jet Propulsion Laboratory into a cell format. (See Table I.)
This format represents a 595-km wide swath on the surface as a
mosaic of square cells. Because of the wide range of wavelengths, the minimum cell size differs from channel to channel.

TABLE I. Data Format

Frequency (GHz)	Wavelength (cm)	Cell size (km)	No. of cells across swath, N/2
6.6	4.55	149	4
10.7	2.81	85	7
18.0	1.67	54	11
21.0	1.43	54	11
37.0	0.81	27	22

The interim data set has been corrected only for contributions from sidelobes that fall off the earth, not for those that
intercept the earth. The antenna temperatures measured by the
instrument are mixtures of vertical and horizontal polarizations
due to two effects: one is the rotation of the antenna
reflector with respect to its feed, and the other is the cross-
polarized component of the antenna gain. The first of these
was compensated in the JPL data processing; for the second an
additional correction was made.

$$T'_{AV \atop H} = \frac{1}{2}\left[T_{AV} + T_{AH} \pm \frac{T_{AV} - T_{AH}}{1 - 2\alpha}\right] \qquad (1)$$

where T_{AV} and T_{AH} are antenna temperatures from the interim data
set, and T'_{AV} and T'_{AH} are the corrected antenna temperatures.
A value of $\alpha = 0.05$ (at all frequencies) was derived empirically
from data measured in the vicinity of ocean weather station
Papa on SeaSat orbit 1178.

Three-dimensional arrays are defined: a--antenna temperatures, b--brightness temperatures, e--measurement errors, and
g--antenna gains. The first index of these arrays denotes
channel number, the second index denotes position of the cell
from left to right across the swath (x-direction), and the
third index denotes position in the y-direction, parallel to
the satellite velocity vector. In the case of a and b, baseline
values are subtracted, e.g.

$$a = T'_A - T_o - \left(\frac{\partial T_o}{\partial \theta}\right)(\theta - 50°) \qquad (2)$$

as listed in Table II. The antenna temperature baseline values
were established empirically from orbit 1178 data so as to
correspond to the geophysical parameter baseline values in

Table II. Baseline Values of Antenna Temperatures or
Brightness Temperatures

Channel	T_o (K)	$\left(\dfrac{\partial T_o}{\partial \theta}\right)_{\theta = 50°}$ (K/deg)
6.6 V	155	2.03
6.6 H	83 (left), 81, 79, 77 (right)	-1.11
10.7 V	162	2.05
10.7 H	91	-1.10
18.0 V	173	2.09
18.0 H	105	-0.81
21.0 V	195	2.05
21.0 H	139	-0.12
37.0 V	200	2.01
37.0 H	140	-0.47

TABLE III. Baseline Values of Parameters

Parameter	Baseline value
Surface T	15 °C
Water vapor	20 kg/m^2
Rain water	0 kg/m^2
Nonprecipitating liquid water	0 kg/m^2
Vapor scale height	2 km
Surface wind	5 m/s

Table III. The reason for the variation of the 6.6 GHz horizontal value across the swath is at present unknown.

III. SYSTEM MODEL

The measurement process will be modeled as

$$a_{jmn} = \frac{Q}{N_j^2} \sum_{m',n'} b_{jm'n'} \; g_{j,m'-m,n'-n} + e_{jmn} \tag{3}$$

where Q is the total area of the image and N_j^2 is the number of cells in it. The indices m and n are considered to be cyclical or modulo N_j. The gain g is normalized so that

$$\frac{Q}{N_j^2} \sum_{m,n} g_{jmn} = 1 \tag{4}$$

which results from the fact that the antenna temperatures have previously been corrected for sidelobes external to the earth. However, Eq. (3) does not accurately describe the far sidelobes of the antenna even on the earth, since they receive emission from the earth at a different angle and possibly a different polarization than the main lobe. This fact perhaps accounts for the rather high value of α obtained. Furthermore, the swath does not cover all of the area seen by the antenna sidelobes, so one may anticipate a particularly inaccurate retrieval at the edges. In order to reduce edge effects, the swath width was doubled by adjoining it with its reversed image. In the y-direction, 1190 km of data was taken, twice the swath width. Thus, the image operated on is square with $Q = (1190 \text{ km})^2$ and has sharp transitions at the top and bottom but not the sides.
    A three-dimensional array of parameters defining the state of the geophysical system will now be defined. The first index has ten values: these are the six parameters minus their

baseline values listed in Table III, augmented by the squares of
the first four. The second and third indices correspond to the
x and y directions. Then the relation between the brightness
temperatures and the parameters is modeled by

$$b_{jmn} = \sum_k K_{jk}\, p_{kmn} \qquad\qquad (5)$$

The coefficients K_{jk} are listed in Table IV. These coefficients
were obtained from theoretical calculations based on Refs. 6-11.
Brightness temperatures were computed from a model with U.S.
Standard Atmosphere temperature profile, an exponential water
vapor profile, a cloud layer extending from 0.25 km to 2.25 km,
and rain from the surface to 2.25 km. The rain was assumed to
have a Marshall-Palmer distribution of drop sizes and scattering
was ignored; only absorption was considered in the radiative
transfer equation (10). Equation (5) contains second-order
terms in surface temperature, water vapor, rain, and nonprecipitating liquid, but no cross-product terms. One may therefore anticipate errors in situations of high atmospheric
opacity, e.g., rain $\gtrsim 0.4$ kg/m^2.

IV. INVERSION

Application of a linear filter to the measurements a_{jmn} is
desired to infer parameters $\hat{p}_{kmn}$. Such a filter is defined by
coefficients d_{kjmn} where

$$\hat{p}_{kmn} = \sum_{j,m',n'} \frac{Q}{N_j^2} d_{kjm'n'}\, a_{j,m-m',n-n'} \qquad\qquad (6)$$

Since the continuous case has been discussed in Ref. 12, the
discussion here will be brief. The expected square error in
the estimate $\hat{p}$ is minimized by requiring the errors to be
uncorrelated with the measurements. (If the error contained

TABLE IV. Observation Matrix, K

		Parameter								
	T	T^2	W	V	V^2	R	R^2	L	L^2	H
6.6 V	0.520	0.0051	0.63	0.030	0.0000	4.0	5.3	3.0	0.0	-0.07
6.6 H	0.260	0.0031	1.09	0.045	0.0000	6.1	8.1	4.7	-0.1	-0.10
10.7 V	0.400	0.0093	0.70	0.084	0.0000	12.9	15.6	7.7	-0.3	-0.20
10.7 H	0.180	0.0059	1.24	0.132	0.0000	20.0	24.6	11.9	-0.4	-0.29
18.0 V	0.160	0.0122	0.69	0.464	-0.0009	39.5	25.3	18.5	-1.5	-0.97
18.0 H	0.020	0.0084	1.44	0.755	-0.0014	63.8	40.9	29.9	-2.4	-1.45
21.0 V	0.110	0.0104	0.68	1.202	-0.0098	44.7	19.4	19.6	-2.1	-0.45
21.0 H	-0.005	0.0073	1.47	2.007	-0.0161	73.6	32.1	32.3	-3.4	-0.15
37.0 V	-0.130	0.0109	0.51	0.540	-0.0019	135.1	-72.9	51.2	-13.9	-1.32
37.0 H	-0.210	0.0093	1.40	0.986	-0.0033	244.1	-129.4	92.8	-24.9	-2.15

Frequency and polarization

a component correlated with the measurements, then its variance
could be reduced by subtracting the correlated component.)

$$< (\hat{p}_{km'n'} - p_{km'n'}) \; a_{j,m'-m,n'-n} > = 0 \qquad (7)$$

The angle brackets denote expectation values of the quantity
inside. Since the search is for a shift-invariant filter, it
must be assumed that the statistics of the variables are
stationary. This assumption implies that the lefthand side of
Eq. (7) is a function of j, k, m, and n only. Substitution of
Eq. (6) gives

$$\sum_{\ell,m'',n''} \frac{Q}{N_j^2} d_{k\ell m''n''} \; <a_{\ell,m'-m'',n'-n''} \; a_{j,m'-m,n'-n}>$$

$$= <p_{km'n'} \; a_{j,m'-m,n'-n}> \qquad (8)$$

Since Eq. (8) is a discrete convolution in two dimensions,
a two-dimensional discrete Fourier series will be used to solve
it. For example, the Fourier components of a_{jmn} are

$$A_{j\mu\nu} = \frac{Q}{N_j^2} \sum_{m=0,n=0}^{N_j - 1} a_{jmn} \; e^{- i\frac{2\pi}{N_j} [(m + \frac{1}{2})\mu + (n + \frac{1}{2})\nu]} \qquad (9)$$

The terms of $\frac{1}{2}$ in the exponent are introduced in order to put
the origins of all the images in the same place. Spectral power
densities of the parameters are also defined

$$S_{pk\ell\mu\nu} = \frac{Q}{N_k^2} \sum_{m,n} <p_{km'n'} \; p_{\ell,m'-m,n'-n}> e^{-i\frac{2\pi}{N_k}(m\mu + n\nu)} \qquad (10)$$

and spectral power densities of the measurement errors,
$S_{ek\ell\mu\nu}$; these are analogous to Eq. (10) with p replaced by e.
Substituting Eqs. (3) and (5) in Eq. (8), transforming into μ, ν
space and solving for the Fourier coefficients of d_{kjmn} yields

$$D_{kj\mu\nu} = \sum_{u',\ell} S_{pk\ell\mu\nu} \, K_{j'\ell} \, G_{j'\mu\nu}$$

$$\left[\sum_{k'\ell'} G_{j'\mu\nu} \, K_{j'k'} \, S_{pk'\ell'\mu\nu} \, K_{j\ell'} \, G_{i\mu\nu} + S_{ej'j\mu\nu} \right]^{-1} \tag{11}$$

where the inverse is an ordinary matrix inverse with respect to indices j and j', and

$$G_{j\mu\nu} = \frac{\Omega}{N_j^2} \sum_{m,n} g_{jmn} \, e^{-i \frac{2\pi}{N_j}(m\mu + n\nu)} \tag{12}$$

The antenna gains are modeled as symmetric functions and the parameter cross-correlations as zero, so G, S_p and D will be real arrays.

The parameter estimate Fourier coefficients are obtained by

$$\hat{P}_{k\mu\nu} = \sum_j D_{kj\mu\nu} \, A_{j\mu\nu} \tag{13}$$

Note that Fourier transformation preserves the x-direction symmetry of the doubled image; also, sign reversal of the indices μ and ν conjugates the transform quantities inasmuch as a and $\hat{P}$ are real. These symmetries together with the cyclical properties of the discrete Fourier series imply that

$$\hat{P}_{k,N_k-\mu,\nu} = \hat{P}_{k\mu\nu} \tag{14a}$$

$$\hat{P}_{k,\mu,N_k-\nu} = \hat{P}^*_{k\mu\nu} \tag{14b}$$

$$\hat{P}_{k,N_k-\mu,N_k-\nu} = \hat{P}^*_{k\mu\nu} \tag{14c}$$

Consequently, $\hat{P}$ needs to be computed in only one quandrant of the μ,ν plane, i.e., for $0 \le \mu < N_k/2$ and $0 < \nu \le N_k/2$. Inverse transformation of $\hat{P}$ gives the parameter estimates

$$\hat{P}_{kmn} = \frac{1}{Q} \sum_{\mu=0,\nu=0}^{N_k-1} \hat{P}_{k\mu\nu} \; e^{i \frac{2\pi}{N_k}(m\mu + n\nu)} \tag{15}$$

V. ANTENNA GAIN

An analytic function is used to model the antenna gain in transform space

$$G_j''(u,v) = \exp\{-[8.31 \cdot 10^{-4} r^2 (\psi_j/\psi_{37})^2 (u^2 + v^2/\cos^2\theta)]^{0.866}\} \tag{16}$$

where u and v are (continuous) spatial frequencies corresponding to x and y; r = 1129 km is the distance from the satellite to the center of the beam spot on the earth; ψ_j/ψ_{37} is the ratio of antenna beamwidth to the beamwidth at 37 GHz, as listed in Table V ; and $\cos^2\theta$ = 0.435 because the earth is viewed obliquely.

Figures 2 and 3 show the inverse transform of Eq. (16) plotted along with measured antenna patterns. Although the constants in Eq. (16) were fitted to the measured gain at 37 GHz, it fits the 6.6 GHz pattern fairly well also. The detailed sidelobe structure is not reproduced by Eq. (16) but the differences correspond to high spatial frequencies which are not too important since their signal-to-noise ratio is low. Equation (16) represents the instantaneous gain function; however, the data are integrated for a certain time period

TABLE V. Beamwidth

Frequency (GHz)	6.60	10.70	18.00	21.00	37
ψ_j/ψ_{37}	4.91	3.17	1.96	1.61	1

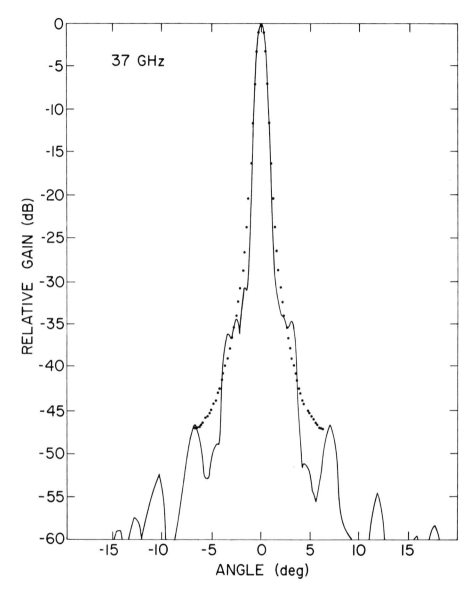

FIGURE 2. Measured antenna gain at 37 GHz (solid line) and
analytic model (dots). The measurements were made in a plane
perpendicular to the axis of symmetry of the reflector. In the
orthogonal plane, the sidelobe pattern shows less symmetry.

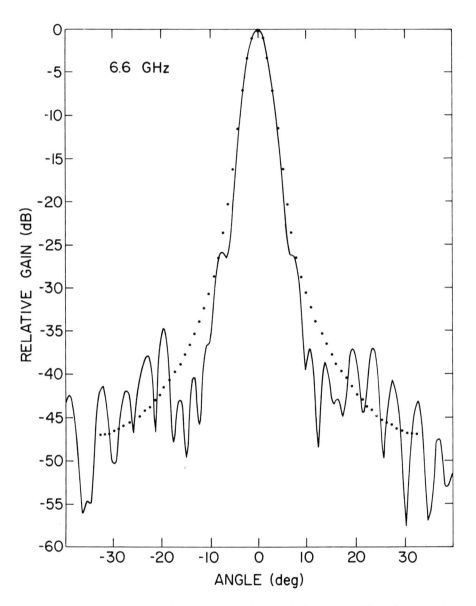

FIGURE 3.   Measured antenna gain at 6.6 GHz (solid line) and
analytic model (dots).   The measurements were made in a plane
perpendicular to the axis of symmetry of the reflector.   In the
orthogonal plane, the sidelobe pattern shows less symmetry.

while the antenna is scanning, which smears the antenna pattern in the x direction and makes its transform narrower in u. For convenience, the smearing is accounted for by arbitrarily making the gain circularly symmetric. The measured antenna temperatures are again averaged into square cells; this is accounted for by considering the antenna pattern to be convolved with a square pulse in x and y. Then the gain function transform becomes

$$G_j'(u,v) = \exp\{-[2443(\psi_j/\psi_{37})^2(u^2 + v^2)]^{0.866}\}$$

$$\frac{\sin(\pi u \Delta x)}{\pi u \Delta x} \frac{\sin(\pi v \Delta y)}{\pi v \Delta y} \qquad (17)$$

where $\Delta x = \Delta y$ is the cell size.

The discrete antenna temperatures operated on constitute a sampled version, at increments Δx and Δy, of a continuous field. The discrete gain function Fourier coefficients defined by Eq. (12) are not equal to the values of Eq. (17) at the corresponding values of $u = \mu \Delta u$ and $v = \nu \Delta V$ for the reason that the fields are aliased by the sampling; i.e., in transform space the function given by Eq. (17) is replicated at intervals of $1/\Delta x$ and $1/\Delta y$. In order to write an analytic expression for the discrete Fourier coefficients (Eq. (12)), there should be an infinite sum of terms of the form given by the righthand side of Eq. (17). However, since the exponential factor decreases very rapidly, only four terms will be used.

$$G_{j\mu\nu} = G_j'(u,v) + G_j'(1/\Delta x - u,v) + G_j'(u, 1/\Delta y - v)$$

$$+ G_j'(1/\Delta x - u, 1/\Delta y - v) \qquad (18)$$

where

$$u = \mu \Delta u$$
$$v = \nu \Delta v$$
$$\Delta u = \Delta v = (N_j\Delta x)^{-1} = Q^{-\frac{1}{2}} = 8.4 \cdot 10^{-4} \text{ cycles/km}$$

Recall that Eq. (18) is to be evaluated only for $0 \leq \mu \leq N_j/2$
and $0 \leq \nu \leq N_j/2$.

VI. SPECTRAL POWER DENSITY FUNCTIONS

The remaining undefined factors in Eq. (11) are S_e and S_p, spectral power densities for the measurement errors and the geophysical parameters. If a radiometer has a 1-second noise variance $_1\sigma_j^2$ then the noise variance when averaged in cell format varies across the scan because the scan angle is a sinusoidal function of time, but near the center it is

$$\sigma_j^2 = 1.6 \cdot 2 \cdot \frac{_1\sigma_j^2}{T} \left(\frac{N_j}{2} \right)^2$$

The factor 1.6 accounts for the fact that less integration time is spent in the middle than at the ends of the scan; the factor of 2 arises because the polarization of the radiometers is switched at the ends of the scan so a given polarization is measured only half the time; and $T = 90$ s is the time required to measure a square block containing $(N_j/2)^2$ cells. For all channels $_1\sigma_j = 0.3$ K was used and the variation of noise with scan angle was ignored, and the cell noise factors in Table VI result.

The spectral power density of measurement noise is then

$$S_{ej\ell\mu\nu} = \delta_{j\ell} \frac{Q}{N_j^2} \sigma_j^2 = 0.8 \, \delta_{j\ell} \, Q \, _1\sigma_j^2/T = 1162 \ (Kkm)^2 \qquad (20)$$

TABLE VI. Cell Noise Factors

Frequency (GHz)	6.6	10.7	18.0	21.0	37.0
σ_j (K)	0.3	0.5	0.7	0.7	1.3

where $\delta_{j\ell}$ is the Kronecker delta. Calibration errors are not considered here because empirically determined baseline antenna temperature values have been used (see Table II).

For the parameter spectral power densities, a function used by Ogura was adopted which has the characteristic of approaching an isotropic turbulence spectrum in the limit of high spatial frequency

$$S_{pk\ell\mu\nu} = \frac{4}{3}\pi\ \delta_{k\ell}\ L_k^2\ <p_k^2>\left[(2\pi L_k)^2(u^2+v^2)+1\right]^{-4/3} \tag{21}$$

where L_k is a characteristic length at which the autocorrelation function drops to 0.26 times its value at zero lag, which is $<p_k^2>$ (13).[2] The values used for these coefficients are listed in Table VII. These values are believed to represent the "worst case" in the sense of as much high-frequency spatial variation of the parameters as one might reasonably expect to find in an image. The effect on the estimated parameters of changing these values is discussed.

One could add aliasing terms to the righthand side of Eq. (21) as in Eq. (18), but there would be little point in doing so since the actual spectra may be different from Eq. (21). This fact is not as ominous as it may seem, because in Eq. (11) S_p occurs in both the numerator and denominator: in the high signal-to-noise ratio regime (where S_e is negligible) the filter coefficients are insensitive to S_p. The need for S_p is primarily to determine the spatial frequency regime for which signal-to-noise ratio is low; these frequencies will be filtered out of the estimate.

---

[2] *The spectrum given by Ogura is in one-dimensional wave number form.*

TABLE VII. Parameter Statistical Coefficients and System
Transfer Characteristics.

Parameter	$<p^2>^{\frac{1}{2}}$	L (km)	RMS noise	Resolution (km)	H_{kk00}
Surface					
temperature	5 °C	120	1.100	88	0.99
Wind speed	5 m/s	30	1.300	46	0.98
Water vapor	15 kg/m^2	30	1.200	51	0.95
Rain water	0.5 kg/m^2	15	0.019	29	0.85
Nonprecipitating					
liquid water	0.5 kg/m^2	15	0.026	32	0.17
Vapor scale					
height	0.5 km^2	30	0.052	118	0.10
$(T - 15)^2$	43.0	120	18.000	169	0.73
$(V - 20)^2$	390.0	30	73.000	85	0.42
R^2	0.43	15	0.028	43	0.97
L^2	0.43	15	0.006	35	0.01

VII. SYSTEM TRANSFER MATRIX

When D has been evaluated, the remote sensing system is
described by the block diagram in the upper half of Fig. 4,
which represents Eqs. (3), (5) and (6). The array $D_{kj\mu\nu}$ is
plotted in Fig. 5 with $\nu = 0$. (Only six of the ten rows (index
k) of D are plotted.) Since Eq. (18) defines $G_{j\mu\nu}$ only for both
μ and $\nu \leq N_j/2$, G is set to zero for higher spatial frequencies;
therefore, $D_{kj\mu\nu} = 0$ at these frequencies also.

The system can also be represented by a single block, as
in the lower half of Fig. 4 where the Fourier coefficients
of the filter are given by

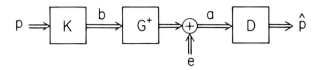

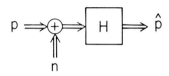

FIGURE 4.   Two equivalent representations of the remote sensing system (12).   In the upper diagram, e represents white noise and in the lower diagram n represents colored noise.

$$H_{k\ell\mu\nu} = \sum_j D_{kj\mu\nu} \, G_{j\mu\nu} \, K_{j\ell} \qquad (22)$$

The array $H_{k\ell\mu\nu}$ describes the deterministic (i.e., nonnoise) part of the response $\hat{P}_{k\mu\nu}$ to an input $P_{\ell\mu\nu}$. In Fig. 6, a normalized, i.e., dimensionless, version of H is plotted

$$H'_{k\ell\mu\nu} = [<p_\ell^2>/<p_k^2>]^{\frac{1}{2}} \, H_{k\ell\mu\nu} \qquad (23)$$

Each row of small plots represents the normalized response of a parameter to inputs of the ten parameters. For example, the upper left plot shows the transfer function for temperature as a first-order input to temperature as output. However, it will be recalled that the nonlinearities in the real system being observed are modeled by including dummy parameters such as $(T - 15°C)^2$, denoted by T2. The dummy parameters are, of course, not independent of the real parameters, although (in Eq. (21)) no linear correlation between them has been included. The reason for not doing so is that a statistical correlation between two parameters tends to increase the amount of crosstalk, i.e., off-diagonal entries in Fig. 6. Crosstalk in the model system between

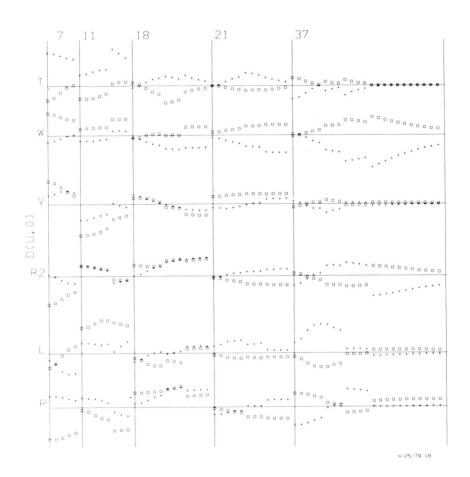

FIGURE 5. Determination coefficients in the v = 0 plane.
Plus signs denote vertical channels, squares denote horizontal
channels. The horizontal scale corresponds to spatial frequency,
with each vertical line representing zero frequency. The
vertical scale differs for each parameter. Maximum values of
the D coefficients in each row are: T, 2.4 °C/K; W, 1.3 m/sK;
V, 2.3 kg/m^{2}K; R2, 0.028 kg^2/m^{4}K; L, 0.042 kg/m^{2}K; R.
0.016 kg/m^{2}K.

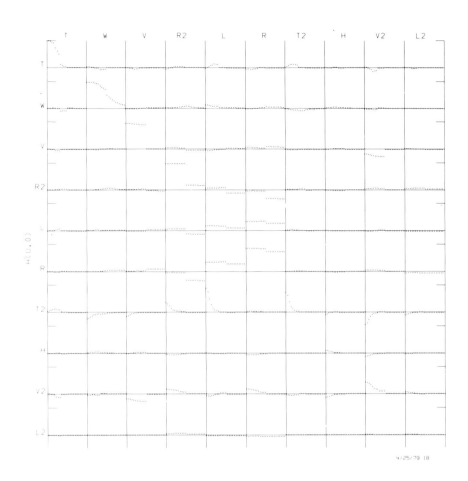

FIGURE 6. Normalized system transfer coefficients in the v = 0 plane. Letters along the top denote input (actual) parameters; letters along the left side denote output (estimated) parameters. Continuous lines are drawn across the figure for H' = 0 and tic marks on either side at H' = 1. Maximum spatial frequency in each small graph is 0.018 cycles/km.

a dummy and a real parameter corresponds to a residual non-linearity in the retrieval.

For the most part, the off-diagonal entries in Fig. 6 are fairly small. The main exceptions are in the crosstalk among nonprecipitating liquid water (L), rain (R) and rain squared (R2). In fact, the crosstalk contribution to nonprecipitating liquid water is greater than the diagonal contribution, whose transfer function is less than 0.2 at all spatial frequencies. It is concluded from these results that a retrieved value of nonprecipitating liquid would be useless. A similar conclusion is reached concerning water-vapor scale height.

This conclusion about nonprecipitating liquid water contradicts results from previous satellite experiments (14,15). The difference apparently results from the fact that these experiments did not attempt to make a distinction between precipitating and nonprecipitating liquid and, in fact, did not include the former in simulations. The absorption spectra of rain and non-precipitating liquid droplets is very similar in shape, as one may judge from their first-order coefficients in Table IV, but rain is more effective as an absorber. In this paper, equal *a priori* variances of rain and nonprecipitating liquid are assumed, with the result that rain is used more than non-precipitating liquid to explain the variance of the observables (the antenna temperatures).

Although the transfer function array H is a complete description of the deterministic aspects of the system, it is convenient for purposes of discussion and comparison to characterize the response in the two spatial dimensions by a single number for each parameter: the resolution. One measure of resolution of an observing system is the integral of the impulse response divided by its peak height. The result is an area, of which the square root may be taken to express resolution in km. (In the present system, resolution is the

same in the x- and y-directions.)   If the impulse response
array is $h_{k\ell mn}$, then for parameter k,

$$\text{Resolution} = \left[\frac{Q}{N_j^2} \sum_{m,n} h_{kkmn} / h_{kk00} \right]^{\frac{1}{2}}$$

$$= \left[Q \, H_{kk00} / \sum_{\mu,\nu} H_{kk\mu\nu} \right]^{\frac{1}{2}} \tag{24}$$

In the case of a boxcar transfer function, the resolution as
defined is the spacing of the coarsest grid that could represent
the impulse response without loss of information.   The resolution
for each parameter is listed in Table VII.   This is a small-
signal resolution, since the second-order response is not
considered to contribute to the first-order resolution.

The nondeterministic part of the system response is due to
the propagation of instrument noise through the D array

$$\text{rms noise} = \left[\sum_{j,\mu,\nu} D_{kj\mu\nu}^2 \, S_{ejj\mu\nu} \right]^{\frac{1}{2}} \tag{25}$$

The total error in the estimated value of a parameter has,
under the assumption that the system has been correctly described,
three components: error due to noise, error due to finite
resolution, and crosstalk.   The last can also be described as
finite resolution in the parameter dimension.   As the a priori
statistics of the parameters change, the relative contributions
of these three components to the total error are found to change.
This behavior of the remote sensing system is very complicated
since there are ten parameters whose statistics are to be
specified and 100 transfer functions (diagonal and crosstalk)
which are affected.   However, it is found that the resolution of
a parameter is much more strongly dependent on its own statistics
than on the statistics of other parameters.

As examples, in Fig. 7 trade-off curves of noise against resolution for sea surface temperature and wind are plotted with correlation length of the parameters as the independent variable.

In computing these curves, the correlation length of the second-order term in sea surface temperature was varied along with the first-order term. Also, as correlation length was increased, the associated parameter rms was decreased in order to keep the spectral power density at $u = v = 0$ constant. (See Eq. (21).)

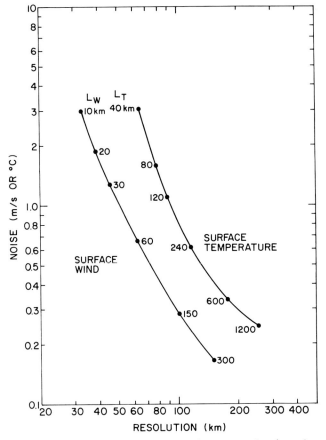

FIGURE 7. *Trade-off of noise against resolution for sea surface temperature and wind as correlation lengths (L_T, L_W) of the parameters vary.*

Thus, the trade-off curves represent the effect of changing the high-frequency response of the system only. These curves exhibit inelasticity of resolution in the sense that although the correlation lengths of the *a priori* statistics change by a factor of 30, the resolution of the remote sensing system changes by a factor of 4 or 5. Corresponding curves for water vapor and rain are even less elastic.

VIII. RESULTS

A. *Sea Surface Temperature*

Figure 8 compares a sea surface temperature retrieval in the Gulf of Alaska with an analysis of conventional data prepared by the SeaSat workshop (16,17). The overall north-to-south increase in temperature is reproduced in both. The righthand and lefthand sides of the retrieval image are colder than the center. This may be an edge effect due to the absence of data from outside the image, or it may be an inadequacy of the linear correction for instrument bias which was introduced into the baseline of the 6.6 GHz data (see Table II). (The first line at the bottom of the retrieval images is not plotted, since it corresponds to the points where the inverse Fourier transform wraps around.) The retrieval contours exhibit a noise level approximately consistent with Table VII. The comparison conventional data has been smoothed.

B. *Surface Wind Speed*

The wind speed retrieval and workshop analysis of conventional surface measurements are plotted in Fig. 9 for the same orbit as Fig. 8. This orbit had the highest winds observed in the region. Some disagreement between the two occurs at higher wind speeds, where the retrieved wind is consistently a few meters per second less than the conventional analysis.

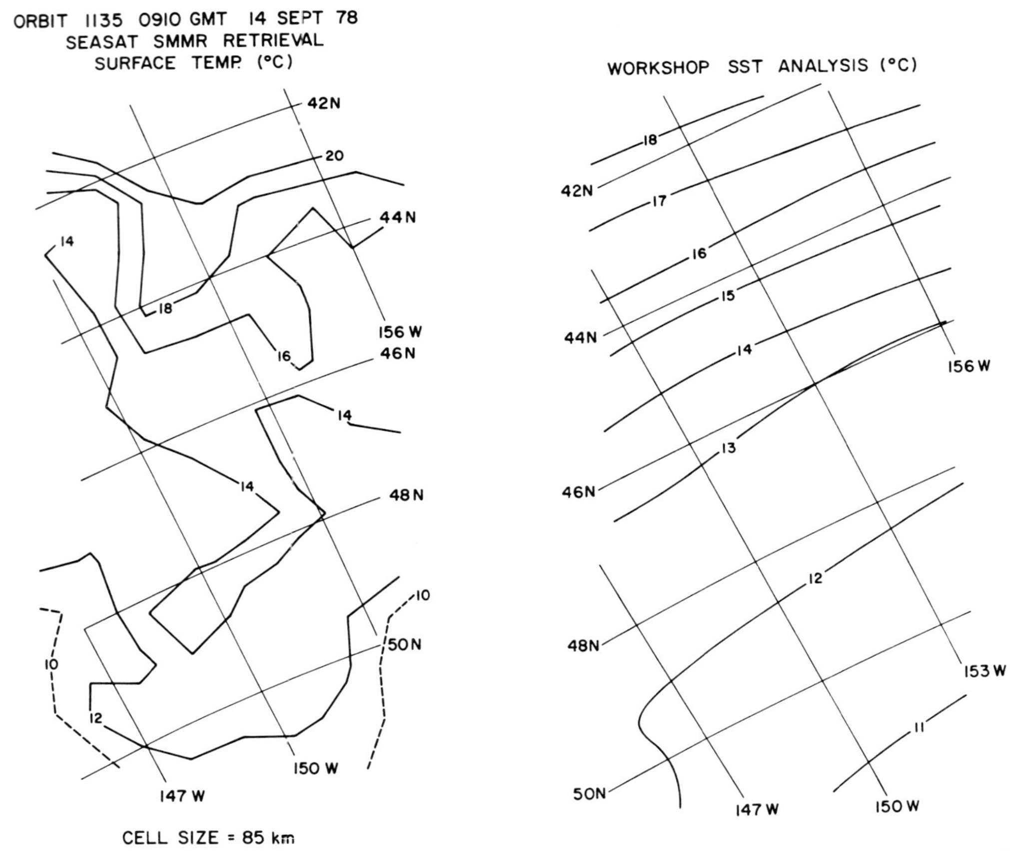

ORBIT 1135 0910 GMT 14 SEPT 78
SEASAT SMMR RETRIEVAL
SURFACE TEMP. (°C)

WORKSHOP SST ANALYSIS (°C)

CELL SIZE = 85 km

Figure 8. Estimated sea surface temperature (left). Conventional analysis from Ref. 16 (right).

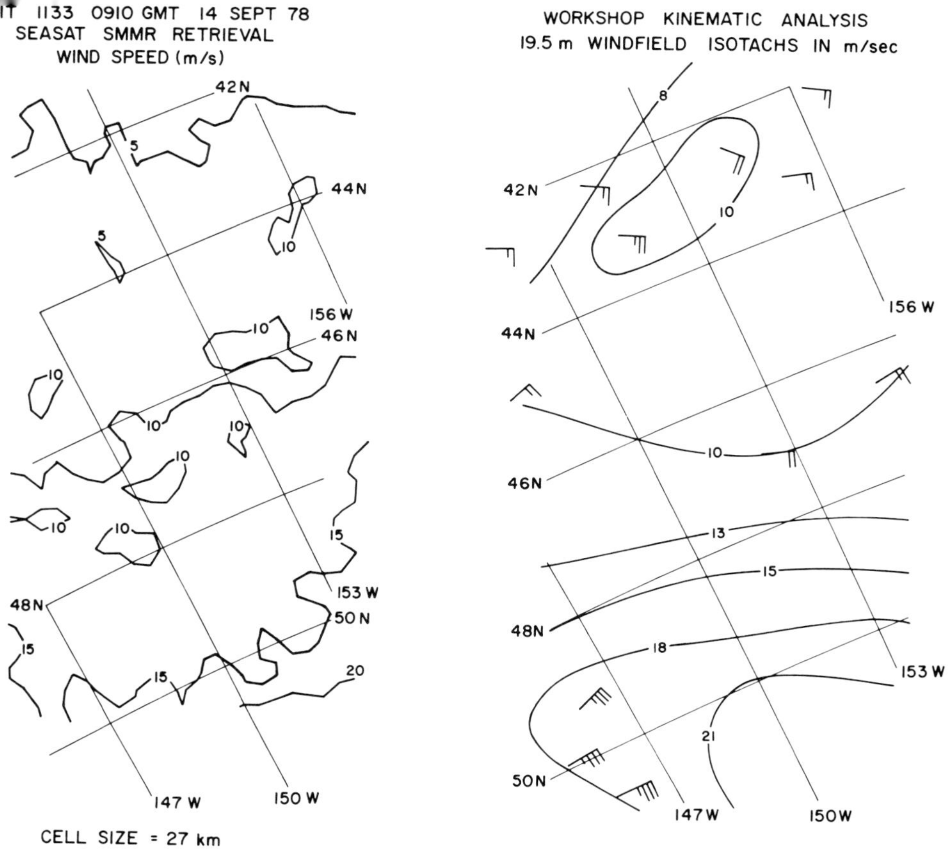
ORBIT 1133 0910 GMT 14 SEPT 78
SEASAT SMMR RETRIEVAL
WIND SPEED (m/s)
42N
44N
46N
48N
50N
156W
153W
150W
147W
5
10
15
20
CELL SIZE = 27 km
WORKSHOP KINEMATIC ANALYSIS
19.5 m WINDFIELD ISOTACHS IN m/sec
42N
44N
46N
48N
50N
156W
153W
150W
147W
8
10
13
15
18
21

Figure 9. Estimated wind speed (left). Kinematic analysis of surface reports from Ref. 16 (right). Surface reports are shown by arrows (each full length feather = 10 knots = 5.1 m/s).

C. Water Vapor and Rain

Figure 10 shows contours of water vapor and rain from a
frontal system (not the same orbit as Figs. 8 and 9). If the
thickness of the rain layer is known or assumed, mean rain
water density can be determined from rain water content. Rain
rate can then be estimated by use of a statistical formula (10).
For example, by assuming a rain height of 2.25 km and the
Marshall-Palmer relation between rain rate and density, a rain
content of 0.6 kg/m^2 corresponds to a rain rate of 4 mm/hr. No
surface measurements of rain are available.

Some crosstalk from rain to the other parameters, especially
water vapor, is noticeable. The retrieved value of water vapor
tends to dip at the peaks of rain cells. This crosstalk is not
surprising since the amount of rain here exceeds the amounts for
which the coefficients in Table IV were computed; higher order
nonlinearities are probably effective here. Furthermore, the
brightness temperature spectrum of rain depends on the drop size
distribution, which is not constant.

Only three radiosonde measurements of water vapor were
available (in addition to the one used to establish the baseline) for verification. These three are listed in Table VIII.
Of these three orbits, 1212 gave very questionable results,
such as negative values of rain. This particular section of
data, unlike the others, was very close to land. The 6.6 GHz

TABLE VIII. Integrated Water Vapor

Orbit	Radiosonde (kg/m^2)	SMMR (kg/m^2)
1135	13	12
1212	30	41
1292	19	25

4. Wiener, N., "Extrapolation, Interpolation, and Smoothing of
    Stationary Time Series, with Engineering Applications."
    M.I.T. Press, Cambridge, Massachusetts (1949).

5. Gloersen, P., and Barath, F. T., *IEEE J. Oceanic Eng.*
    OE-2, 172 (1977).

6. Klein, L. A., and Swift, C. T., *IEEE Trans. Antenn. Propag.*
    AP-25, 104 (1977).

7. Webster, W. J., Wilheit, T. T., Ross, D. B., and Gloersen,
    P., *J. Geophys. Res. 81,* 3095 (1976).

8. Waters, J. W., *in* "Methods of Experimental Physics, vol. 12:
    Astrophysics, Part B" (M. L. Meeks, ed.), p. 142.   Academic
    Press, New York (1976).

9. Liebe, H. J., Gimmestad, G. G., and Hopponen, J. D., *IEEE
    Trans. Antenn. Propag. AP-25,* 327 (1977).

10. Savage, R. C., *J. App. Meteor. 17,* 904 (1978).

11. Goldstein, H., *in* "Propagation of Short Radio Waves"
    (D. E. Kerr, ed.), p. 671.   Dover Pub., New York (1965).

12. Rosenkranz, P. W., *Radio Science, 13,* 1003 (1978).

13. Ogura, Y., *J. Meteor. Soc. Japan, 31,* 355 (1953).

14. Staelin, D. H., Kunzi, K. F., Pettyjohn, R. L., Poon,
    R. K. L., and Wilcox, R. W., *J. App. Meteor. 15,* 1204
    (1976).

15. Rosenkranz, P. W., Staelin, D. H., and Grody, N. C.,
    J. Geophys. Res. 83, 1857 (1978).

16. Born, G. H., Lame, D. B., and Wilkerson, J. C. (eds.),
    "SeaSat Gulf of Alaska Workshop Report" (2 vols.),
    Report 622-101, Jet Propulsion Laboratory, Pasadena,
    California (1979).

17. Wilkerson, J. C., Brown, R. A., Cardone, V. J., Coons, R. E.,
    Loomis, A. A., Overland, J. E., Peteherych, S., Pierson,
    W. J., Woiceshyn, P. M., and Wurtele, M. G., *Science,*
    204, 1408 (1979).

18. Libes, R. G., Bernstein, R. L., Cordone, V. J., Katsaros,
     K. B., Njoku, E. J., Riley, A. L., Ross, D. B., Swift, C. T.,
     and Wentz, F. J., *Science, 204,* 1415 (1979).

DISCUSSION

Staelin: This is interesting. These are the first satellite
data retrievals for some of these parameters.

Twitty: Is this a three-dimensional deconvolution you are
talking about?

Rosenkranz: It is a deconvolution in two dimensions and
inversion in the third dimension, which is the channel number.

Twitty: So the two dimensions are "x" and "y" and you did a
reflection in one dimension?

Rosenkranz: Yes, in x-dimension. Maybe I can explain this
with the Figure D-1. Start with an observed image composed of
two squares and you may have some stuff like rain in one of
them and we replace the rectangular image with the square
image (which consists of the observed image plus reflected
image). So that the function wraps around on itself smoothly
in these two directions. We could also have done it in the
other direction.

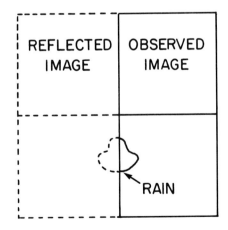

Fig. D-1

Twitty: But normal procedures were to extend the function in some way so that it was continuous, so that you did not get aliasing problems at the edges of the sample space that you are in. If the sharp edge is dropping off on any of those corners you are going to get aliasing and you will get sin x/x in the transform which you can see in your result. It is artificial.

Rosenkranz: Yes, I did not do it. There is inevitably some aliasing. If you do a discrete Fourier series, it is a complete representation of whatever you started with. I should say first that the beam spot on the earth is actually elliptical instantaneously because we are looking at an angle, but then it is smeared because the beam is moving so it is primarily circular. So, we replace this with the function that is symmetrical with respect to interchange of U and V, multiplied by sin x/x and sin y/y because the data has been averaged into cells which effectively convolve it with the square box car function in two dimensions. It also replicates the gain function it aliases it at points where the grid repeats itself in frequency space. So the discrete transform would actually have an infinite number of terms each of which has this function. You know, but I use four terms because the farther away you get, the less significant all this is. The data is also aliased. But the amount of it depends on the actual spectral distribution of the data which is not necessarily the same as what we have assumed in the simulation for the spectral part. Let me just say that the different frequencies is different and therefore, in the Fourier series the N depends on frequency. But if the fields are sufficiently smoothly varying so that the grid is an accurate representation, then it goes over into the limit of band limited functions in the frequency domain which are representable on a finite grid.

Staelin: I would like to recommend that we continue these discussions in the open discussion session. There is a lot in this paper, and I do not think any small number of questions can focus on some of these interesting things, such as, the issue of the horizontal resolution versus accuracy.

DESIGN OF A GROUND-BASED REMOTE SENSING SYSTEM USING RADIO WAVELENGTHS TO PROFILE LOWER ATMOSPHERIC WINDS, TEMPERATURE, AND HUMIDITY

D. C. Hogg
C. G. Little
M. T. Decker

F. O. Guiraud
R. G. Strauch
E. R. Westwater

Environmental Research Laboratories
National Oceanic and Atmospheric Administration
Boulder, Colorado

An essential requirement of a regional (mesoscale) observing and forecasting service is continuous monitoring of temperature, moisture, and winds. To meet these needs, the Wave Propagation Laboratory, NOAA, is constructing a six-channel microwave radiometer (two moisture channels and four temperature channels) and a UHF radar for wind profiling. The radar also aids in the retrieval of temperature profiles by measuring the altitude of significant features of the profile; e.g., the base of and elevated inversion, the height of a nocturnal inversion, or the height of the tropopause. Moisture, wind and temperature sounding capabilities are analyzed and results are presented of field experiments and of computer simulations. Retrieval algorithms to incorporate active and passive measurements are presented. Finally, improvement in profile retrieval using combined ground-based and satellite soundings is discussed.

I.   INTRODUCTION

Atmospheric observations form the essential base for almost
all atmospheric research and services. Since the atmosphere is
a highly variable three-dimensional fluid, these observations
must be obtained in all three spatial dimensions. Ideally, such
data sets should be continuous in both space and time; in
practice, this has not been possible. Because existing
observational systems use *in situ* instruments carried aloft by
balloons, the National Weather Service has had to accept a
compromise in which the data sets are neither continuous in time,
nor continuous in space, but instead are taken once every 12
hours at stations spaced roughly 350 km apart across the United
States. This system provides observations of upper air conditions that are suitable for identification and forecasting of
synoptic scale phenomena such as cyclones and anti-cyclones
(which have lifetimes of days and dimensions of 1000 km or more),
but is not adequate for the observation and prediction of
smaller scale, shorter lived phenomena such as thunderstorms,
flash floods, etc. Other disadvantages of the existing system
are that the profiles obtained are not usually vertical, and
that significant manpower is required, totaling approximately
4 manhours per profile.

Thus, the present upper air radiosonde network does not
adequately meet present operational and research needs,
especially on the shorter time and space scales. Nor does it
appear at this time that satellites will soon be able to provide
the necessary data sets, primarily because of their inability
to provide the desired spatial resolutions, and to take
observations below thick clouds.

This paper discusses an alternative approach, in which
ground-based active (radar) and passive microwave (radiometric)
systems are combined to provide profiles of wind, temperature,

and, to a lesser extent, humidity, continuously in time, in an unmanned mode. This mode of operation is possible largely as a consequence of the use of solid-state electronics and the lack of mechanical moving parts in the system design. The value of integrating the measurements from such a system with data from satellite microwave radiometers is discussed in the last section. Background information influencing design of the pro-filer is discussed in the first section, and the specifics of the design in the second section.

II. THE PROFILER SUB-SYSTEMS

A. *Precipitable Water Vapor and Integrated Cloud Liquid*

The amount of water vapor integrated along a path through the atmosphere is quite variable. The amount is relevant not only to meteorology and weather forecasting but also to millimeter-wave astronomy, and to the determination of phase shifts caused by changes in refractivity for long baseline radio interferometry and metrology. In the clear air a measure of precipitable water vapor can be obtained by observing brightness temperatures with a single-frequency microwave radiometer operating on the peak of a vapor absorption line, for example, at 22.2 GHz or 183 GHz. But these are pressure-broadened lines, and since the pressure and amount of vapor vary with height, errors arise in the derived vapor. However, there are frequencies displaced from the peak of the line that are more suitable, for example 20.6 GHz rather than 22.2 GHz, because the absorption (and therefore the derived vapor) is relatively independent of pressure (1).

Precipitable water vapor and integrated cloud liquid are measured more accurately by dual-frequency rather than single-frequency radiometry, especially when the system operates in the presence of liquid-bearing clouds (2). (Ice clouds have

negligible effect because the absorption and, therefore, the
radiation by ice, is insignificant.) Using frequencies of
20.6 GHz and 31.6 GHz in a zenith-pointing mode, the precipitable
water vapor (V) and integrated liquid (L) are written in linear
relationship to the respective measured brightness temperatures
(in K) T_2 and T_3

$$V = a_1 + a_2 T_2 + a_3 T_3 \text{ cm}$$

$$L = b_1 + b_2 T_2 + b_3 T_3 \text{ cm} \tag{1}$$

where the a and b terms are constants, derived by statistical
inversion, for a given climate.[1] The 20.6 GHz brightness is
most sensitive to vapor, and the 31.6 GHz, to cloud liquid. The
constants, a_1 and b_1, primarily offset the effect of oxygen
absorption and the cosmic background radiation.

Equations (1) for the climatology of Denver, Colorado, for
example, are

$$V = -0.19 + 0.12 T_2 - 0.056 T_3 \text{ cm},$$

$$L = -0.018 - 0.0011 T_2 + 0.0028 T_3 \text{ cm} \tag{2}$$

A dual-channel system has been in operation for more than
6 months at a Weather Service Forecast Office (WSFO) radiosonde
launch site, and considerable experience on accuracy and
reliability has been gained. A typical analog plot of total

---

[1]*Although a linear relationship between precipitable water
vapor and brightness temperature is used here, it is known that
a linear relationship with absorption is more accurate; however,
even that is not strictly correct because the absorption has
been found to increase quadratically with precipitable water
vapor at 31 GHz (3).*

precipitable water vapor above the Denver, Colorado, WSFO, taken over a 2-week period, is shown in Fig. 1. The solid curve is the radiometrically measured vapor and the triangles are total vapor from integrated radiosonde data, which are obtained at 12-hour intervals. The agreement is quite good; of the differences that do exist on individual readings, it is not known precisely how much is due to the radiometric system and how much error stems from the radiosonde measurements. It is clear, however, that during certain 12-hour intervals (e.g., 8 August p.m. to 9 August a.m.) considerable amounts of vapor passed overhead unobserved by the sondes.

Further verification of performance is obtained by plotting precipitable water vapor amounts measured by the radiometer and the radiosonde directly against one another as in Fig. 2. In this 6-month data sample, the root-mean-square (rms) difference

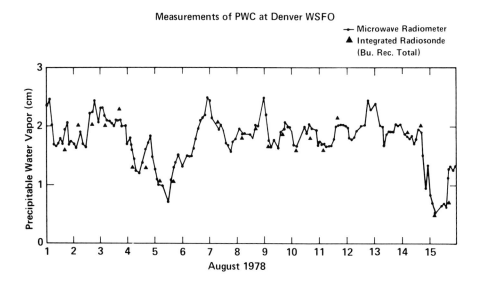

FIGURE 1. *Time series of radiometrically retrieved and radiosonde- measured precipitable water vapor at Denver, Colorado.*

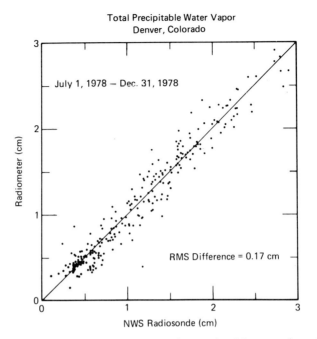

FIGURE 2. *Scatter plot of radiometrically retrieved and*
radiosonde-measured precipitable water vapor at Denver,
Colorado.

between the two is 1.7 mm, some part of this being attributable
to each measuring device. Although establishment of the error
in operational radiosonde measurements is difficult, estimates
by experienced operators and scientists, for typical mid-
range humidity and temperature, place

$$\left[\overline{(V_S - V_A)^2} \right]^{1/2}$$

at about 1.5 mm, where V_S is the sonde-measured precipitable
water and V_A the actual value. Thus, by assuming no correlation
between radiosonde and radiometric errors, one obtains

$$\overline{(V - V_A)^2} = 1.7^2 - 1.5^2 = 0.64 \text{ mm}^2$$

or 0.8 mm for the rms error of the radiometric measurement. It
is believed, therefore, that the dual-channel radiometer is a
reliable instrument for providing continuous data on precipitable
water vapor for the weather services.

In Fig. 2 all data have been included for the 6 months,
regardless of weather conditions, for those times when total
precipitable vapor from the radiosonde was available. Many
cases involving clouds are included. The error in vapor that is
associated with a nonprecipitating cloud of given integrated
liquid can be calculated; computations for the 20.6/31.6 GHz
system are shown in Fig. 3 (1). The upper abscissa gives the
31 GHz attenuation corresponding to the integrated cloud liquid

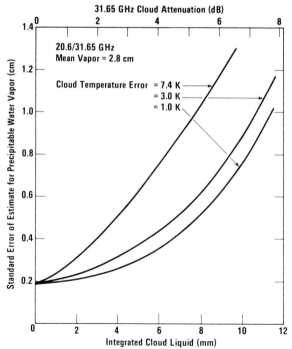

FIGURE 3. *Theoretical error in dual-frequency measurement of
precipitable water vapor as a function of cloud liquid.*

on the lower abscissa. The particular set of curves in Fig. 3,
calculated by assuming 0.5 K rms noise level in the radiometric
system, is carried to cloud liquid of about 10 mm, although most
nonprecipitating clouds observed at Denver do not contain more
than about 3 mm.[2] The three curves of the set correspond to
various errors in knowledge of the temperature of the liquid in
the cloud; the most pessimistic, 7.4 K, is the estimate of
climatological error in the temperature of cloud liquid. In that
case, a cloud producing a line integral of 3 mm of liquid
(31 GHz attenuation of about 2 dB) results in an error of 4 mm
in the vapor, which is about 15 percent of the mean (28 mm)
of the data used for Fig. 3.

   Determination of cloud liquid (L), by using the second of
Eqs. (2), has proven useful in sensing super-cooled liquids in
clouds. This function is of importance in seeding of clouds
during weather-modification experiments and is also valuable to
the local weather forecaster in connection with icing of
aircraft. An example of correlation with icing measured at
Stapleton Airport, Denver, Colorado, on February 15, 1979, is
shown in Fig. 4. As shown in that figure, on the morning of
that day, cloud cover moved in accompanied by a decrease in
surface temperature to well below freezing. Simultaneously,
significant amounts of cloud liquid water were indicated by
the radiometric system. Reports of icing on aircraft in flight
over the airport were then received as indicated on the bottom
trace in Fig. 4.

   Algorithms for retrieving precipitable water vapor and
liquid during rain are at present being developed; the radiometric system operates satisfactorily (remains below saturation)

[2] *1 mm of liquid corresponds to 1 g/m^3 over a cloud
thickness of 1 km.*

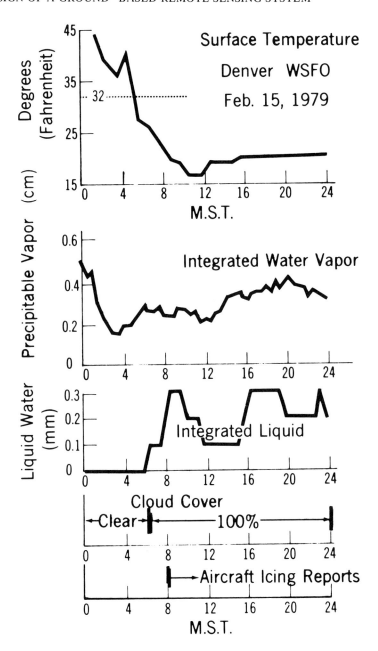

FIGURE 4. Radiometric measurements of precipitable water vapor and cloud liquid during reports of aircraft icing.

up to rain attenuations of at least 6 dB at 30 GHz. This
amount of attenuation is produced by a rain rate of about
10 mm/hr falling from a height of 3 km above the site. For
rain rates and heights exceeding these values, retrieval of the
vapor and liquid is not possible with this system.

   This dual-channel sub-system forms part of a multichannel radiometer for temperature profiling as discussed in
the following section. In that role, it is also used to enable
retrieval of accurate profiles during the presence of nonraining
clouds.

B. Temperature Profiling

   Unlike water vapor, the oxygen content of the atmosphere is
constant, and, therefore, it affords a means of passively
measuring temperature; it has characteristic absorption lines
at millimeter wavelengths. Temperature profiling has been
achieved by operating at three frequencies on the side of an
absorption band near 60 GHz. The theory of ground-based microwave thermal sounding is reviewed by Westwater and Decker (4).

   Microwave temperature profilers have been designed
primarily for a downward-looking mode, from a satellite. But
several instruments have been operated by NOAA and the Jet
Propulsion Laboratory upward looking from the surface (5,6).
In particular, data have been taken at Pt. Mugu, California,
and on the Canadian Ship QUADRA in the Gulf of Alaska for
analysis in an ocean data buoy feasibility study. Statistical
inversion was applied to the brightness temperatures measured
by the radiometers at the various frequencies to retrieve both
temperature and humidity profiles.

   A typical set of measurements using a 5-channel radiometer
at Pt. Mugu, California, were inverted to yield the temperature

profiles on the right-hand plots in Fig. 5.[3] The dashed lines
are the radiometric retrievals; they are compared with radiosonde profiles (the solid lines). Two of the examples were
measured with liquid-bearing clouds in the antenna beam. The
effect of clouds does not degrade the profile since it can be
corrected with simultaneous measurements using the water vapor
and liquid (20 GHz and 30 GHz) radiometers discussed in the
preceding section. An immediate criticism of these retrieved
profiles is that they have smoothed rapid changes in temperature
with height, for example, the inversion about 200 mb above the
surface in the profile of 11:15, 3/11/76 at top center of
Fig. 5. But hybrid methods which obviate this problem are now
under development as discussed later in this section.

There are several ways of presenting data measured radiometrically for comparison with data measured by radiosonde;
here only a variable useful in the forecasting process, pressure
thickness, is presented. Figure 6 shows thicknesses of four
layers which were obtained from data taken at both Pt. Mugu and
the Gulf of Alaska by using a 5-channel radiometer. The
thickness intervals range from 1000 mb to 800 mb level to the
500 mb to 300 mb level. The agreement is excellent for the
lowest interval, rms difference is 5 m, but degrades to 41 m at
the 500 mb to 300 mb level. Although in plotting the
differences the sonde values are taken as reference, it is not
actually known how much of the differences in Fig. 6 are
introduced by errors in the sonde.

[3]*Although approximations to the humidity profiles are also
shown (retrieved) in Fig. 5, it must be emphasized that this
radiometric system is not designed specifically to profile
water vapor.*

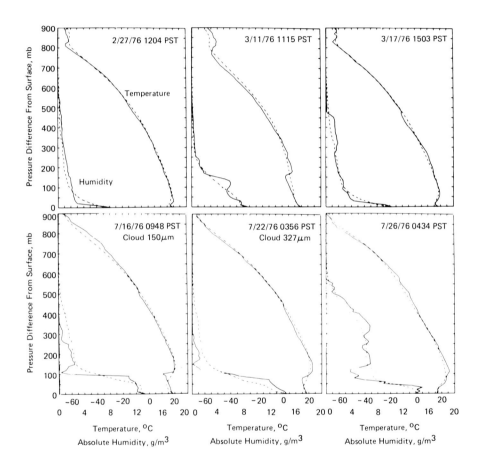

FIGURE 5.   *Examples of retrievals of temperature and water vapor profiles at Pt. Mugu, California; solid lines, rawinsonde; dashed lines, radiometer.*

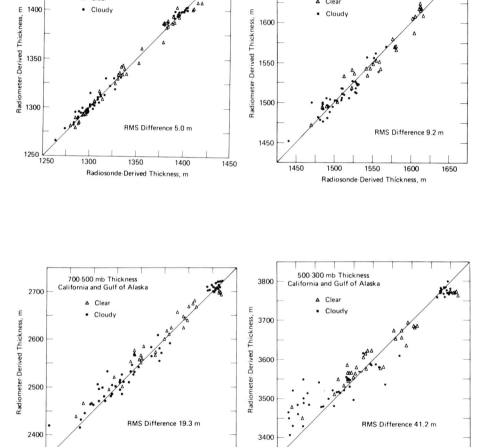

FIGURE 6. Comparison of layer thicknesses as derived from rawinsonde and from radiometer.

Radiometric retrievals of the 500 mb height, using data
from both Pt. Mugu, California, and Gulf of Alaska, are shown
in Fig. 7; the rms difference between radiometric and radiosonde
measurements is about 25 m. Again, the question of how much of
this difference is attributable to the sondes arises. Estimates
have been made that about 10 m could be attributed to instrumental and procedural errors in the sonde measurement of
temperature, and a few meters to measurement of vapor. The
radiometric retrievals shown in Figs. 6 and 7 are subject to
further improvement when refinements such as the addition of
satellite radiation measurements are included (to be discussed
in a subsequent section).

As mentioned earlier, these measurements were made with an
instrument designed for use on a satellite. For ground-based
profiling, the 60-GHz radiometric system will undergo

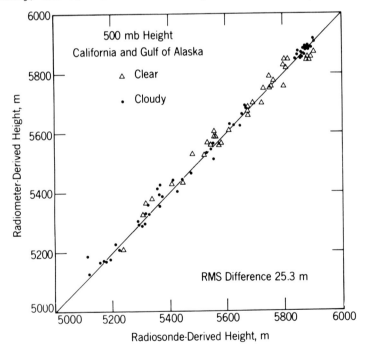

FIGURE 7. *Comparison of 500 mb heights as derived from*
rawinsonde and from radiometer.

considerable redesign as discussed in the next section. In
addition, other types of sounding information can be incorporated
into the profiling system. As discussed below, both radar and
satellite information can supplement the ground-based passive
system.

    As measurements have shown, clear air radar returns from
layers in the atmosphere are frequently associated with the
heights of significant levels, such as elevated thermal
inversions (7). If the heights of a significant level in a
temperature profile can be measured, this information can be
used to improve the accuracy of profile retrieval. A simple
extension of statistical inversion can be used for incorporating
this information. In the method of statistical inversion,
profiles are retrieved from brightness temperature observations
by minimum variance estimation over a representative ensemble,
usually a climatological history, of temperature profiles and
radiance observations. In many cases, the radiance observations
are computer-simulated. Previously, for a given location,
ensembles were chosen strictly on a climatological basis. If,
however, an active sounder can measure an inversion at a
certain altitude level, then this profile can be retrieved
radiometrically by using coefficients derived from an ensemble
consisting of profiles that contain inversion heights within
the appropriate interval of height. The width of the interval
should be somewhat larger than the resolution of the radar that
measures the inversion, and, in addition, should be large
enough to include a sufficient number of profiles in the
ensemble; thus, the retrieval coefficients can be constructed
with a good degree of statistical confidence.

    The improvement in temperature retrieval accuracy by the
addition of knowledge of inversion heights can be evaluated as
follows: If the rms retrieval accuracy for profiles whose
inversions are contained in an interval i is σ_i and P_i is

the frequency of occurence of i, then the total mean square
error σ^2 is

$$\sigma^2 = \Sigma_i P_i \sigma_i^2 \tag{3}$$

An example of the estimated reduction in error is shown in
Fig. 8 for Pt. Mugu, California, during the summer months.   The
simulated radiometer data are at the frequencies of 22.235 GHz,
31.65 GHz, 52.85 GHz, 53.85 GHz, and 55.45 GHz; these are the
frequencies of the Scanning Microwave Spectrometer (SCAMS)
radiometer designed for the Nimbus 6 satellite.   The joint
retrieval process results in an RMS error of less than about
1.5 K to 10 km altitude; at the height of the persistent marine
inversion (just below 1 km), the reduction in error is almost
1°, a 30 percent improvement.

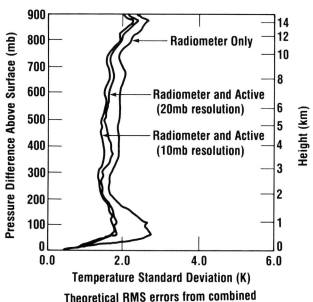

FIGURE 8.   *Theoretical rms errors from combined active-*
passive (radar and radiometric) system.

In addition to this theoretical result on the improvement
in retrieval accuracy by measurement of inversion heights,
field experiments have been recently conducted with both active
and passive sounders. These experiments provide an example of
the use of clear air returns obtained by an FM-CW radar (8)
in improving a temperature profile retrieval. (See Fig. 9.)
These data were obtained at the Boulder Atmospheric Observatory
(BAO) near Boulder, Colorado, on September 6, 1978.   The
radiosonde profile is from a conventional system operated by the
National Center for Atmospheric Research (NCAR) at the BAO
during this experiment. Radiometric data were measured with a
ground-based version of the SCAMS radiometer which was operated
by personnel from the Jet Propulsion Laboratory.   Radiation
measurements used here include zenith radiation at the five

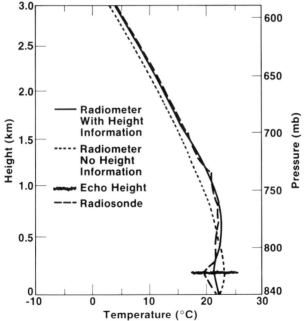

FIGURE 9.   Temperature profile derived by a radiometer with
and without radar-derived height information; data were measured
at the Boulder Atmospheric Observatory, Boulder, Colorado,
9/6/78.

frequencies plus 55.45 GHz measurements at a zenith angle of
59.3°.[4] With the use of these six radiometric measurements and
measurements of surface temperature, pressure, and relative
humidity the profile designated "Radiometer, No Height
Information" of Fig. 9 was obtained. As expected, this profile
smooths the elevated temperature inversion shown in the
radiosonde profile.

At the time of the radiometric measurement discussed above,
a persistent echo was observed at a height of 189 m above the
surface with the 10 cm wavelength FM-CW radar. This information
is now included in the temperature profile retrieval process by
assuming that the radar echo represents the base of an elevated
temperature inversion. The retrieval method discussed above is
applied by deriving statistical coefficients from a set of
a priori profiles that contain elevated temperature inversion
bases between 10 mb and 30 mb pressure difference from the
surface. The resulting profile designated "Radiometer With
Height Information" of Fig. 9 now exhibits the structure of the
radiosonde profile.

Although the improvement in retrieval of temperature profiles
with the combined technique is evident, work remains before this
technique may be applied in an operational mode. It is noted
that echoes from an active sounder, whether radio or acoustic,
may arise from atmospheric properties different from those
assumed here and hence further experience must be gained before
this technique is applied with confidence. The method of
using conditional statistics does, however, show promise for
improving profile retrievals.

[4]*This sixth observation is taken to simulate use of an
additional (58.8 GHz) zenith channel, as discussed subsequently.*

C. Wind Profiling

The vertical profile of horizontal wind can be measured in nearly all weather conditions with a sensitive Doppler radar. In the optically clear atmosphere the radar signals result from scattering from refractive irregularities at scales of half the radar wavelength; scattering from hydrometeors enhances the radar return during precipitation. The scatterers move with the mean wind and the Doppler frequency shift of the back-scattered signal determines their radial velocity. The feasibility of measuring wind profiles has been demonstrated with research radar systems with wavelengths of 0.1 m (micro-waves) to 10 m (VHF). The background information needed to design a radar wind profiler is reviewed in this section.

Microwave meteorological Doppler radars have been used to measure winds in precipitation for more than a decade. The technology has been sufficiently developed so that plans are being made for Doppler capability in the next generation of operational weather radars (9). In the past decade it has also been known that microwave radars can detect scattering from refractive index fluctuations in the optically clear air (10) but even high quality radars used in meteorological research lack the sensitivity needed to utilize clear air scattering for routine wind profile measurements. However, during the past several years very sensitive radars designed primarily to study ionospheric scattering have shown that atmospheric winds can be measured from a few kilometers altitude (where the radar receiver has recovered from the transmitted pulse and short-range clutter) to an altitude greater than 20 km. This result has led to the development of several long wavelength (6 m to 7 m) radar systems for atmospheric research; these systems have shown that routine wind profile measurements to the mid-latitude tropopause (or higher) are feasible (11).

Although microwave meteorological radars have demonstrated
clear-air wind profile measurements in the boundary layer (8,12)
only a few have the sensitivity for routine measurements in
the mid- and upper-atmosphere.

The minimum detectable reflectivity of a radar system can
be expressed as

$$\eta_{min} = \frac{KR^2}{\Delta R} \frac{T_{sys}}{\lambda^{1/2}} \frac{1}{\bar{P}_t A_e} \tag{4}$$

where K is a constant that includes loss terms and signal
processing gain for atmospheric echoes observed with long
averaging times, R is the range, ΔR is the range resolution,
T_{sys} is the radar system noise temperature, λ is the radar
wavelength, $\bar{P}_t$ is the average transmitted power, and A_e is the
effective area of the antenna. The radar reflectivity of
spherical raindrops is given by

$$\eta = \frac{\pi^5}{\lambda^4} \left[\frac{m^2 - 1}{m^2 + 2} \right]^2 \Sigma D^6 \tag{5}$$

where m is the complex dielectric constant of water and D is
the drop diameter which is assumed to be small relative to the
radar wavelength (13). The sum of the 6th power of drop
diameters per unit volume is the so-called radar reflectivity
factor, Z. The Z values typically range from 10^{-4} mm^6/m^3 in
clouds to 10^{+5} mm^6/m^3 in severe storms (-40 dBZ to +50 dBZ).
Radar reflectivity in the optically clear air is given by (14)

$$\eta = 0.38 \, c_n^2 \, \lambda^{-1/3} \tag{6}$$

where c_n^2 is the refractive index structure constant,

$$C_n^2 = \frac{\overline{\{n(x) - n(x + r)\}^2}}{r^{2/3}} \qquad (7)$$

Here, n is the refractive index and r is a separation distance
in a Kolmogorov inertial subrange. The radar detects that
component of the refractive turbulence spatial spectrum whose
scale size is one-half the radar wavelength. Typical values
for C_n^2 range from 10^{-13} $m^{-2/3}$ near the surface to 10^{-18} $m^{-2/3}$
at an altitude of 10 km. Figure 10 shows the radar reflectivity
for precipitation and for refractive turbulence at wavelengths
of 0.1 m to 10 m. Radars designed for clear air detection
also detect particle scattering; for microwaves, particle
scattering starts to dominate clear-air scattering in some
clouds whereas at longer wavelengths clear-air scattering
dominates until there is moderate precipitation. Thus, a radar

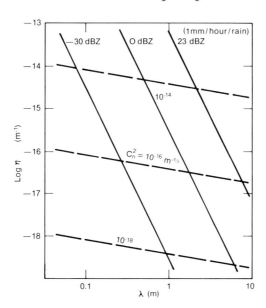

FIGURE 10.   Radar reflectivity (η) for spherical raindrops
(solid lines) and refractive turbulence (dashed lines) at wavelengths considered for wind-profiling radar.

wind profiler is designed to have sufficient sensitivity for
clear air detection, but it must also operate in precipitation
where particle fall speeds of 0.1 m/s to 9 m/s must be accounted
for.

The sensitivity needed for a wind profiling radar can only
be obtained with systems having high average power and/or large
antenna apertures. Figure 11 shows the average power-aperture
product needed to detect refractive turbulence with C_n^2 of

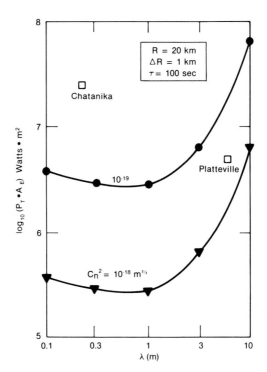

FIGURE 11.   *Average power* (P_T) *--effective antenna aperture*
(A_e) *product needed for radar detection of refractive turbulence*
in the clear atmosphere at a range of 20 km.   Range resolution
is 1 km and total observation time is 100 sec.   Realistic
efficiency factors and system noise temperatures have been
included.   Also indicated are the power-aperture products of
radars that measured the wind profiles in Figs. 12 and 13.

$10^{-18} m^{-2/3}$ and $10^{-19} m^{-2/3}$ at a range of 20 km for radar wavelengths of 0.1 m to 10 m. The range resolution is 1 km and
wavelength independent loss factors have been assumed.
Equations (4) and (6) result in a very weak explicit wavelength
dependence $(\lambda^{-1/6})$ for clear air detection; the large increase
in power aperture at long wavelengths is required to offset
increased galactic noise (15). The transition from a system
noise temperature dominated by receiver noise to one dominated
by cosmic noise occurs at about 1 m. Also shown in Fig. 11
are the power-aperture products of two research radar systems
that have been used for wind measurements. Since both power
and aperture costs increase with decreasing wavelength,
economic considerations weigh against wavelengths much less
than 1 m. The aperture needed to define a given angular
resolution increases as λ^2 and the need for acres of land at
long wavelengths poses economic and logistical problems for
some applications. Another factor in wavelength selection
is the difficulty in obtaining frequency allocations for
unrestricted continental United States operation, especially
at longer wavelengths.

Examples of vertical profiles of the horizontal wind
measured by radar in the optically clear air are shown in
Figs. 12 and 13 (11). The data in Fig. 12 were acquired with
a VAD (Velocity Azimuth Display) scan where the radial velocity
is measured at each resolution height as a function of azimuth
angle for a complete antenna rotation (16). Fourier analysis
of the measured radial velocity yields not only the mean wind
but the shearing and stretching deformation and convergence
(The latter if vertical air motion can be neglected or measured
with a zenith-pointing antenna) (17). In Fig. 12, the radar
data obtained with the 23 cm Chatanika radar facility near
Fairbanks, Alaska (dots and crosses), are compared with winds

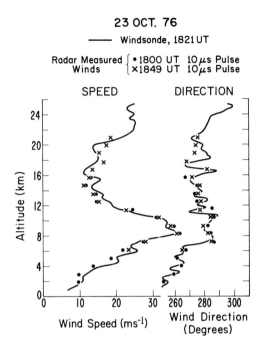

23 OCT. 76
—— Windsonde, 1821 UT

Radar Measured {• 1800 UT 10 μs Pulse
      Winds   {× 1849 UT 10 μs Pulse

FIGURE 12. *Wind profile measured by the 23-cm wavelength Chatanika radar near Fairbanks, Alaska. Data were obtained with the VAD scan method.*

derived from a radar-tracked balloon (solid line) launched a few km away. The 10 μ-sec radar pulse width gives 1 km height resolution for an antenna elevation angle of 45°. (The antenna is fully steerable.) Both wind speed and direction show good agreement for the two measurements.

Figure 13 shows hourly averages of the N-S and E-W wind components with approximately 2 km height resolution measured with a 6 m wavelength radar at Platteville, Colorado. In this case the data are obtained with just two antenna pointing positions. The antenna consists of two planar phased arrays (100 m x 100 m) of crossed dipoles. The arrows denote the times of synoptic radiosonde ascents which can obviously miss much of the temporal variability of the wind profile.

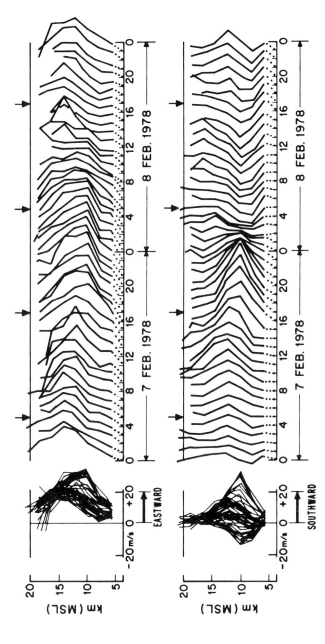

FIGURE 13. *North-south and east-west wind profiles measured with a 6 m wavelength radar near Platteville, Colorado; radar beams are fixed in position; separate antennas are used for each of wind components.*

In precipitation, the vertical motion of the scatterers cannot be neglected since raindrop fall speeds can be as high as 9 m/s. The fall speed does not affect the VAD measurements of horizontal wind, but a third measurement is needed, in addition to the N-S and E-W measurements shown in Fig. 13, to account for the vertical velocity. By using the VAD technique, a wind field which varies linearly with horizontal distance can be measured whereas the 3-point method assumes a constant wind. However, considerations of reliability and design simplicity dictate against large-aperture physically steerable antennas, and additional steering also complicates a phased array. Thus, for mean wind measurements, a simple 3-point scan is planned. Measurements will be made at the zenith and toward the north and east at an antenna elevation angle θ. Then the radar radial velocities are

$$V_Z = w - V_t$$

$$V_E = u \cos \theta + (w - V_t) \sin \theta$$

$$V_N = v \cos \theta + (w - V_t) \sin \theta \qquad (8)$$

where u, v, and w are the x, y, and z wind components and V_t is the mean fall speed of the scatterers. Errors in u and v will result if $(w - V_t)$, which is measured directly above the radar, differs from $(w - V_t)$ at the E and N measurement points which are displaced by $(h \cot \theta)$ from the Z measurement point. These errors are

$$\Delta u = h \, \frac{\partial (w - V_t)}{\partial x}$$

$$\Delta v = h \, \frac{\partial (w - V_t)}{\partial y} \qquad (9)$$

where h is the height. Horizontal gradients of u and v do not
affect the measurement directly; however, when the data are
assumed to represent a profile directly above the radar, a
displacement error of

$$\Delta u = \frac{\partial u}{\partial x} \, h \, \cotan \, \theta$$

and

$$\Delta v = \frac{\partial v}{\partial y} \, h \, \cotan \, \theta$$

occurs. The usual assumption is that these errors are negligible
for long averaging times but horizontal wind profile measurements
during convective precipitation, for example, could have large
errors.

In addition to mean wind profiles, the radar may be able to
supply the following:

1. profiles of C_n^2--the profile of radar reflectivity in the
clear air is a measure of the C_n^2 profile. These profiles
frequently show a layered structure associated with inversions;
the heights of these inversions are useful for radiometric
retrieval of temperature profiles.

2. tropopause height--radar detection of the tropopause
height was reported using centimeter wavelength radars at
Wallops Island in 1966 by detection of a turbulent layer which
caused enhanced C_n^2 (18). Recently, longer wavelength radars
have reported tropopause detection by specular reflection (19).
Knowledge of the tropopause height can greatly aid in retrieval
of temperature profiles and is also useful in local prediction
of severe weather.

3. profiles of turbulence--this information is potentially
available from (a) the width of the Doppler spectrum which
represents the variability of the radial velocity of the
scatterers in the radar resolution volume, (b) the temporal

variations in the mean wind which represents the variability of the motion of the scatterers on scale sizes larger than the radar resolution volume, or (c) the measured profile of radar reflectivity in the clear air (20).

Although considerable research has been performed, a number of unanswered questions remain:

1. What is the temporal and spatial variability of C_n^2 throughout the atmosphere? That is, given a radar system with known sensitivity, what fraction of the time will it be able to measure the wind profile to various altitudes? This information is needed to design an operational wind sounding radar system.

2. What is the interpretation of the layers of enhanced C_n^2 and can layer heights be correlated with temperature structure in radiosonde profiles and therefore be used to improve radiometric temperature retrievals?

3. What is the optimum method of using radar to detect tropopause height?

4. Are there layers of enhanced C_n^2 that exhibit specular reflections throughout the height of the atmosphere and what is their importance?

5. What are the important scales of motion at the various heights in the atmosphere? This information needs to be known before optimum averaging schemes can be employed and before optimum use can be made of continuous profile data.

Answers to some of these questions are needed before radar can measure continuous wind profiles for operational meteorology. Research conducted with the proposed prototype system described in the next section will provide some of the answers. Existing research systems will also supply data that will assist in development of an operational version of a radar wind profiler.

III. DESIGN OF AN INTEGRATED PROFILING SYSTEM

A. *General*

The background material discussed in the preceding sections
serves as a basis for design of a prototype operational system
useful for weather-forecasting services and research. In this
section, details of the equipment and their basic functions will
be given; perhaps more importantly, interrelationships between
the sub-systems will be discussed.

The physical layout of the system as presently conceived is
shown in Fig. 14. Two buildings are sketched; the upper one in
the figure contains the microwave radiometers and the radar
receiver, both of which have some data preprocessing. A
separate computer accepts data from both systems where processing
for internal and external display is carried out. Control
voltages are taken by land line to the radar transmitter which
is in a separate housing shown in the lower part of Fig. 14.
The six radiometer beams (upper right) are all reflected by a
45° flat plate which directs them to the zenith; the beamwidths
(2.5°) are all equal.[5] This arrangement is similar to the
antenna in the presently operational dual-channel system (2).
Beams of the UHF radar (beamwidth 2.5°), sequentially switched
for zenith, 15° east, and 15° south, are shown schematically in
the lower right side of the figure.

[5]*Equality of the beamwidths of the antenna radiation patterns
is necessary to accomplish the same beam-filling at all
frequencies when liquid-bearing clouds enter the beams.*

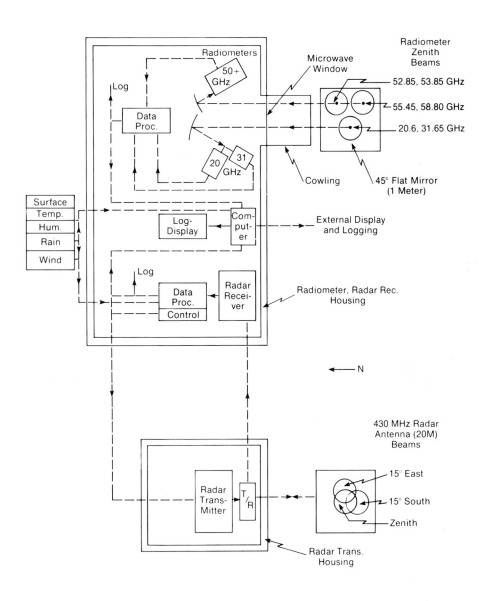

Plan View of Profiler System Layout

FIGURE 14.   Plan view of the profiler system layout.

B.   Temperature-Humidity Profiler

The 20.6 GHz and 31.65 GHz radiometers which constitute two of the six channels of the temperature-humidity profiler are identical, except for data processing, to those discussed previously and in Ref. 2.   The four 50 GHz channels are at center frequencies of 52.85 GHz, 53.85 GHz, 55.45 GHz and 58.80 GHz which are chosen to provide surface-based weighting functions suitable for temperature sensing to the 500 mb pressure level.   The design of these four channels is similar to that of the 20.6 GHz and 31.65 GHz channels.   All radiometers use a three-step Dicke-switching sequence which samples the antenna, reference, and calibration sources.   Comparison of the latter two sources provides a continuous monitor of any gain fluctuations that occur in the system; this establishes an automatic gain control that is implemented in the computer. Offset parabolic reflectors are used in the antenna system with two of the frequencies received in each antenna feed.

C.   Wind-Profiling Radar Design

The prototype wind profiling radar is designed to meet the following goals:

1.   continuous, automated, and unattended operation;

2.   measurement of wind profiles from 100 m above the surface to the tropopause or higher in nearly all weather conditions;

3.   height resolution of 100 m in the boundary layer, 300 m at mid-levels, and 1 km at high levels;

4.   wind component accuracy of 1 m/s or better;

5.   temporal resolution of 15 min or less; and

6.   operation in clutter and interference expected near urban areas.

The Doppler radar will use a mechanically fixed antenna with three pointing positions: zenith, 15° off zenith toward N or S and 15° off zenith to E or W.[6] At each pointing position there will be three operating modes for measurements at low-, mid-, and high-altitudes. Each pointing position will use a dwell time of less than 5 min. Radar control and data processing will be performed with a minicomputer.

A radar wavelength of 0.71 m was initially selected for the following reasons:

1. The power-aperture product required to measure wind profiles to a given height decreases with wavelength (Fig. 11) until it reaches a broad minimum at about this wavelength. At shorter wavelengths the required power-aperture product remains nearly constant but both the transmitter and antenna become more expensive.

2. The frequency band 420 MHz to 450 MHz was the lowest frequency band where frequency allocation with sufficient bandwidth for 100 m height resolution seemed possible.

3. The required antenna characteristics could be achieved with either a reflector antenna or a phased array at this wavelength.

4. Solid-state transmitters for radars have been developed at this wavelength.

5. This wavelength allows for tradeoffs between antenna size and transmitted power for various applications. A 0.71 m wavelength radar could be built with both the sensitivity needed for high-altitude measurements and the resolution needed for low-altitude measurements. The same basic radar design, with

[6]*Lack of moving parts is mandated because of maintenance and reliability consideration.*

smaller aperture and/or less transmitted power, could be used
for applications that do not require profiles to extend to the
tropopause.

A conceptual design was made for a 0.71 m wavelength with a
power-aperture product of 5 x 10^5 W-m^2. The sensitivity of such
a radar should be approximately equivalent to that of the
Platteville radar which achieved the results shown in Fig. 13.
An antenna diameter of 20 m (2.5° beamwidth) and an average
power of 1.5 kW were chosen. The antenna beam would be
steered by selecting one of three feeds of an offset paraboloid
reflector. The transmitter would be solid sate and would use
high duty cycle to obtain the required average power with
relatively low peak power. Pulse compression by pseudo-random
binary phase coding the transmitted pulses achieves the required
height resolution with wide pulses (21). A pulse compression
factor of 8 was chosen for mid- and high-altitude modes and a
narrow (uncoded) pulse for the low-altitude mode. Range
sidelobes typically found in pulse-compressed radars are reduced
by using a complementary code pair so that sidelobes of one
code (transmitted with every other pulse) are cancelled by
those of the complement (22). This sidelobe cancellation
technique is possible because the signal remains highly
correlated from pulse to pulse. Integration of 10^5 to 10^6
radar pulses would be used to measure the radial velocity at
each height. Height resolutions are 90 m, 300 m, and 900 m
for low-, mid-, and high-altitude operation, respectively.

Frequency allocation for 2 MHz bandwidth in the bandwidth
in the 420 MHz to 450 MHz band has been requested, but it does
not seem likely that this band will be available for operational
wind-profiler systems. Allocation of a lower frequency with
the required bandwidth is also unlikely, but experimental
authorization in the 902 MHz to 928 MHz band has been granted.

Therefore, the radar concept and design has had to be
reexamined. Some of the possible alternatives are:

1. Develop a complete profiler system in the 902 MHz to
928 MHz band.

2. Utilize a dual-radar concept whereby a 902 MHz to
928 MHz radar with high resolution and low sensitivity would
complement a low resolution and high sensitivity radar at long
wavelengths. Frequency allocation may be possible for the
latter radar because the bandwidth required would be an order of
magnitude less than that needed for high resolution measurements.

3. Develop a profiler with reduced sensitivity in the
902 MHz to 928 MHz band and evaluate its performance before
deciding whether a profiler with the total capability desired
is technically and economically feasible.

Present plans are to pursue the third alternative. The
design concepts used for the 420 MHz to 450 MHz radar will be
retained as much as possible. A power-aperture product of
10^5 W-m^2 (10 m diameter antenna with 1.5 kW average power)
would be used. The antenna would be a scaled version of the
one designed for the 420 MHz to 450 MHz band. A solid-state
transmitter, with pulse compression as described above, could
be used although solid-state technology with the required power
level may be too expensive at this time. Three operating
modes with different height resolution would be used for low-,
mid-, and high-altitude measurements. A minicomputer with an
array processor would calculate the Doppler spectrum of the
scattered signal at each height and estimate the radial
velocity, signal power, and spectrum width. The radial
velocities are used to generate a mean wind profile. Profiles
of mean wind, signal power and spectrum width are sent to the
profiler data system.

D. *Data Processing*

As indicated in Fig. 14, data processors are associated with
both the radiometric and Doppler radar sub-systems. In the
former, voltages from the radiometers are converted to atmospheric brightness temperatures and then routed in digital form
to the computer. Data from the radar receiver, processed in
an array processor to determine the speed and direction of the
wind, and the intensity of backscattered power at various
height intervals, are forwarded to the same computer. The
surface temperature, wind, humidity, and rain rate are also
entered. Some of these data are interrelated in the computer;
for example, the intensity of backscattered power is associated
with temperature-profile retrieval (discussed previously), and
the surface rain rate may be used for decision making in choice
of algorithm for profile retrieval (that is, whether liquid
water is in the form of rain or cloud).

The computer in Fig. 14 provides data in suitable format
for three users: the synoptic forecaster, the local forecaster,
and the research forecaster. In simplest terms, the synoptic
mode entails availability of data in the same form as from a
radiosonde that can be called every 12 hours, or more frequently
if necessary. The local forecasting (PROFS) mode, made available
at the local Weather Service Forecast Office, is comprised of
current time series of significant variables such as wind at
given heights, thickness of the surface to 500 mb layer, total
precipitable water vapor, etc.[7] On the other hand, in the
research mode, the basic meteorological variables are available
in highest time resolution; if desired, they can be stored by
the user. It is planned that the most suitable variables and

[7] *PROFS is a Prototype Regional Observing and Forecasting
Service.*

display format for the local (WSFO) mode be chosen and
implemented some time after the initial stages of operation
of the system, following advice from both operational and
research forecasters.

IV. JOINT SATELLITE AND GROUND-BASED PROFILING

Temperature profiling from orbiting satellites using
microwave radiometers is of great interest because of potential
wide application to synoptic-scale forecasting. Although these
downward-looking profilers can operate in the presence of cloud,
there is a difficulty in obtaining accurate data, especially
at altitudes near the earth's surface. The difficulty stems
from coarse spatial resolution and from uncertain background
brightness temperatures seen by the antenna beam of the radiometer; that is, spatial and temporal variations in ground
emission cause problems, as do variations in sea state. These
effects also lead to uncertainties in retrieved values of
integrated vapor and liquid when using a satellite platform.

Because of the continuous observational capability of both
satellite and ground-based profilers, it seems plausible that
with suitable interpolation techniques, the ground-based
observations could be combined with those of a satellite as it
passes near the zenith of the ground-based sounder to provide
useful profiles for mesoscale applications. In addition, the
satellite profiler could be provided with accurate measurements
of downwelling radiance from the ground-based profiler and
estimates of surface temperature and atmospheric transmission.
During clear conditions, such combined measurements could
yield the surface emissivity.

Experience has already been gained using the SCAMS radiometer
to measure temperature profiles from space and from the ground
(see section "Temperature-Humidity Profiler) (23). To extend

the vertical coverage of this instrument above 300 mb and in the first 100 mb above the ground, in subsequent simulations, a 58.8 GHz channel was added.[8] Temperature weighting functions for the ground-based and satellite systems, normalized to a unit maxima, for these channels are shown in Fig. 15. The exponential-like ground-based functions, which have poor resolution above 500 mb, are supplemented above this altitude with the broad Gaussian-like functions of the satellite radiometers to yield an extended range of profiling capability. For comparison of magnitudes, two unnormalized weighting functions are shown in Fig. 16. Here, the integrals with respect to height of these nearly opaque channels are both close to unity.

[8] This channel was used from the Nimbus 5 satellite platform as well as from the earth's surface.

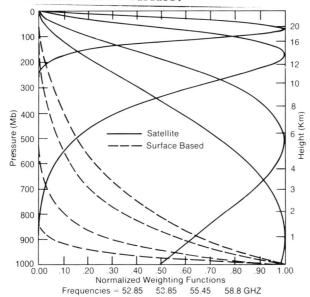

FIGURE 15. Weighting functions, normalized to unit maxima, for surface-based and satellite-based microwave temperature profiling.

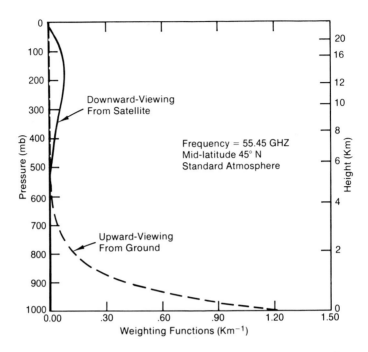

FIGURE 16. Weighting functions for surface-based and satellite-based microwave temperature profiling.

The temperature retrieval accuracy was simulated for the combined system using two highly variable climatologies, each with strongly contrasting surface emission characteristics-- Ocean station P in the Gulf of Alaska during February, March and April; and Washington, DC (Dulles airport) during February. For these climatologies, the effects of nonraining clouds are included by following the modeling techniques in Ref. 6.   The surface emissivities ε_s of the sea and ground were simulated as random variables with $\varepsilon_s = 0.4 + 0.1u$ and $\varepsilon_s = 0.9 + 0.1u$, respectively, where u is a random variable that is uniformly distributed between 0 and 1. Noise in the radiometers was assumed to be Gaussian with standard deviation of 0.5 K. The results are shown in Fig. 17. Surface measurements of tempera- ture, pressure, and humidity were assumed to be available for

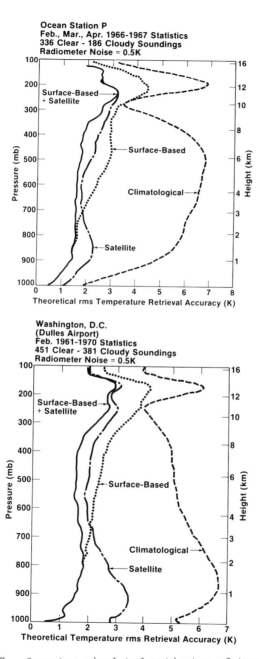

FIGURE 17. *Computer-simulated estimates of temperature pro-*
file retrieval accuracy using dependent sample linear regression
analysis. Assumed radiometric noise = 0.5 K.

the ground-based observations. As expected, the additional
information contributed by the satellite sounder improves
markedly the upper air retrieval accuracies; similarly, lower
altitude soundings of the satellite are much improved with the
surface-based observations. Retrieval accuracies of the combined
system are generally predicted to be around 1.5 K below about
300 mb.

The accuracies for both climatologies are degraded considerably above 300 mb, presumably because of the variation in
the height and intensity of the tropopause. However, recent
experiments suggest that the height of the tropopause can be
routinely measured by ground-based radar (19). Given that this
is achievable, the tropopause height can be used in a retrieval
algorithm as described in an earlier section. Under the
assumption that the tropopause height can be measured with a
resolution of 30 mb ($\approx$ 1 km), again the retrieval accuracies for
the two climatologies mentioned were simulated. As is evident
from Fig. 18, this single piece of height information yields a
substantial improvement in accuracy of retrieval above 300 mb.

Although these simulations suggest the combined ground- and
satellite-based temperature sounding system yields promising
results for mesoscale applications, some problems are anticipated.
The largest uncertainty in evaluating the capability of a
combined system is amalgamation of the two sets of data which
represent widely different spatial scales. With satellite
footprints of the order of tens of kilometers and the much
smaller spatial resolution of a ground-based system, difficulties
in interpretation arise. The degree such difficulties can be
reduced by data processing techniques awaits experimental
evaluation.

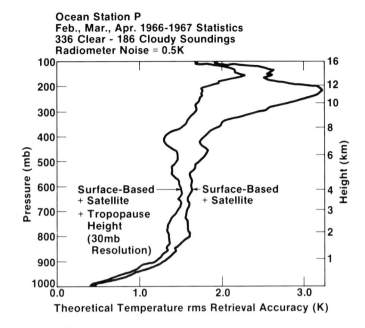

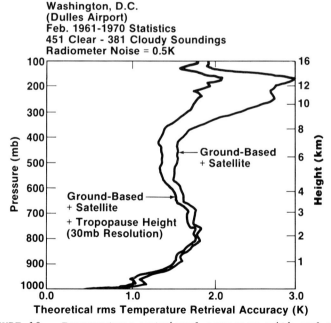

FIGURE 18. Temperature-retrieval accuracy with and without knowledge of tropopause height. Same conditions as in Fig. 17.

V.   POTENTIAL IMPACT

The advent ⟨f a new geophysical measurement capability
typically opens up new research opportunities, followed somewhat
later by opportunities for improved or broadened services.
The continuous profiler described represents such a capability;
it is therefore appropriate to begin to consider possible new
research and service opportunities.

A.   *Research*

The uniqueness of the profiler system lies in its ability
to measure meteorological variables *continuously in time*, as
compared with radiosondes which operationally obtain data only
once per 12 hours, and for research purposes are rarely launched
more frequently than once per 3 hours to 6 hours.

A continuous vertical observing capability will make it
possible to extend studies of the temporal and spatial spectra
of atmospheric variability at least two orders of magnitude
beyond the one cycle per day, one cycle per 700 km of the
existing operational radiosonde network. This extension of the
spectrum of the measurements includes a wide range of what might
be called "significant weather" (e.g., fronts, thunderstorms,
flash floods, chinook winds, land-sea breezes, etc.) whose
study and modeling has been handicapped by lack of suitable
observational data. Many of these phenomena could be investigated by using small arrays of continuous profilers whose
essentially instantaneous, continuous, vertical profiling would
be much more appropriate than the intermittent, slow ascents of
radiosondes. For example, the temporal continuity of the data
would be of great value in identifying the presence of gravity
waves, which are believed to be responsible for triggering some
severe thunderstorms (24). Similarly, the monitoring of
moisture convergence (both vapor and liquid) is much better

achieved with the proposed continuous profiler than by intermittent radiosondes. It is also possible that the Taylor
hypothesis will be applicable to the continuous data and
thereby result in information relevant to the volume surrounding
a profiler.

B. *Operations*

Numerical weather predictions (NWP) by the National
Meteorological Center (NMC) are currently tied to the 12-hour
cycle of the upper air (radiosonde) network. The impact on NWP
of having vertical profiles available continuously rather than
once per 12 hours has not been analyzed, but could well be
significant. Possibilities include (a) improved analyses of
current weather patterns (resulting from continuous monitoring
of profiles) leading to better initialization of the NMC models;
(b) incorporation of time derivatives in the initialization
process, as well as the instantaneous or time-averaged values;
(c) more frequent updating of NWP model inputs, for example,
once per 3 hours or 6 hours rather than once per 12 hours; and
(d) improvements to the NWP models and parameterization, through
improved ability to monitor continuously the developing
differences between the real atmosphere and the NWP models.

The continuous profiler will also play a significant role in
PROFS. Three profilers at the corner of a triangle some 150 km
to 200 km on a side surrounding an urban area with its central
profiler site, are being considered in order to give continuous
information on local meteorological conditions. In areas where
PROFS-type coverage might eventually be essentially contiguous
(such as much of the eastern half of the nation, the Great Lakes
region, and part of the West Coast), such a network would provide
continuous data at twice the spatial frequency of the present
upper air network.

The profiler is also useful to the local forecaster for monitoring conditions in his immediate area. Sudden changes in upper-air winds, existence of super-cooled liquid (relevant to aircraft icing), and changes in persistent inversions (relevant to capping of air pollution) are examples of such useful observations. Application of the observations to forecasting of local precipitation is also anticipated.

SYMBOLS

a_1, a_2, a_3	constants relating measured brightness to water vapor (Eq. (1))
A_e	effective antenna area, m^2 (Eq. (4) and Fig. 11)
b_1, b_2, b_3	constants relating measured brightness to liquid water (Eq. (1))
c_n^2	refractive index structure constant, $m^{-2/3}$ (Eqs. (6) and (7) and Figs. 10 and 11)
dBZ	$= 10 \log_{10} Z$
D	diameter of raindrops, m (Eq. (5))
h	height of radar measurement, m
K	radar constant (Eq. (4))
L	radiometrically measured liquid water, mm (Eqs. (1) and (2))
m	complex dielectric constant of water (Eq. (5))
n	refractive index (Eq. (7))
P_i	frequency of occurrence (Eq. (3))
$\bar{P}_t$	average transmitted power, w (Eq. (4) and Fig. 11)
r	separation distance, m (Eq. (7))
R	range, m (Eq. (4) and Fig. 11)
T_2, T_3	brightness temperatures of 20.6 GHz and 31.6 GHz radiometers, K (Eqs. (1) and (2))
T_{sys}	system noise temperature, K (Eq. (4))
u, v, w	Cartesian wind components, m/s (Eqs. (8) and (9))

V	radiometrically measured precipitable water vapor, mm (Eqs. (1) and (2))
V_A	actual precipitable water vapor, mm
V_E, V_N, V_Z	radar-measured radial velocity, m/s (Eq. (8))
V_s	radiosonde-measured precipitable water vapor, mm
V_t	mean fall speed of raindrops, m/s (Eqs. (8) and (9))
x, y, z	Cartesian coordinates (Eqs. (8) and (9))
Z	radar reflectivity factor
ΔR	range resolution, m (Eq. (4) and Fig. 11)
Δu, Δv	error in measurement of u, v, m/s (Eq. (9))
ε_s	surface emissivity
η	radar reflectivity, m^{-1} (Eqs. (5) and (6) and Fig. 10)
η_{min}	minimum detectable radar reflectivity, m^{-1} (Eq. (4))
θ	elevation angle (Eq. (8))
σ^2	mean square error of temperature retrieval, K^2 (Eq. (3))
σ_i^2	mean square error of temperature retrieval for profiles with inversions in i, K^2 (Eq. (3))
λ	radar wavelength, m (Eqs. (4), (5), and (6), and Figs. 10 and 11)
τ	observation time, s (Fig. (11))

ACKNOWLEDGMENTS

We thank R. B. Chadwick of WPL/ERL/NOAA for the FM/CW radar data used in the section "Temperature Profiling," N. C. Grody of NESS/NOAA for collaboration on the satellite analysis in section "Joint Satellite and Ground-Based Profiling," B. B. Balsley and associates at the Aeronomy Laboratory/ERL/NOAA for data presented in several sections, E. B. Burton and staff of the NWS/WSFO, Denver, Colorado, for help and advice on operational considerations, and J. B. Snider of WPL/ERL/NOAA for a critical reading of the manuscript.

REFERENCES

1. Westwater, E. R., *Radio Sci. 13(4)*, 677-685 (1978).
2. Guiraud, F. O., Howard, J., and Hogg, D. C., *IEEE Trans. on Geosci. Elect. GE17(4)*, 129-136 (1979).
3. Hogg, D. C., and Guiraud, F. O., *Nature, 279,* 408 (1979).
4. Westwater, E. R., and Decker, M. T., *in* "Inversion Methods in Atmospheric Remote Sounding" (A. Deepak, ed.), pp. 395-427. Academic Press, New York (1977).
5. Snider, J. B., *J. Appl. Meteor. 11,* 958-967 (1972).
6. Decker, M. T., Westwater, E. R., and Guiraud, F. O., *J. Appl Meteor. 17,* 1788-1795 (1978).
7. Gossard, E. E., Jensen, D. R., and Richter, J. H., *J. Atmos. Sci. 28,* 794-807 (1971).
8. Chadwick, R. B., Moran, K. P., Strauch, R. G., Morrison, G. E., and Campbell, W. C., *Radio Sci. 11,* 795-802 (1976).
9. Johannessen, K., and Kessler, K. E., *in* "17th Conference on Radar Meteorology," pp. 560-561. American Meteorological Society, Boston, Massachusetts (1976).
10. Hardy, K. R., and Katz, I., *Proc. IEEE, 57,* 466-480 (1968).
11. Gage, K. S., and Balsley, B. B., *Bull. Amer. Meteor. Sci. 59,* 1074-1093 (1978).
12. Hennington, L., Doviak, R. J., Sirmans, D., Zrnic, D., and Strauch, R. G., *in* "17th Conference on Radar Meteorology," pp. 342-348 (1976). American Meteorological Society, Boston, Massachusetts (1976).
13. Battan, L. J., "Radar Observations of the Atmosphere," University of Chicago Press, Chicago, Illinois (1973).
14. Ottersten, H., *Radio Sci. 4,* 1251-1255 (1969).

15. Hogg, D. C., and Mumford, W. W., *Microwave J. 3*, 80-84 (1960).

16. Lhermitte, R. M., and Atlas, D., *in* "Proceedings of 9th Weather Radar Conference," pp. 218-223. American Meteorological Society, Boston, Massachusetts (1961).

17. Browning, K. A., and Wexler, R., *J. Appl. Meteor. 7*, 105-113 (1968).

18. Atlas, D., Hardy, K. R., Glover, K. M., Katz, I., and Konrad, T. G., *Sci. 153*, 1110-1112 (1966).

19. Gage, K. S., and Green, J. L., *Sci. 203*, 1238-1240 (1979).

20. Van Zandt, T. E., Green, J. L., Gage, K. S., and Clark, W. L., *Radio Sci. 13*, 819-829 (1978).

21. Nathanson, F. E., "Radar Design Principles," McGraw-Hill, New York (1969).

22. Golay, M. J. E., *IRE Trans Infor. Theo. IT-7(2)*, 82-87 (1961).

23. Grody, N. C., *IEEE Trans. Ant. Prop., AP-24*, 155-162 (1976).

24. Chimonas, G., Einaudi, F., and Lalas, D. P., A Wave Theory for the Onset and Initial Growth of Condensation in the Atmosphere, *J. Atmos. Sci. 37(4), n.p.*, (1980).

DISCUSSION

Green: Would this then entail having lots of radiometers distributed over the globe for global purposes?

Westwater: For global purposes it would. For local mesoscale
forecasting, a radiometer could be placed at an observing site at
a weather station and it would continuously observe weather
changes that would probably be representative of that region as
a whole and it would help to interpolate in time between radiosonde and records but as far as a synoptic tool, I don't believe
this is really what it is geared for. The satellite temperature
soundings would help the mesoscale forecasting, but I don't really
think that the ground-based system would help the satellites
that much because they would just be at one place in space.

Staelin: Could you elaborate on how you include the decreased
tropopause height in a retrieval with temperature sounding data.
Is that simply a linear regression where you treat the height
as a number?

Westwater: A regression does not work at all.

Staelin: I did not think so. How did you do it?

Westwater: The way to do it, or the way I did it was to average
over an ensemble of profiles which contained inversions within
the specified resolution interval, so that for the tropopause I
retrieved profiles over an ensemble which contained tropopause
height within a 30 mb resolution interval. If you use the
height in the algorithm as a simple regression predictor, there
is very small reduction in variance at all. But the other
technique does work.

Reagan: In your discussion of the water vapor comparison between
the radiometer and the radiosonde, there was some scatter in the
results. I was curious, indeed, really about how much of that
scatter might be radiosonde as opposed to the radiometer. Would
you talk a little bit about the relative distribution of
uncertainty there?

Westwater: Well, that's a number that is very difficult to
determine. One hint that we have that maybe the radiometer is
working even better than those records indicate is that there
was a time in which the humidity sensor element was changed. It
was upgraded to a more recent element, and the Weather Bureau
thought that this was an improvement. And the correspondence
between our radiometrically inferred and the radiosonde measured
humidities were closer to the order of 10 percent closer. So

that there are always problems in the interpretation of the radiosonde, particularly in the humidity element. And quantitatively, it is really difficult to put a number on that.

Staelin: Let me elaborate on that. Our data and other studies of the water vapor distribution in the atmosphere suggest that a large part of the radiosonde error in water vapor sensing is actually aliasing error due to the inhomogenity of the water vapor; this is in addition to whatever errors there may be in the sensor. I have not seen an exact analysis of how much that contributes, although in principle it could be done. It could be very interesting because there is a significant advantage in remote sensing from space or the ground because of the averaging that takes place over those local inhomogeneities.

Westwater: I would like to make another comment on that point. One way that this might be studied would be with two or more ground-based radiometers separated by a fixed horizontal distance. The resultant time series of inferred precipitable water vapor, coupled with measurements of horizontal wind, would yield information on the spatial inhomogeneity.

Grody: I would like to know how you do the stratification for the low level inversions? Is this done in the same manner as you do the tropopause inversion?

Westwater: For the low level inversions, I picked out the first discontinuity in the slope of the profile within a 20 mb pressure interval and then did retrievals over that ensemble. Now with the complicated structure, sometimes, it becomes a little difficult about what height you would choose, and I always chose the first height assuming the radar would have no difficulty in interpreting the first inversion height.

Grody: Also, what mechanism do you use for compensating for cloud effects?

Westwater: We use the five channel data to estimate whether conditions are clear or cloudy. If it is clear, we use ordinary statistical inversion. If cloudy, an equivalent clear set of brightness temperatures are derived and again used with statistical inversion, but with an appropriately larger noise level.

Staelin: I should like to conclude with the observation that remote sensing appears to be alive and well in terms of the innovations that have occurred over the past two and a half years since the last meeting on inversion methods held in Williamsburg in 1976.

Note: The following questions were addressed to Dr. R. G. Strauch.

Deepak: Dr. Strauch, you mentioned the horizontal scan. Does it give the velocity component in the x-y direction?

Strauch: The radar measures the radial velocity, the component of motion toward or away from the radar. At 15 degrees off the vertical we get approximately one-fourth of the horizontal wind vector projecting in the radial direction. If we look at the variance of the horizontal wind measurement, it becomes 16 times the variance of the actual radial measurement. But with the long integration time we are talking about, we can tolerate that much uncertainty. I would not want to go much higher than 75° elevation angle since the uncertainty increases as the reciprocal of the cosine of the elevation angle and it starts going up very rapidly.

Rosenkranz: You mentioned the measurement of vertical winds. Could you elaborate on that with respect to accuracy?

Strauch: In the clear atmosphere, we expect a very low mean vertical velocity unless we were in a location such as the lee of the mountains or something like this, and certainly you would want to pick your site so that you would not have that situation. Where vertical velocity is important is in precipitation where the fall speed of the particles, that is the fall speed of rain-drops, varies from about a meter per second to about nine meters per second depending on their size. We can measure this particle speed with very good accuracy; probably to a few centimeters per second. We have to take the fall speed and correct the measurement of the horizontal velocity for the projection of that vertical velocity, which is almost the entire value of the fall speed, on to the radial; so you have to make that correction in precipitation. We expect that measurement of particle fall speed to be very accurate because the signal noise ratio will be high even at the 70 cm wavelength we are talking about. Is that what you are alluding to?

Rosenkranz: Well, I would think that if you wanted to measure vertical winds, you would look vertically.

Strauch: We have one measurement mode looking vertical, one to the north, and one to the east, but in the clear air, we may not even use the vertical mode because we would expect over the time that we are averaging, that we would get near zero vertical velocity. It would only be centimeters per second, and the uncertainty would be as large as that or more.

Westwater: What will you be using for ground-truth once the instrument is constructed and you are inferring winds? Will the radiosonde be used as ground-truth?

Strauch: The plans are to build the first unit at Denver Stapleton Airport, and there we would be within a hundred meters of the radiosonde launch site; the radiometers will also be there. The radiosondes will be ground-truth. Now, obviously, we could be 20 km away and at some times still be close to the radiosonde, so you really would not have to be right at that site. But the radiosonde will be the ground-truth.

Deepak: You said that the reflection of the scattering is from the C_n^2. Could you amplify on that?

Strauch: Reflections occur because of the inhomogeneities in the medium. The resolution is determined, since it is an active system, by the pulse length and the beamwidth. At a 75° elevation angle, the resolution is primarily governed by the pulse length. In the vertical, it is strictly governed by the pulse length. We are talking about a 3° beam width so that at the very highest altitudes, the 3° beam width starts to cause a slight increase in the resolution over that given by the pulse width. At the low altitude, we plan on using pulse widths of 90 meters; middle altitudes-300 meters; highest altitude-1 kilometer. And what we are probably going to be measuring is not the average wind through that depth, but the wind that actually exists at certain layers in that depth where the C_n^2 is high.

Westwater: I would like to make one more quick comment. The minimum height at which C_n^2 could be inferred from these was at the order of 100 meters. And for many meteorological situations, particularly in locations like Denver, a very high percentage of ground based inversions and trapped pollutants are below 100 meters. And this radar may not have the sensing capabilities that would really help us in this regime. For that probably the acoustic sounder or some related device would be needed to really supplement the closest boundary layer for sensing of inversions relevant to pollutant studies.

Strauch: There is another radar device that could observe C_n^2 at a minimum altitude of 10 to 15 meters and that is the Fm-Cw radar which is a very high resolution boundary layer radar. One could envision using a Fm-Cw radar for high resolution measurements at low altitudes and a VHF radar, for example, the one at Platville, for upper altitude measurements. But we are not trying to build another research device--we are trying to build what might be viewed as operationally attractive. By the same token, consider the radiometers. If we had, as Ed Westwater pointed out, difficulties with some of the radiometric channels, one could simply shift the frequency and overcome that with a multi-frequency device. Here, we could have several radars, but we would rapidly approach the situation where the operational weather service would not consider it to be a practical device. We are developing an operationally attractive tool in our first step.

Unidentified Speaker: In the operational context, what kind of cost are you talking about for one of these stations, and how does that compare to a radiosonde station?

Strauch: I am not totally familiar with the radiosonde facility but I believe that the hundred radiosonde sites in the United States cost of the order of 15 million dollars per year. Now another thing that enters here is that the present radiosonde network is expected to be obsolete mechanically by 1990. So that it must be replaced. The reason why we must start on this now is not just the development of the equipment and the measurement techniques, it is the transition to accepting and using such a device, too, that takes time. We are talking on the order of half a million dollars per site, which is about the cost for 5 years of operation of a radiosonde site.

CO_2 LASER RADAR FOR ATMOSPHERIC
ENERGY MEASUREMENTS[1]

Charles A. DiMarzio
Albert V. Jelalian
Douglas W. Toomey

Raytheon Company
Equipment Division
Wayland, Massachusetts

Measurements have been made of the atmospheric velocity turbulence spectrum using a mobile, CO_2, continuous wave, Doppler LIDAR System. The LIDAR was used in a heterodyne configuration with a 30 cm output opted to produce a conical scan. Velocity data from this scan were Fourier analyzed to obtain velocity power density spectrum. The -5/3 power law of Kolmogorov was observed, along with variations in total energy which could be correlated with storm activity. In addition, departures from the Kolmogorov spectrum were observed under some conditions. It is hoped that the total energy and departure from the -5/3 power law may prove useful in short-term meteorological forecasting.

[1]*This work was performed under contract number N00014-75-C-0282, for the Office of Naval Research. The authors gratefully acknowledge the assistance of the contract monitors, Drs. J. Hughes and V. Nicholai.*

I.   INTRODUCTION

In recent years, a number of significant advances have been
made in the use of CO_2 laser Doppler radars for measurement of
atmospheric wind flow fields.   The simplest measurement is that
of the line-of-sight wind velocity along a transmitted laser
beam.   This eventually led to the concept of a conical scanning
system to make several measurements in different directions
resulting in a measurement of the three-dimensional wind
velocity vector.   The raw data obtained from such a system
consists of three components of the wind velocity for each
revolution of the scanner, and as such, is too large a body of
data to be easily assimilated.   To reduce this mass of data to
a meaningful data set, various manipulations may be used to
display significant characteristics of the velocity time history.
One of these characteristics is the power density spectrum of
the magnitude of the velocity vector.   This spectrum may be
related to the theoretically predicted spectrum of Kolmogorov
by using Taylor's hypothesis to relate a temporal correlation
function to a spatial one.   It will be demonstrated here that a
CO_2 laser Doppler radar has been used successfully to measure
the kinetic energy density spectrum of the atmosphere.

II.   OPERATING PRINCIPLES

A brief description of the operating principles of the
system including the principles of line-of-sight velocity
measurement, conical scan principles, and construction of the
power density spectrum is presented.   This will be followed by
a description of the system used to collect the data and
samples of actual data obtained from the system.

The first and simplest consideration is the measurement of a line-of-sight velocity component using a CO_2 laser radar. This can be understood with the aid of Fig. 1 which shows the paths of energy within the laser system. A plane wave Gaussian beam transmitted from the laser is separated into two components: the larger component going to the target, and the smaller component falling on the detector. Energy backscattered from the target returns along the path shown in Fig. 1b and is incident on the detector, superimposed with the small portion of the transmitter energy. If the target is moving, the returning energy will be Doppler shifted in frequency by an amount proportional to the velocity component along the system line of sight, as illustrated in Fig. 2. There are two beams incident on the detector with different amplitudes and frequencies. The larger amount of energy generally comes from the reference beam, and the smaller amount of energy is that

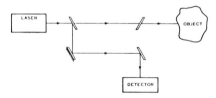

A) PATHS OF TRANSMITTED BEAM

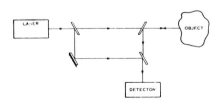

B) PATHS OF TRANSMITTED
AND RECEIVED BEAMS

FIGURE 1. Doppler LIDAR transmission and reception.
(a) Paths of transmitted beam. (b) Paths of transmitted and received beams.

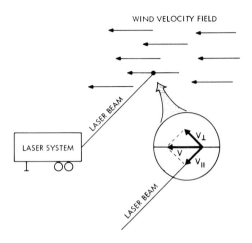

FIGURE 2. Velocity component sensed by the LIDAR.

scattered from the target and Doppler shifted in frequency.
These two frequency components are illustrated in Fig. 3. The
output of the detector is an electrical signal at the difference
between these two frequencies as shown in Fig. 4. This figure
represents an actual atmospheric signal measured by a CO_2 laser
radar. The target consists of aerosols naturally suspended in
the atmosphere. This figure was associated with a line-of-sight
velocity component of approximately 3 m/s.

The next concept in developing an understanding of the system
is that of the upward looking conical scan, often referred to as
a velocity azimuth display (VAD). In this concept, the laser
beam is scanned from a fixed source in a number of different
directions. It is assumed that the wind flow field is uniform
over the area being scanned. In fact, this restriction can be
relaxed to say that the only velocities of interest are those
having spatial variations on a size scale larger than the scan

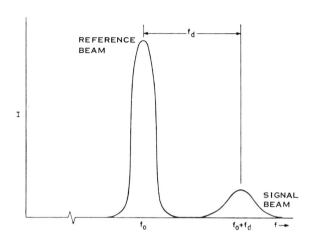

FIGURE 3. Spectral distribution of power on the detector.

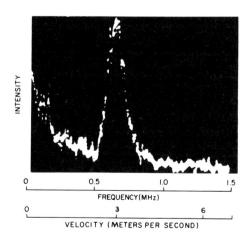

FIGURE 4. Typical LIDAR Doppler Signal Spectrum.

area. It will be observed that if the laser beam is directed either upwind or downwind, a Doppler shift will occur because of the existence of a line-of-sight velocity component. In one case, the shift will be toward increasing frequency, and in the other case, it will be toward decreasing frequency. If the beam is directed at 90° to either of these positions, the Doppler shift will be zero. A simple geometric analysis of Fig. 5 will show that if the velocity is plotted as a function of the azimuth angle of the scan, and if the velocity field is uniform over the scan area and over the time required to complete the scan, the resulting plot will be a sinusoid with the peak-to-peak excursion indicating the horizontal wind velocity magnitude, the phase indicating its direction, and the offset of the sine wave from zero indicating the vertical velocity component.

It will be noted at this time that no discrimination exists between positive and negative Doppler frequency shifts. Therefore, it is not possible with this system in its present

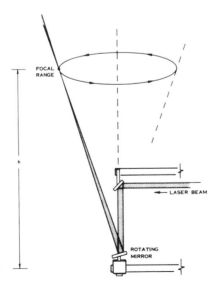

FIGURE 5. *Upward looking VAD scan concept.*

form to determine the sign of the velocity components, and there results a 180° ambiguity in wind direction. For the purpose of measuring the magnitude of the wind vector, this ambiguity is of no significance, and the sign of the components may be arbitrarily chosen. This is most readily accomplished by determining the highest velocity component measured in the scan and arbitrarily selecting it to be positive. All velocities within 90° either side of this point are also assumed positive, and the remaining velocities are assumed negative. Clearly, in cases of a vertical wind component, this assumption results in some errors. Careful analysis of a large number of atmospheric measurements has shown that these errors can be removed, but amount to only an error of about 2 percent in the actual wind velocity component. A sample plot of velocity against azimuth is shown in Fig. 6. The rough curve in this figure shows the raw data with the sign reconstituted as described above, and the smooth curve shows a least-squares fit, performed in closed

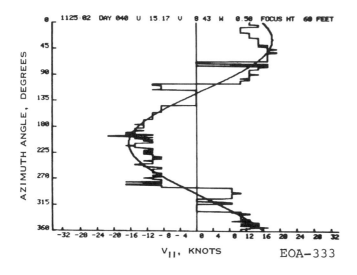

FIGURE 6.   Typical VAD scan data.

form in real time, to the data. The U, V, and W components of
the wind velocity are indicated at the top of the plot as calculated from the least-squares fit. The U component is defined
as being from the north, V from the west, and W upward. By
using the least-squares-fit technique, a time history of U, V,
and W can be constructed as shown in the sample of Fig. 7.

III. DATA COLLECTION SYSTEM

The power density spectrum can be constructed by Fourier
transforming the time history of the magnitude of the wind
velocity vector and squaring. To reduce storage requirements,
an autocorrelation function of this velocity time history is
performed first. Then the Fourier transform of this autocorrelation function is taken to determine directly the power
density spectrum. By using Taylor's hypothesis, the frequency

1136:11	DAY 094	U	9.56	V	8.03	W	2.92
1136:18	DAY 094	U	10.53	V	8.05	W	1.61
1136:25	DAY 094	U	11.00	V	-5.78	W	-0.85
1136:32	DAY 094	U	10.04	V	9.37	W	1.68
1136:40	DAY 094	U	10.02	V	9.03	W	1.74
1136:49	DAY 094	U	9.81	V	9.30	W	1.83
1136:57	DAY 094	U	4.09	V	11.04	W	0.61
1137:04	DAY 094	U	7.63	V	10.49	W	1.46
1137:11	DAY 094	U	6.08	V	11.61	W	1.09
1137:18	DAY 094	U	10.89	V	8.43	W	1.72
1137:25	DAY 094	U	0.29	V	-12.23	W	-0.40
1137:33	DAY 094	U	10.68	V	8.58	W	1.70
1137:40	DAY 094	U	12.72	V	-10.36	W	-0.52
1137:47	DAY 094	U	9.26	V	10.02	W	1.74
1137:54	DAY 094	U	10.68	V	- 9.05	W	-0.15
1138:02	DAY 094	U	13.78	V	5.72	W	-0.74
1138:09	DAY 094	U	10.96	V	-12.88	W	1.26
1138:16	DAY 094	U	2.35	V	12.02	W	1.34
1138:23	DAY 094	U	12.44	V	5.06	W	-0.01
1138:30	DAY 094	U	9.90	V	- 8.81	W	0.19
1138:38	DAY 094	U	7.54	V	- 7.25	W	-0.45

(VELOCITIES IN KNOTS)

FIGURE 7. U, V, and W wind components as function of time.

scale is related to a spatial frequency scale using the average
wind velocity over the time of measurement. Data are plotted
on a logarithmic scale to show the anticipated negative 5/3
power law of Kolmogorov.

A block diagram of the system used for data collection is
shown in Fig. 8. The laser is a Raytheon Model LS-10A, 20 watt,
CO_2 laser with an output beam which is Gaussian, and has a
diameter at $1/e^2$ points of approximately 7 mm. The laser energy
enters the interferometer and optical system, which includes a
beam expander to produce a 1-foot-diameter beam which is
focused in the atmosphere at the altitude of interest. Most of
the experiments were performed at an altitude of 500 m. In this
configuration, if the aerosol distribution of the atmosphere is
uniform, half of the signal will be received from an altitude
segment 150 m thick covered at 500 m. The detector is a liquid
nitrogen cooled mercury-cadmium telluride photodetector operated

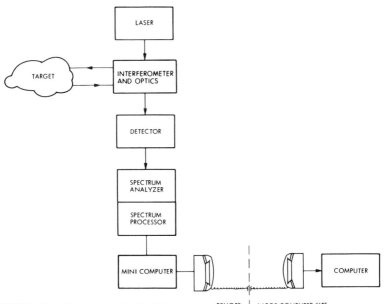

FIGURE 8. System block diagram.

in the photovoltaic mode. Signals from this detector are
frequency analyzed by using a surface acoustic wave delay line
spectrum analyzer followed by a digital signal processor which
obtains the significant parameters of the spectrum. The digital
output of the processor, along with the azimuth data from the
scanner and from a time code generator, are input to a Raytheon
706 mini-computer which constructs the autocorrelation function.
A scan is performed approximately once per second and autocorrelation functions are performed on the complete data set
after an initial 15-minute interval. Data collection intervals
are typically 30 minutes to 1 hour. At the end of a data
collection interval, the autocorrelation function is transferred
from the Raytheon 706 mini-computer to a larger computer: a
CDC 6700 at Raytheon's Missile System Division in Bedford,
Massachusetts, by using a telephone line with modems. The large
computer is responsible for processing the correlation function
into a form suitable for display and for further evaluation, for
correctly scaling the velocity data, for performing a fast
Fourier transform of 1024 points to produce the power density
spectrum, and for constructing suitable displays of the data.
The data are stored on magnetic tapes and on disks so that
algorithm changes may be incorporated and old data sets rerun.
The laser system and real time mini-computer and all supporting
hardware are mounted in a self-contained vehicle at Raytheon's
Equipment Development Laboratory in Sudbury, Massachusetts.
The scanner is mounted on a small aluminum tower immediately
behind the van. The entire system is self-contained and mobile
and may be moved from one site to another. Recently, similar
systems have been developed in self-contained motor vehicles
which include a self-contained scanner, power sources, air-
conditioning, and more complete data processing facilities.

IV.   SAMPLE DATA

A sample of a typical power density spectrum is shown in
Fig. 9. The spectrum is plotted on logarithmic scales of both
spatial frequency and power so that the -5/3 slope may be
readily verified. The resolution limit of this system, as
determined by the size of the circle generated by the conically
scanned, focused spot, is shown on the plot. It may be seen
that meaningful data are obtained at all spatial frequencies
corresponding to size scales longer than this limit. Data was
collected by use of this system during Hurricane Belle in August
1976. The first run was made while the hurricane was between
Long Island in New York and the state line between Massachusetts
and Vermont. These results are shown in run 1 in Fig. 10. The
most significant feature of this data is the presence of two very
sharp discontinuities in the spectrum. A second run was made
later, after the hurricane had been over land for a long period
of time, and shows a much reduced structure. By this time,
the storm was no longer defined as a hurricane. For comparison,
a third run was made several days later, and is identified as
run 3. In this case, very little structure is visible.

V.   CONCLUDING REMARKS

A system has been developed for measuring the distribution
of kinetic energy in the atmosphere for comparison to the theory
of Kolmogorov. The system is capable of measurements at remote
sites at a variety of altitudes, and data can be processed from
each run within a few hours time. Indications of changes in
the structure of the spectral distribution of energy related to

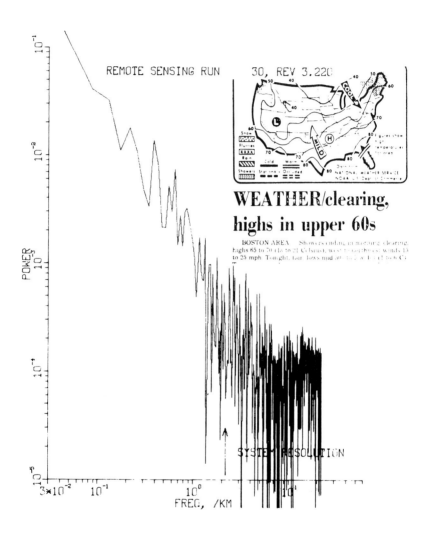

FIGURE 9.   Energy density spectrum.

nearby meteorological conditions have been observed. This
program illustrates the versatility for atmospheric velocity
measurements which is possible by use of CO_2 laser Doppler
radar combined with real-time data-processing capability.

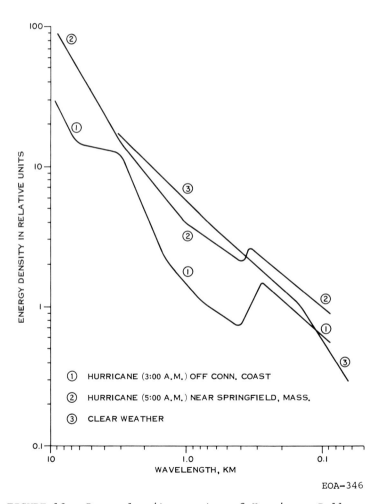

EOA–346

FIGURE 10. Power density spectra of Hurricane Belle.

DISCUSSION

Westwater: Do you have any editing procedure for your data? If there are partial clouds, how do you edit those?

Toomey: We usually just focus down below the cloud cover.

Westwater: Do you have to know that the clouds are there before you can edit the data?

Toomey: Yes, you can see them by how the system performs. High amplitudes of signal make it very apparent when you are hitting clouds.

Westwater: So you cannot edit the effects of clouds from the signal alone without having an operator looking at the sky?

Toomey: Yes. You can make an algorithm, based on S/N versus azimuth.

Rosenkranz: The last slide you showed with the three curves was relative to power, but were the three curves in the true ratio?

Toomey: No. What happened was, when we originally built the system and started the data, we were not keeping track of all the constants that were necessary. We were just looking at the relative spectrum. Later on we went back and did all that was necessary to have an absolute calibration. The relative position of the curves to each other was absolute.

Rosenkranz: Oh, so in fact there was less turbulence in that hurricane nearby?

Toomey: There was less total energy in the atmosphere at that time.

Rosenkranz: What is the explanation for that?

Toomey: I don't know.

Strauch: You said your turbulence spectra were from vertical Doppler velocities—that is when you are vertically pointing.

Toomey: The velocity that we used for the spectral measurements was the magnitude of the vector sum of three components.

DePriest: I have two questions. The first question concerns the relationship between the measured energy and the theoretically derived energy shown in the graph. Were you able to establish a positive correlation?

Toomey: Like I said, about all that we can really state
conclusively was that there would be this kind of break in the
spectrum when a storm was in the immediate area of the system.
That's the phenomena that we usually saw--that the spectra would
come down and then there would be a rise in it and then it would
continue down at something close to a five-third slope.

DePriest: And my second question concerns the new facility that
you developed for EPA. How is that system performing?

Toomey: That system was not used to make this kind of measurement, although it could be. It is used to measure pollutants
out of smoke stacks and the reason that they have the conical
scan is so that they can get an idea of what the atmosphere is
doing around a power plant as well.

Deepak: The EPA LDV system, I believe, was developed for measuring and monitoring the stack effluents by measuring the effluent
velocity and concentration, so that one can calculate the total
amount of effluents put out by the smoke stack within a certain
time period. And often these measurements have to be made from
outside the boundary of the power plant.

Toomey: When we tell the companies that we are coming in, they
clean up everything and there is all of a sudden--no smoke!
Amazing.

Deepak: Can you explain why you use the CO_2 10.6 μm wavelength
for the LDV system?

Toomey: I think that one reason why we use CO_2 10.6 μm wavelength is that we have most experience with CO_2 lasers. We have
developed these lasers for electro-optic systems for a long time,
and they are capable of high power with good coherence for
velocity measurements. CO_2 10.6 μm wavelength is very good in
smoke. I cannot think of any other reason.

SATELLITE-BASED MICROWAVE RETRIEVALS
OF TEMPERATURE AND THERMAL WINDS:
EFFECTS OF CHANNEL SELECTION AND
A PRIORI MEAN ON RETRIEVAL
ACCURACY

Norman C. Grody

National Oceanic and Atmospheric Administration
National Environmental Satellite Service
Washington, D.C.

Temperatures and thermal winds are derived by using three channels from the Microwave Sounding Unit (MSU) aboard Tiros-N for a frontal system over the United States on April 6, 1979. The MSU retrievals are compared with radiosonde measurements and simulations of the MSU and a five-channel sounder using radiosonde computed brightness temperatures. Increasing the number of channels significantly improves the retrieval accuracy, although specific features such as boundary-layer inversions and middle tropospheric changes in lapse rate appear to be averaged out by either three- or five-channel instruments. There is, however, less vertical smoothing in the five-channel case, as shown by the larger, more accurate thermal winds derived in the frontal band compared with the MSU results. It also appears that the temperature retrieval errors at a particular pressure level are strongly dependent on the local differences in curvature between the true temperature and the a priori mean profile, *and is not substantially effected by the absolute difference between these two temperatures. These error characteristics are shown to be in agreement with a general relationship between retrieval error and the departure of the true sounding from the* a priori mean.

I.   INTRODUCTION

Vertical temperature profiles are an important parameter for
observing and forecasting synoptic scale disturbances.   These
must be monitored at least every 12 hours for distances less than
500 km apart, over an area about the size of the United States.
The global network of fixed radiosonde stations provides much of
the required data over the continental areas, but additional
data are needed over the oceans and remote land areas.   Microwave
temperature soundings from satellites can play an important role
in filling this data gap.

The Scanning Microwave Spectrometer (SCAMS) on the Nimbus 6
satellite demonstrated the ability of the three oxygen channels
(52.85, 53.85, 55.45 GHz) for retrieving temperatures between
700 mb to 300 mb with an accuracy less than 3K (root mean
square) under clear and cloudy situations (1).   Over the central
Pacific, the SCAMS-retrieved 1000 mb to 500 mb thickness displayed the development of a short wave, which was initially
undetected from conventional data analysis by the National
Meteorological Center (NMC) (2).   These two investigations showed
the potential of microwave sounders for complementing conventional data and supplying additional information over the ocean.
However, the latter study also exhibited thickness errors
exceeding 90 meters over Europe as a result of extreme liquid
water attenuation in the 52.85 GHz channel due to precipitating-
type clouds.   In the absence of precipitation conditions, the
vertical smoothing in the retrieved profiles depends primarily on
the number of sounding channels, the instrumental noise, the
uncertainties resulting from cloud and surface emissivity variations, and the *a priori* statistical information.

This paper examines the influence of increased sounding
channels and statistical information on the retrieval accuracy
for a frontal system over the United States on April 6, 1979.

Temperatures and thermal winds are first derived by using three channels (53.74, 54.96, 57.95 GHz) from the Microwave Sounding Unit (MSU) aboard Tiros-N and are compared with radiosonde measurements. It should be noted that the MSU instrument also contains a 50.3 GHz channel which lies on the edge of the oxygen band (centered at 60 GHz). However, it requires "window" channels to correct for cloud and surface effects in order to extract temperature information. Its application will be discussed later. The radiosonde measurements are also compared with simulations of the MSU and a five-channel sounder using radiosonde-computed brightness temperatures. The simulated five-channel sounder contains a blend between the SCAMS and MSU channels, having frequencies of 53.40, 54.35, 54.96, 55.50, and 57.95 GHz, and results in a more complete sounder. These five channels are part of the complement of 20 channels being considered by the National Oceanic and Atmospheric Administration (NOAA) for the Advanced Microwave Sounding Unit (AMSU) to be placed aboard a new series of weather satellites to be launched in the mid 1980's. Also examined for this case study is the influence of the *a priori* mean profile on retrieval accuracy, that is, the dependence of retrieval accuracy on the departure of the true sounding from the *a priori* mean temperature.

II. RETRIEVAL ACCURACY DEPENDENCE ON CHANNEL SELECTION AND THE *A PRIORI* MEAN TEMPERATURE PROFILE.

Before examining the measurements and simulated results for the frontal system, it is appropriate to outline some of the basic relationships between brightness temperature measurements and retrieved temperatures. Included are retrieval error statistics obtained for the MSU and five-channel sounder using a simulated climatological data set. Also discussed is the relationship between retrieval error and the departure of the

a priori mean from the true temperature profile.   These results
will be referred to in later sections.

The brightness temperature measurement $T_B(\nu)$ at frequency ν
is linearly related to the atmospheric temperature profile
$T(p)$ by an effective transmittance function $\tau_\nu(p)$ as

$$T_B(\nu) \; = \; \int_{\ln p_s}^{-\infty} T(p) \; \frac{d\tau_\nu(p)}{d \ln p} \; d \ln p \; + \; \tau_\nu(p_s) \; T_s \qquad (1a)$$

where

$$\tau_\nu(p) \; = \; \left[1 - \left(\frac{\hat{\tau}_\nu(p_s)}{\hat{\tau}_\nu(p)} \right)^{2 \sec \theta} \left(1 - \varepsilon_s(\nu,\theta) \right) \right] \hat{\tau}_\nu(p)^{\sec \theta} \qquad (1b)$$

$\tau_\nu(p)$ includes the zenith transmittance $\hat{\tau}_\nu(p)$ at pressure p.
T_s and p_s are the surface temperature and pressure, θ is the
local zenith angle, and $\varepsilon_s(\nu,\theta)$ is the surface emissivity.

The kernel $- \dfrac{d\tau_\nu(p)}{d \ln p}$ in Eq. (1a) is called the temperature
weighting function, which for frequencies in the 50 to 70 GHz
oxygen band is a unique function of pressure for a particular
frequency, local zenith angle, and surface emissivity.   For
nadir and a unity surface emissivity (approximately true for
dry land), Fig. 1 shows the temperature weighting functions
computed for the three MSU channels and the five-channel
sounder, based on the theoretical transmittance model of
Rosenkranz (3).   Note that the functions generally keep the same
shape, but increase in height as the frequency approaches the
center of the oxygen band at 60 GHz.

A pseudoinverse relationship between $T(p)$ and $T_B(\nu)$ is
obtained by correlating temperatures at different pressures
with brightness temperatures at different frequencies, and using
a least-squares error criteria (e.g., Rodgers (4)).   The
regression solution can be written in the form

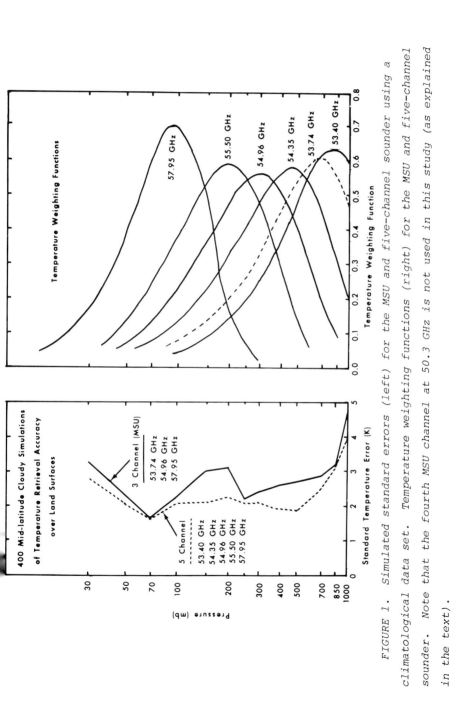

FIGURE 1. Simulated standard errors (left) for the MSU and five-channel sounder using a climatological data set. Temperature weighting functions (right) for the MSU and five-channel sounder. Note that the fourth MSU channel at 50.3 GHz is not used in this study (as explained in the text).

$$\hat{T}(P) = \bar{T}(p) + \sum_{n=1}^{N} a_n(p) \left[T_B(\nu_n) - \bar{T}_B(\nu_n) \right] \qquad (2)$$

where $\hat{T}$ is the estimated (retrieved) temperature, and a_n are the regression coefficients based on the statistical correlation between temperatures and the "N" brightness temperature measurements. In this study the parameters for the sounders are determined through simulations by using a historical data set consisting of 400 mid-latitude radiosonde soundings and corresponding brightness temperatures computed by using Eq. (1). Effects due to cloud and surface emissivity variations over land and a 0.3 K (root mean square (rms)) instrumental noise are included in the simulations as described by Grody and Pellegrino (1). The data sample also provides the *a priori* mean temperature $\bar{T}(p)$ and mean brightness temperatures $\bar{T}_B(\nu_n)$ in addition to the regression coefficients a_n contained in Eq. (2). In the operational use of the MSU data, the National Environmental Satellite Service (NESS) of NOAA obtains the parameters in Eq. (2) from global match-up statistics between nearly coincident radiosonde temperature soundings and the corresponding MSU brightness temperature measurements. The empirically determined parameters are updated weekly and stratified according to latitude zone (tropics, mid-latitude, polar). The purpose of the operational procedure is to remove any uncertainties in the oxygen transmittance model, to adjust for any possible instrumental changes, and to introduce seasonal and climatological statistical information into the parameters.

Equation (1a) is substituted into Eq. (2) to eliminate the brightness temperatures T_B and $\bar{T}_B$ and to obtain a relationship between the retrieval error $\hat{T} - T$ in terms of the departure of the *a priori* mean from the true temperature $\bar{T} - T$ and the "averaging kernel" $-d\rho/d\ln p'$; thus,

$$\Delta \hat{T}(p) = \Delta \bar{T}(p) - \int_{\ln p_s}^{-\infty} \Delta \bar{T}(p') \, \frac{d\rho(p,p')}{d \ln p'} \, d \ln p'$$

$$- \Delta \bar{T}_s \, \rho(p,p_s) \qquad\qquad\qquad (3a)$$

where

$$\rho(p,p') = \sum_{n=1}^{N} a_n(p) \, \tau_n(p') \qquad\qquad\qquad (3b)$$

where $\Delta \hat{T} = \hat{T} - T$, $\Delta \bar{T} = \bar{T} - T$, and $\Delta \bar{T}_{,s} = \bar{T}_s - T_s$.

The averaging kernel $-d\rho/d \ln p$ is a linear combination of single channel temperature weighting functions multiplied by the regression coefficients and generally peaks around pressure p. This relationship is similar to that derived by Conrath and is particularly useful for examining the influence of increased sounding channels and *a priori* statistics on retrieval accuracy (5).

It can be seen from Eq. (3a) that the atmospheric contribution of retrieval error vanishes in the limit as the averaging kernel approaches a Dirac delta function. In actual practice, the retrieval error is minimized by using broad averaging kernels whose shape depends on the temperature weighting functions and the statistically derived regression coefficients according to Eq. (3b). The shape of the averaging kernel is therefore a function of the number of channels, instrumental noise, statistical correlations between channel measurements and level temperatures, and uncertainties arising from cloud and surface emissivity variations. In general, the averaging kernel narrows for increasing channels. This effect is shown in Fig. 2 where the functions are plotted for the 850-mb, 500-mb, 250-mb, and 150-mb levels using the regression coefficients obtained from the mid-latitude data set for the three- and five-channel sounders. Observe that the 500-mb

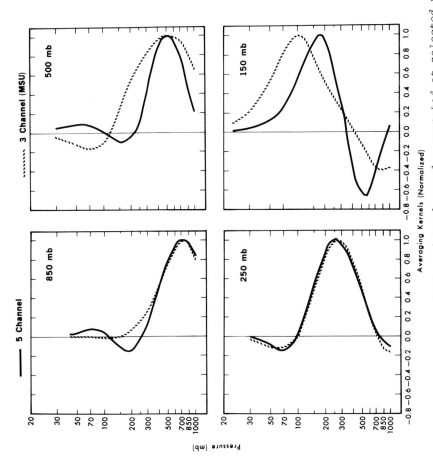

FIGURE 2. Averaging kernels for the MSU and five-channel sounder computed at selected pressure levels.

function is significantly reduced in width by the additional channels, although little improvement occurs for the 850-mb and 250-mb levels. The 150-mb averaging kernel is bimodal because of the large negative correlation between this level and the lower tropospheric sounding channels, and it also shows a sharper positive and negative mode for the five-channel sounder. There is less of such interlevel correlation between the stratospheric and tropospheric temperatures for the other averaging kernels, as evident by their unimodal appearance. Also, the 70-mb averaging kernel (not shown) is mainly defined by the 57.95 GHz temperature weighting function for both the three- and five-channel sounder. In summary, except for the region between 150 mb and 100 mb, it is found that the averaging kernels are basically unimodal, peaking near pressure p, and generally become narrower for increasing channels (particularly near 500 mb).

Figure 1 shows the standard errors obtained for the three- and five-channel sounders. These results were determined from simulations by using the same 400 mid-latitude data sample that was used for obtaining the regression coefficients. The smallest differences between the two standard error curves occur below 700 mb, near 250 mb, and above 100 mb where the averaging kernels are approximately the same for the two sounders. The region below 700 mb also corresponds to the largest errors because of the lack of sharp temperature weighting functions and the atmosphere being most variable in the boundary layer. Large errors occur near the tropopause region around 250 mb where compensating temperatures produce small changes in brightness temperature. However, around 500 mb and near 150 mb, the sharper more optimal averaging kernels for the five-channel sounder results in large improvements in retrieval accuracy

compared with the MSU. Above 100 mb the errors could be substantially reduced if additional stratospheric channels were
included.

Equation (3a) also reveals the dependence of the retrieval
error on the variations between the *a priori* mean and the true
temperature. The error characteristics can be illustrated by
representing $\Delta\bar{T}$ as a Taylor expansion about $\ln p' = \ln p$ in the
integral of Eq. (3a) so that

$$\Delta\hat{T}(p) = (1 - S_o(p)) \, \Delta\bar{T}(p) - \sum_{m=1}^{\infty} S_m(p) \, \frac{d^m \Delta\bar{T}(p)}{d \ln p^m} \qquad (4a)$$

where

$$S_m(p) = \frac{1}{m!} \int_{\ln p_s}^{-\infty} (\ln \frac{p'}{p})^m \, \frac{d\rho(p,p')}{d \ln p'} \, d \ln p' \qquad (4b)$$

where the small surface term contribution in Eq. (3a) is omitted
from Eq. (4a). Although convergence of the Taylor expansion is
implicitly assumed, the argument presented below is not particularly dependent on the representation of $\Delta\bar{T}$, but only on the
fact that it contains different degrees of vertical variation.

Equation (4a) expresses the retrieval error by two different
physical quantities: the first term contains the difference
between the *a priori* mean and true temperature, and the remaining
series of terms contains local derivatives of this difference
temperature. The shape of the averaging kernel is contained in
the integrals $S_m(p)$ which weight the various terms.

From Eqs. (3b) and (4b),

$$S_o(p) = \sum_{n=1}^{N} a_n(p) \, (1 - \tau_n(p_s)) \qquad (5)$$

which is found to differ from unity by about 15 percent for all
the standard pressure levels below 50 mb, excluding the 1000-mb,
150-mb and 100-mb levels. For these levels, the difference

from unity is about 30 percent. These computations were carried out for the three- and five-channel sounder by using the regression coefficients derived from the 400 mid-latitude data set; these results appear to be consistent with results obtained by using other climatological data sets. Based on these results, and by considering an extreme case where the *a priori* temperature is 10 K different from the true temperature, the first term in Eq. (4a) would then contribute a 1.5 K error in the middle troposphere. This is generally much smaller then the remaining terms so that to a good approximation

$$\Delta \hat{T}(p) \simeq - \sum_{m=1}^{\infty} S_m(p) \frac{d^m \Delta \bar{T}(p)}{d \ln p^m} \qquad (6)$$

This result is based on the Taylor series expansion of $\Delta \bar{T}$, and the fact that the area under the averaging kernel is close to unity (i.e., $S_o \simeq 1$) for a wide range of pressures. Alternate techniques for obtaining a more rigorous result are being considered, such as that of using transform analysis to obtain the minimum-mean-square error solution (6).

Equation (6) shows the strong dependence of the retrieval error on the vertical structure of $\Delta \bar{T}$. Hence, *large errors result due to local differential changes between the a priori mean and true temperature,* whereas the absolute difference between the two temperatures only weakly contributes to the error. This point will be dramatized later in comparisons between MSU retrievals with radiosonde temperatures along a cross section through a frontal system. It is shown that although the radiosonde departure from the *a priori* mean is larger for certain cases, the maximum errors occur where the radiosonde temperatures contain the largest structural differences relative to the *a priori* mean profile.

III.   MSU RETRIEVED TEMPERATURES AND THERMAL WINDS

A component of the NESS operational sounding system is the
microwave sounder on the Tiros-N satellite. The MSU is a four
channel (50.30, 53.74, 54.96, and 57.95 GHz) temperature sounder
based on the SCAMS design and provides nearly global coverage
every 12 hours with a 110-km field of view at nadir (increasing
to about 250 km at the extreme scan position). Temperature
sounding information is contained in the three highest frequency
channels which lie on one side of the oxygen band centered at
60 GHz. As mentioned earlier, the 50 GHz channel cannot be used
for providing reliable temperature information without adequate
atmospheric and surface emissivity corrections. It is presently
used to sense the emissivity differences between land and ocean
in order to correct the lowest sounding channel (53.74 GHz) for
the less than 2 K change resulting from the different surface
emissivities. Since the 50 GHz channel is also a sensitive
indicator of liquid water, the data may eventually be used to
edit or possibly estimate the degradations of the 53.74 GHz
measurements due to the large amounts of attenuation encountered
under precipitation conditions (7).

The vertical structure retrieved by the MSU channels is
examined for the frontal system over the United States on
April 6, 1979. Figure 3 displays the brightness temperature
fields determined from the NESS operationally processed MSU
measurements. Unlike the actual MSU measurements, the processed
data is corrected for viewing angle (i.e., limb corrected) and
surface emissivity (the brightness temperatures are normalized
to a unity surface emissivity), in addition to being smoothed
by averaging a number of individual measurements to the observation points shown in Fig. 3. For details on the processing and
utilization of the MSU in operational soundings, the reader is

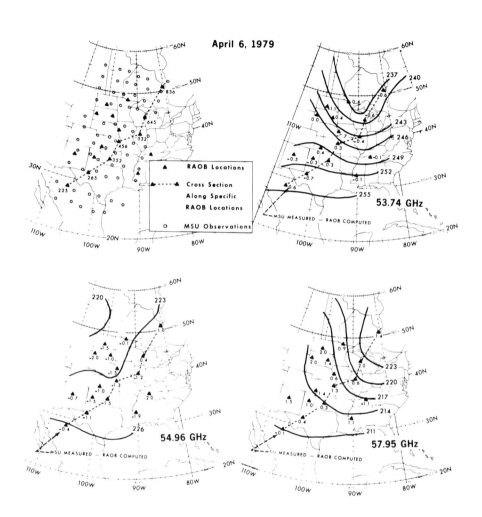

FIGURE 3. Display of radiosonde coverage (1200 GMT) and the available MSU observations (1000 GMT) (upper left corner). Brightness temperature (K) fields for the MSU channels; inserted are the differences relative to radiosonde-computed values (K).

referred to the paper by Smith *et al.* (8). The brightness
temperatures are contoured every 3 K in Fig. 3 and display the
thermal pattern of the frontal system. The 53.74-, 54.96-, and
57.95-GHz measurements represent mean temperatures weighted
around 700 mb, 300 mb, and 90 mb, according to the temperature-
weighting functions shown in Fig. 1. Also shown in Fig. 3 are
the differences between the measured brightness temperatures
(interpolated to radiosonde locations) and that computed from
radiosonde temperatures using Eq. (1), and employing the
theoretical oxygen transmissions model of Rosenkranz (3).
Figure 4 shows the scatter diagram between the 17 measured and
computed brightness temperatures for the MSU channels and also
indicates the mean differences and standard errors. The largest
differences which occur for the two highest frequency channels
are presently unexplainable.

A sample plot of the measured and computed brightness
temperatures are displayed in Fig. 5 for the cross section shown
in Fig. 3 (upper left corner). The 54.96- and 57.95-GHz
channels exhibit the biases found from the complete set of 17
radiosonde comparisons. Figure 6 shows the vertical cross
section analyses of potential temperature and thermal winds[1]
based on the radiosonde (RAOB) temperature measurements as well
as from the MSU retrievals. The retrievals are obtained by
using the measurements in Fig. 5 with the regression parameters
in Eq. (2) generated from the historical data sample discussed
earlier. It is observed that the MSU generally produces a
smoother temperature structure with weaker thermal winds than

---

[1] *Thermal winds are computed from gradients of thickness,
where the lower boundary is set to 1000 mb in this study. As
such, the winds are geostrophic with zero surface wind assumed.*

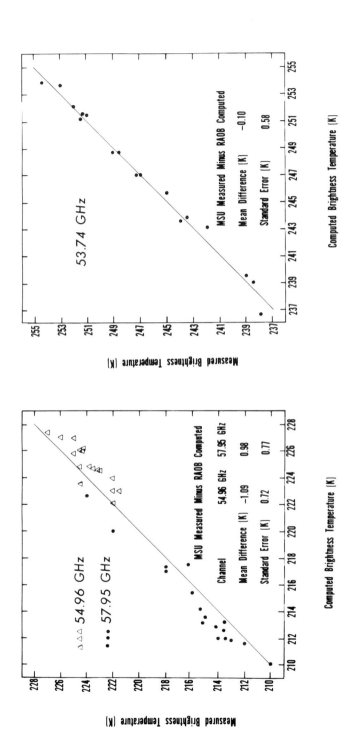

FIGURE 4. Comparisons between MSU measured brightness temperatures and calculated values using radiosonde data. Also indicated are the mean differences and standard errors relative to calculated values.

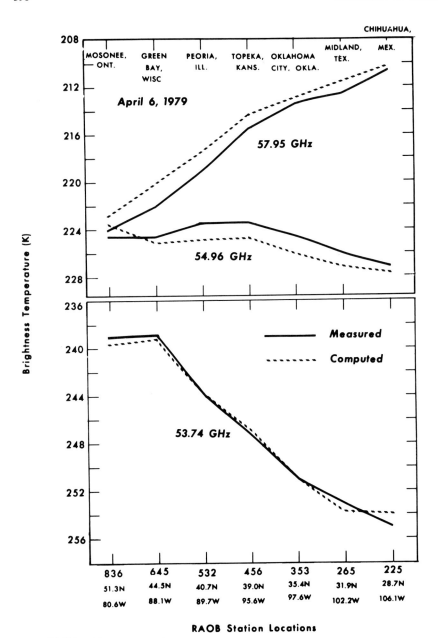

RAOB Station Locations

FIGURE 5.   Comparisons between MSU measured brightness
temperatures and radiosonde-computed values along a cross section
through the frontal system.   (See Fig. 3, upper left corner.)

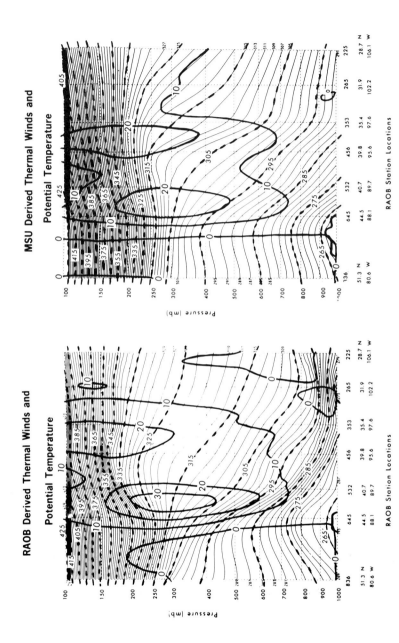

FIGURE 6. Vertical cross sections of thermal winds (heavy lines) (m/sec) and potential temperature (K) determined from radiosonde soundings and MSU retrievals. Cross section is that used in Fig. 5.

that obtained from the radiosonde data. Notice the lack of
vertical resolution for pressures greater than 800 mb, particularly south of 40° latitude where the boundary-layer inversions
produce large vertical gradients in the potential temperature
that are smoothed out in the MSU retrievals. The cold dome in
the frontal band (around 44.5° latitude) is reduced in amplitude
at pressures greater than 700 mb by the MSU results. Vertical
smoothing in the troposphere is particularly revealing in the
frontal zone where the jet level thermal winds are weakened.
The reduced vertical resolution around the tropopause region is
evidenced by comparing the 325 K isotherm with the radiosonde
result, and noticing the reduction in amplitude of the cold
dome around 250 mb near the 40° latitude. Some of this vertical
smoothing could be reduced if more sounding channels were available, as shown by the following simulations.

IV. THREE CHANNEL (MSU) AND FIVE CHANNEL SIMULATIONS

    As indicated in Fig. 1, large improvements in retrieval
accuracy are generally possible by adding more channels. To
examine this for the frontal case, simulated retrievals are
obtained using radiosonde-computed brightness temperatures
for the three MSU channels and the five-channel sounder. The
retrieval algorithms are the regression solutions based on the
historical data sample previously described. Figure 7 shows
the comparisons between the vertical cross sections based on the
three- and five-channel simulated retrievals. Comparing them
with the radiosonde cross section (Fig. 6), the five-channel
results generally indicate more accurate temperatures and thermal
winds than that of the MSU. The temperature structure is not
improved in the boundary layer by the additional channels,
although there are improvements in the troposphere. This is
particularly evident in the frontal band where the five-channel

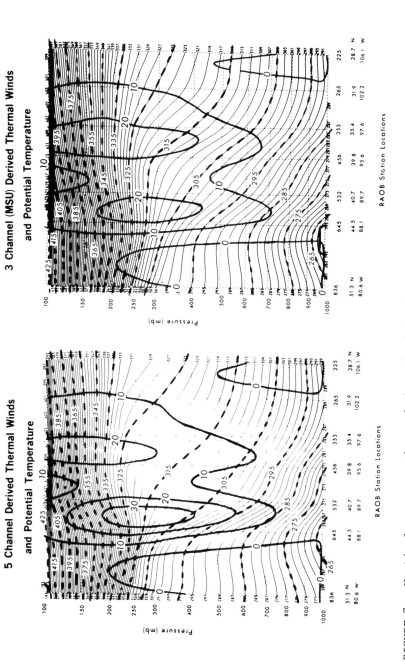

FIGURE 7. Vertical cross sections of thermal winds (heavy lines) (m/sec) and potential temperature (K) determined using three-channel and five-channel retrievals simulated using RAOB data.

results show larger, more accurate thermal winds at the jet
level in comparison with the MSU simulations. Also, the cold
dome in the upper troposphere is better defined by the five-
channel retrievals, as seen by comparing the 325 K isotherm
with the radiosonde and simulated MSU cross sections. In general,
the five-channel temperature retrievals are substantially more
accurate than the three-channel results for the region between
700 mb and 400 mb and between 200 mb and 150 mb, as will be
more clearly shown from detailed analysis presented later.

It is gratifying to note that the MSU retrievals based on
computed and measured brightness temperatures show similar
vertical cross sections of temperature and thermal winds
because much of the differences between the two sets of brightness temperatures is in the form of a bias (see Fig. 5) which
does not affect the derivation of thermal winds. The most
noticeable differences in the potential temperature occur at
pressures less than 300 mb, and are due to the differences
between the measured and computed brightness temperatures for
the 54.96 and 57.95 GHz channels. For a more detailed comparison,
the individual temperature profiles used in generating the
vertical cross sections in Figs. 6 and 7 are considered.

Figure 8 shows the radiosonde (RAOB) temperature profiles
(solid line), retrievals based on the MSU brightness temperature
measurements (dashed line), and the simulated retrievals from
the MSU (Δ) and the five-channel sounder (o) which are based on
the radiosonde computed brightness temperatures. The MSU
retrievals based on measurements generally appear as a smoothed
version of the radiosonde profiles, and average out the boundary-
layer inversions and the mid-tropospheric change in lapse rate
seen in the frontal band (station 645). Notice, however, that
the mid-latitude data set (used in obtaining the regression
coeeficients and a priori means) appears to be adequate for

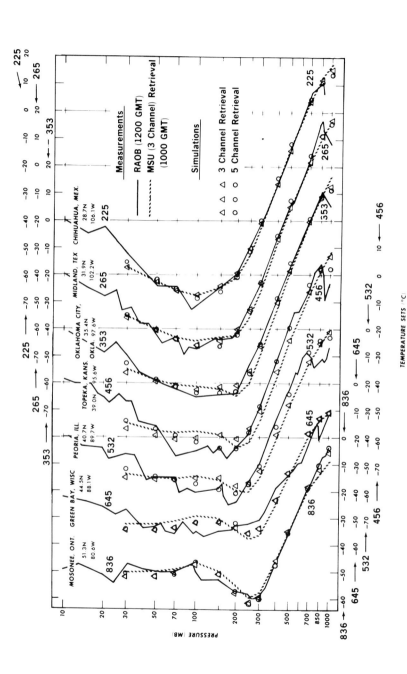

FIGURE 8. Comparisons between radiosonde soundings, MSU retrievals, and simulated MSU and five-channel retrievals. The data corresponds to that used in generating Figs. 6 and 7.

retrieving both the extreme northern and southern soundings. The
largest retrieval errors occur within the frontal zone where
there are large-amplitude small-scale fluctuations that are
unresolved by the three MSU channels. It is clear from the small
differences between the "measured" and simulated MSU retrievals
that the problems encountered in the frontal region are not a
result of instrumental errors or cloud effects; the stations 836
and 645 were, in fact, cloud covered. In general, the largest
differences between the two MSU retrievals occur for pressures
less than 300 mb and are due to the differences in the measured
and computed brightness temperatures mentioned previously.

It is evident from Fig. 8 that the simulated five-channel
retrievals follow the radiosonde profiles with greater accuracy
(see stations 532 and 456) than the MSU retrievals (simulated or
measured). However, even the five-channel sounder appears to be
ineffective in resolving the boundary-layer inversions or the
lapse-rate change in the frontal band. The vertical distribution
of errors for the two simulated sounders are summarized in Fig. 9
by displaying the rms and mean differences based on the seven
retrieval against radiosonde comparisons. Although these results
are for a localized data set of seven samples, the rms difference
plots have the same characteristics as Fig. 1 which was obtained
by using the 400 mid-latitude dependent data sample. Unlike the
dependent data set, mean differences are found for the frontal
case. The effect of adding more channels is seen to improve the
standard error as well as the mean difference. In summary, the
distribution of weighting functions for the five-channel sounder
produces more accurate retrievals than the three-channel results,
particularly for the regions between 700 mb and 400 mb and
between 200 mb and 150 mb.

From the relationship developed earlier (Eq. 6), it was
deduced that the retrieval error is strongly influenced by the
variability between the true temperature and the *a priori* mean

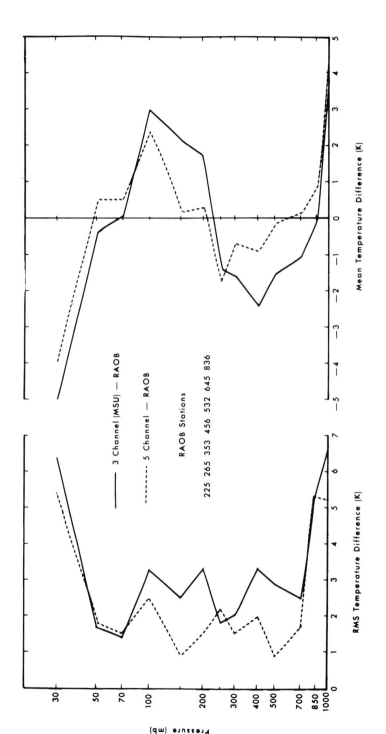

FIGURE 9. Vertical distribution of root-mean-square and mean differences relative to radiosonde for the simulated MSU and five-channel retrievals shown in Fig. 8.

profile. This point is well dramatized for the frontal case by
referring to the difference temperature profiles plotted in
Fig. 10. The simulated three- and five-channel retrieval
errors are individually displayed along with the difference
between the *a priori* mean and radiosonde temperature. Consistent
with the theory, one observes that although the radiosonde
departure from the *a priori* temperature is largest for the
northernmost and southernmost stations, larger MSU errors occur
for the three centrally located stations (532, 456, and 353).
These three middle latitude radiosonde profiles are closest to
the mean temperature of the 400 mid-latitude data sample but
show large variability relative to this *a priori* mean temperature.
It is, however, interesting to note that for these profiles,
the additional channels provide the largest improvements in
retrieval accuracy, and this result is reflected in the difference curves of Fig. 9.

V.   CONCLUSIONS AND FUTURE CONSIDERATIONS

A detailed analysis of a frontal system was performed by
using retrievals based on the three MSU sounding channels and
a simulated five-channel sounder. Retrieval errors of temperature and thermal winds are significantly improved by increasing
the number of channels, although specific features such as
boundary-level inversions and mid-tropospheric changes in lapse
rate appear unresolvable by the two sounders. It is found that
temperature profiles having large-amplitude small-scale
fluctuations relative to the *a priori* mean result in the largest
retrieval errors. This dependency of retrieval error on the
variability between the true temperature and the *a priori* mean
is also expressed analytically, and reveals the insensitivity of

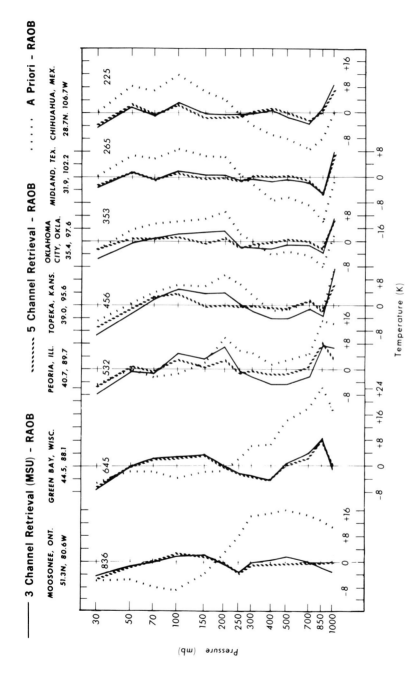

FIGURE 10. *Vertical profiles of the differences between simulated retrievals and radiosonde temperatures for the MSU and five-channel sounder. Also shown are the differences between the a priori mean and radiosonde temperatures. The data corresponds to those in Fig. 8.*

retrieval accuracy to gross changes in statistics, that is, where the a priori value does not contain the small-scale vertical structure associated with the actual synoptic situation.

The NESS operational sounding system employs continuous updating of statistical regression parameters every week to remove any uncertainties in the oxygen transmittance model, to adjust for possible instrumental changes, and to introduce seasonal and climatological statistical information. However, the a priori statistics are obtained from a global data base which is only stratified according to the three basic latitude bands, and not by synoptic situation. The difficult task of introducing localized (synoptic) stratification to improve the vertical structure of the a priori value is still in the future.

ACKNOWLEDGMENTS

The author greatfully acknowledges Mr. Harold Brodrick of NOAA/NESS for developing the many computer programs central to the analysis of the Tiros-N data, and for his many suggestions. Also to be thanked are Dr. Arnold Gruber and Dr. Burt Morse of NESS for their recommendations and critical review of the material.

SYMBOLS

a_n	regression coefficients
$-d\rho/d \ln p$	averaging kernel (sum of temperature weighting functions weighted by the regression coefficients)
$-d\tau/d \ln p$	temperature weighting function
p	atmospheric pressure
p_s	surface pressure
S_m	weighted area of averaging kernel

Grody: The clouds are basically modeled as Rayleigh absorbing media and the emissivity having a random variable component.

Westwater: Do you model the height and thickness and liquid water content distributions of the clouds?

Grody: Yes, I put in all ranges of cloud top-cloud bottom and liquid-water density. Every profile--there are essentially 400 profiles that are used in generating the algorithms--in the ensemble contains some cloudiness. Some atmospheres are less cloudy than others because the top and the bottom pressure of the clouds and liquid-water density are uniformly distributed by a random number generator operating between fixed upper and lower limits of the cloud parameters.

Susskind: In the particular results that you showed[2] for the wind fields for April 6--were those conditions clear or cloudy?

Grody: The first two most northerly soundings were cloudy. The last five were clear.

Susskind: Were the quality of the retrievals basically the same?

Grody: Yes, there was not any obvious difference that I could see between the clear and cloudy results. Also, the simulations were the same as the measurements for the clear as well as cloudy situations.

Westwater: I wonder if there are any infrared people who would comment on deriving thermal waves from the radiance observations in the infrared and how they would compare with the microwave inferences that Norm Grody has shown?

Susskind: I would just like to ask a question with regards to that question. Were you deriving winds from the observations or deriving winds from the retrievals based on the observations?

Grody: In principle, I see no reason why you cannot do the same. Obviously, from the observations themselves in the infrared, if they are cloud contaminated, something has to be done to them. That can be taken into account in the retrieval and the same thing can be done. I like the idea of what you are showing and we are going to begin to try to do winds from our retrievals and see what they look like. Maybe at the next meeting we will report on how they look.

---

[2] See Fig. 6 in this paper.

McMillin: Just a comment. There are some retrievals being done
now that way with infrared retrievals and they look good. I am
not involved directly in it. Particularly, with retrievals that
are in good cloud conditions look very impressive.

INFRARED REMOTE SENSING OF SEA SURFACE TEMPERATURE[1]

M. T. Chahine

Jet Propulsion Laboratory
Pasadena, California

The surface temperature T_S of the ocean and solid earth can be derived from the radiance data measured in the 3.7 μm transparent region between 2700 and 2500 cm^{-1} as well as from the 11 μm water vapor continuum between 960 and 775 cm^{-1}. The effects on the accuracy of the recovered values of T_S of surface emissivity, reflection of solar radiation and variations in the concentration of water vapor in the atmosphere are different in the two bands. In this paper the accuracy of the surface temperature derived from each of these two transparent spectral regions is discussed. The possibility of determining the difference between the air temperature at the surface and the true skin surface temperature is also considered.

[1]*This paper presents the results of one phase of research carried out at the Jet Propulsion Laboratory, California Institute of Technology, and was supported in part by the National Aeronautics and Space Administration under Contract NAS 7-100 and by the National Science Foundation, Office of the International Decade of Ocean Explorations, and the Office of Naval Research in support of the NORPAX Program, under NSF Grant OCE 78-20991.*

I.   INTRODUCTION

   Accurate determination of sea surface temperature is essential
for ocean and climate studies. The thermal structure of sea
surface plays a critical role in regulating the exchanges between
the atmosphere and oceans, and a knowledge of sea surface temperature is important for verifying circulation models. Satellite
measurements of the upwelling radiance from the Earth's surface
and atmosphere have been carried out in a routine manner for many
years. However, the interpretation of such data for accurate
determination of sea surface temperature remains a challenge.

   Because the atmosphere is not completely transparent even in
the least absorbing regions of the infrared spectrum, the outgoing spectral radiance of the Earth is influenced not only by
the Earth's surface but also by the composition and thermal
structure of the atmosphere. The observed radiance is further
modified by the presence of clouds and by scattered solar
radiance. It is therefore necessary to take all these factors
into account if accurate, reliable, sea surface temperatures are
to be obtained.

   In this paper the components of the surface emission are
derived and the sea surface temperature is expressed in terms of
the observed radiance and other related surface and atmosphere
parameters. The advantages and disadvantages of the spectral
regions used for obtaining the surface radiance are examined.
Experimental results will be presented and a brief discussion of
the possibility of determining the difference between the sea
surface temperature and the air temperature just above the interface will be offered.

II.   ELEMENTS OF SURFACE RADIANCE MEASUREMENTS

The total spectral radiance $I(\nu,\theta)$ observed at frequency ν, and zenith angle θ can be expressed in terms of four main components:

$$\overline{I}(\nu,\theta) = I_s(\nu,\theta) + I_a(\nu,\theta) + I_d(\nu,\theta) + I_h(\nu,\theta) \qquad (1)$$

where

$\overline{I}(\nu,\theta)$ clear-column radiance

$I_s(\nu,\theta)$ surface emission

$I_a(\nu,\theta)$ atmospheric emission

$I_d(\nu,\theta)$ reflected thermal downward-flux

$I_h(\nu,\theta)$ reflected solar flux

The geometry associated with observations made when the instrument is viewing the Earth at a zenith angle θ is shown in Fig. 1. The surface is assumed to be horizontal so that the local normal

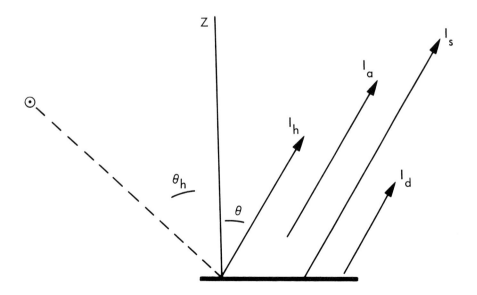

FIGURE 1.   Components of surface emission.

is parallel to the vertical axis. In general, the outgoing radiance has both zenith and azimuth angular dependence; in this section the azimuthal dependence will be suppressed. The cloud effect will be only bri' fly discussed in this paper because it has been given in detail by Chahine (1-4).

A. *Clear-Column Radiance*

It is very difficult to be certain that an observed field of view is cloud free, no matter how small the field of view is. A method was proposed by Chahine to derive the clear-column radiance from observations made in two different spectral bands over adjacent fields of view (1-4). In this case $\bar{I}(\nu,\theta)$ may be written as

$$\bar{I}(\nu,\theta) = \tilde{I}_1(\nu,\theta_1) + \eta\left[\tilde{I}_1(\nu,\theta_1) - \tilde{I}_2(\nu,\theta_2)\right] + \dots . \qquad (2)$$

where $\tilde{I}_1(\nu,\theta_1)$ and $\tilde{I}_2(\nu,\theta_2)$ are the observed radiance in the first and second field of view, respectively. When the two fields are small and contiguous, assume

$$\theta \approx \theta_1 \approx \theta_2$$

and solve for the constant η in a manner described in Refs. 1 to 4.

B. *Surface Emission*

The emitted radiance reaching the instrument from the surface can be written as

$$I_s(\nu,\theta) = \varepsilon_s(\nu,\theta)\, B(\nu,T_s)\, \tau(\nu,\theta,z_s) \qquad (3)$$

where $\varepsilon_s(\nu,\theta)$ is the surface emissivity.

The Planck function $B(\nu,T)$ at frequency ν and temperature T is given by

$$B(\nu,T) = a\nu^3 / (e^{\frac{b\nu}{T}} - 1) \tag{4}$$

where a and b are constants, and $\tau(\nu,\theta,z_s)$ is the atmospheric spectral transmittance between the surface and the instrument.

C. Atmospheric Emission

The atmospheric emission for a plane, parallel, and homogeneous atmosphere in local thermodynamic equilibrium can be expressed as

$$I_a(\nu,\theta) = \int_{z_s}^{\bar{z}} B[\nu,T(z)]\ \frac{\partial\tau(\nu,\theta,z)}{\partial z}\ dz \tag{5}$$

where T(z) is the vertical atmospheric temperature profile as a function of height z, and $\tau(\nu,\theta,z)$ is the clear-column atmospheric transmittance between height z and the observing instrument at $\bar{z}$. For most atmospheric conditions the use of the trapezoidal rule is adequate for evaluating Eq. (5).

D. Reflected Thermal Downward Flux

The reflected thermal downward flux originates from the atmosphere above the surface. In general, it is seen from Fig. 2 that the surface element dA receives radiant energy through an elementary beam of solid angle $d\Omega_i$ from a direction (θ_i,ϕ_i). The radiant energy reflected into the solid $d\Omega_r$ in the direction (θ_r,ϕ_r) comes from all directions above the surface. This radiant energy will be attenuated by the atmosphere when it traverses the atmospheric layers between the surface and the observing system. At the observation point, the reflected thermal downward flux can be expressed in its general form as

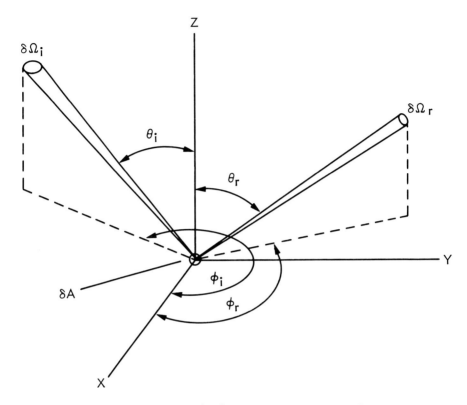

FIGURE 2. Geometry of incident and reflected elementary geometry.

$$I_d(\nu,\theta_r,\phi_r) = \tau(\nu,z_s,\theta_r,\phi_r) \int_{FOV} dA \int_{Hemisphere} d\Omega_i$$

$$I_i(\nu,\theta_i,\phi_i) \; \rho(\nu,\theta_i,\phi_i,\theta_r,\phi_r) \; \cos\theta_i \qquad (6)$$

where $I_i(\nu,\theta_i,\phi_i)$ is the atmospheric downward radiation in the direction (θ_i,ϕ_i); $d\Omega_i = \sin\theta_i \; d\theta_i \; d\phi_i$; and $\rho(\nu,\theta_i,\phi_i,\theta_r,\phi_r)$ is the partial spectral reflectance.

The expression for I_d can be simplified by assuming an optically thin isotropic atmosphere where

$$I_i(\nu,\theta_i,\phi_i) = I_a(\nu,0)/\cos \theta_i \qquad (7)$$

where $I_a(\nu,0)$ is given by Eq. (5) for $\theta = 0$. Furthermore, by assuming that

$$\rho(\nu,\theta_i,\phi_i,\theta_r,\phi_r) = \rho(\nu)$$

Equation (6) for $\theta = \theta_i$, can be written as

$$I_d(\nu,\theta) = 2 \pi\rho(\nu) I_a(\nu,0) \tau(\nu,\theta,z) \qquad (8)$$

where $\tau(\nu,\theta,z)$ is the transmittance of a clear-column of absorbers between the surface and the instrument taken along the direction θ.

Note that for a perfect Lambertian surface, the directional surface reflectivity $\rho_s(\nu)$ is related to $\rho(\nu)$ and $\varepsilon_s(\nu)$ by

$$\rho_s(\nu) = \int_{\text{Hemisphere}} \rho(\nu) \cos \theta \, d\Omega = \pi\rho(\nu) = 1 - \varepsilon_s(\nu) \qquad (9)$$

E. *Reflection of Solar Flux*

During daytime observations, an additional term caused by scattering of solar flux from the surface, $I_h(\nu,\theta)$, should be included. The sun can be considered as a source subtending a small solid angle $d\Omega_h$ in the direction θ_h. The reflected solar flux measured by the instrument can be written as

$$I_h(\nu,\theta) = C H_h(\nu) \cos \theta_h \, \tau(\nu,\theta_h,z_s) \rho_h'(\nu) \tau(\nu,\theta,z_s) \qquad (10)$$

where the solar irradiance for normal incidence outside the atmosphere at $T_h = 5600$ K is

$$H_h(\nu) = 2.16 \pi 10^{-5} B(\nu,T_h) \qquad (11)$$

where θ_h is the direction of sunbeam, $\tau(\nu,\theta_h,z_s)$ is the trans-mittance of entire atmospheric clear column traversed by solar flux, $\tau(\nu,\theta,z_s)$ is as defined in Eq. (5), and C is the fraction of the solid angle $d\Omega_h$ *not* covered by clouds.

III.   DETERMINATION OF SEA SURFACE TEMPERATURE

Substituting Eqs. (3), (4), (8), (9), (10) and (11) into Eq. (1) yields

$$\bar{I}(\nu) = \tau(\nu,z_s) \left\{ \epsilon_s(\nu) \ B(\nu,T_s) + [1 - \epsilon_s(\nu)] \ I_a^{\uparrow}(\nu) \right.$$
$$\left. + CH_h(\nu) \ \cos \theta_h \ \rho_h'(\nu) \ \tau_h(\nu,\theta_h,z_s) \right\} + I_a(\nu) \qquad (12)$$

with

$$I_a(\nu) = \int_{z_s}^{\bar{z}} B[\nu,T(z)] \ \frac{\partial \tau(\nu,\theta,z)}{\partial z} \ dz \qquad (13)$$

$$I_a^{\uparrow}(\nu) = 2 \int_{z_s}^{\bar{z}} B[\nu,T(z)] \ \frac{\partial \tau(\nu,0,z)}{\partial z} \ dz \qquad (14)$$

Now group the terms representing atmospheric attenuation and emission in Eq. (12) and write

$$\frac{\bar{I}(\nu) - I_a(\nu)}{\tau(\nu,z_s)} = \bar{\epsilon}_s \ B(\nu,T_s) + (1 - \bar{\epsilon}_s) \ I_a^{\uparrow}(\nu)$$
$$+ C \ \bar{\rho}_h' \ \tau_h(\nu,\theta_h,z_s) \ H_h(\nu) \ \cos \theta_h \qquad (15)$$

To solve for T_s, it is assumed that $\epsilon_s(\nu)$ and $\bar{\rho}_h'(\nu)$ are indepen-dent of frequency within a range $\nu'' < \nu < \nu'$. This assumption will allow the grouping of the two terms in the reflected solar flux component as

$$\gamma(\nu) = C \, \overline{\rho}'_h(\nu) \tag{16}$$

and this term can be treated as a single unknown parameter independent of frequency in the range $\nu'' < \nu < \nu'$.

The sea surface temperature can now be expressed by

$$B(\nu, T_s) = \frac{1}{\varepsilon_s} \left[\frac{\overline{I}(\nu) - I_a(\nu)}{\tau(\nu, z_s)} - (1 - \varepsilon_s) \, I_a^{\uparrow}(\nu) \right.$$

$$\left. - \gamma \, H_h(\nu) \, \cos \theta_h \, \tau_h(\nu, \theta_h, z_s) \right] \tag{17}$$

T_s can be computed from Eq. (17) provided that information is available concerning clear-column radiance $\overline{I}(\nu)$, atmospheric transmittances $\tau(\nu, \theta, z)$, atmospheric temperature profile $T(z)$, and sun zenith angle θ_h. Furthermore, if it is assumed that the surface is isothermal, the surface is a diffuse reflector, and the atmosphere is optically thin and isotropic, Eq. (17) can be solved for T_s, ε_s, and γ.

The determination of T_s, ε_s and γ requires three sounding frequencies for daytime observations or two sounding frequencies for nighttime ($H_h = 0$) observations. Such a set of frequencies could be obtained from the 3.7 μm part of the spectrum, as shown by Shaw (5).

For observations near the 11 μm or 9 μm water vapor windows of the spectrum, the solar reflection term can be neglected but the dependence of the atmospheric transmittance on the local concentration of water vapor introduces a new unknown factor which should be determined.

IV. METHOD OF SOLUTION

The surface temperature T_2 appears in a nonlinear form in Eq. (17). Its determination is based on the formulation given by

Chahine (10) and can be carried out by iterations using a simple
relaxation equation of the form

$$e^{\dfrac{b\bar{\nu}}{T_s^{(m+1)}}} = e^{\dfrac{b\bar{\nu}}{T_s^{(m)}}} \; F^{(m)}(T_s^{(m)}, \ldots) \tag{18a}$$

or alternatively

$$\dfrac{b\bar{\nu}}{T_s^{(m+1)}} = \dfrac{b\bar{\nu}}{T_s^{(m)}} + \ln F^{(m)} \tag{18b}$$

where m is the order of iteration, $\bar{\nu}$ is an average frequency and
$F^{(m)}$ is derived from Eq. (17) as follows.

In general, the formulation of F requires the use of three
surface sounding frequencies ν_1, ν_2, and ν_3 chosen such that

$$B(\nu_1,T_s) \gg B(\nu_2,T_s) \gg B(\nu_3,T_s) \tag{19}$$

The resulting system of three equations corresponding to ν_1, ν_2,
and ν_3 has three unknowns T_s, ε_s, and γ. One can easily eliminate
the linear terms ε_s and γ and reduce the system of three equations
into one equation. If one groups all $B(\nu_i,T_s)$ terms on the left
hand side and all remaining terms on the right hand side of the
resulting equation, one obtains the expression for $F^{(m)}$ as the
ratio of the left hand side over the right hand side.

For the simple case of nighttime observations, one has
$H_n(\nu) = 0$. The derivation of $F^{(m)}$ will lead to a grouping of the
$B(\nu_i,T)$ terms as

$$B(\nu_1,T_s) - \alpha B(\nu_2,T_s) = I_a^\uparrow(\nu_1) - \alpha I_a^\uparrow(\nu_2) \tag{20}$$

where

$$\alpha = \dfrac{\left[\dfrac{\bar{I}(\nu_1) - I_a(\nu_1)}{\tau(\nu_1,z_s)} - I_a^\uparrow(\nu_1) \right]}{\left[\dfrac{\bar{I}(\nu_2) = I_a(\nu_2)}{\tau(\nu_2,z_s)} - I_a^\uparrow(\nu_2) \right]} \tag{21}$$

Hence

$$F^{(m)} = \frac{B(\nu_1, T_s^{(m)}) - \alpha \, B(\nu_2, T_s^{(m)})}{I_a^{\uparrow}(\nu_1) - \alpha \, I_a^{\uparrow}(\nu_2)} \tag{22}$$

and the value of $\bar{\nu}$ in Eq. (18) becomes

$$\bar{\nu} = \frac{\nu_1 + \nu_2}{2} \tag{23}$$

The iteration process starts with an initial guess (m = 0) taken from Eqs. (20) and (4) as

$$T_s^{(0)} = \frac{-b\nu}{\ln\left[\dfrac{\bar{I}(\nu) - I_a(\nu)}{a\nu^3 \, \tau(\nu, z_s)}\right]} \tag{24}$$

Subsequent iterations are obtained by substituting $T_s^{(m)}$ into Eq. (22) and on the right hand side of Eq. (18) to calculate $T_s^{(m+1)}$. Repetition of this process leads to rapid convergence because F is a monotonic function of T_s. Convergence occurs when $F^{(m)} \to 1$.

The rate of convergence depends on the properties of the selected set of sounding frequencies. In the case of Eq. (22), the optimum set of frequencies should be selected to satisfy Eq. (19) and to ensure that the numerator and denominator of Eq. (22) are different from zero and larger than the magnitude of the effects of uncertainty in the data.

V. PROPERTIES OF THE 11 μm AND 3.7 μm SPECTRAL WINDOWS

The ideal spectral regions for remote sensing of sea surface temperature are those for which a simple direct relationship exist between the clear column radiance $\bar{I}(\nu)$ and the sea surface Planck emission $B(\nu, T_s)$. According to Eq. (17), this would be true if $\tau(\nu, z_s) = 1$, $\varepsilon_s(\nu) = 1$, and $\gamma(\nu) = 0$ (or for night observations when $H_h(\nu) = 0$). Two spectral regions around the 3.7-µm CO_2 window between 2700 cm^{-1} and 2500 cm^{-1} and 11-µm H_2O window between 960 cm^{-1} and 775 cm^{-1} are suitable for remote sensing of sea surface temperature. The accuracy of the sea surface temperature recovered from these two regions depend on the wavelength of observations, the degree and variability of atmospheric opacities, and surface emissivities. The causes of inaccuracy in T_s will now be discussed.

A. *Dependence of Accuracy of T_s on Wavelength of Observations*

According to Eq. (4), the effect of errors (or uncertainties) in the Planck function $\Delta B/B$ on the accuracy ΔT_s of the recovered sea surface temperature can be expressed as

$$\Delta T_s = \left[\frac{T_s^2 \left(1 - e^{-\frac{b\nu}{T}}\right)}{b\nu} \right] \frac{\Delta B}{B} \qquad (25)$$

From Eq. (18) it is clear that for a given error $\frac{\Delta B}{B}$, the resulting temperature error ΔT_s is larger for $\nu = 900\ cm^{-1}$ than for $\nu = 2500\ cm^{-1}$ as illustrated in Fig. 3. This property clearly shows one of the advantages of observations in the short wavelength region of the infrared spectrum.

B. *Sea Surface Emissivity ε_s*

Water is essentially opaque to infrared radiation longer than 3 µm. But, even though $\varepsilon_s(\nu) \approx 1.0$, accurate knowledge of ε_s is required for accurate determination of T_s. Values of ε_s are illustrated by Wolfe and by McCalister (6,7).

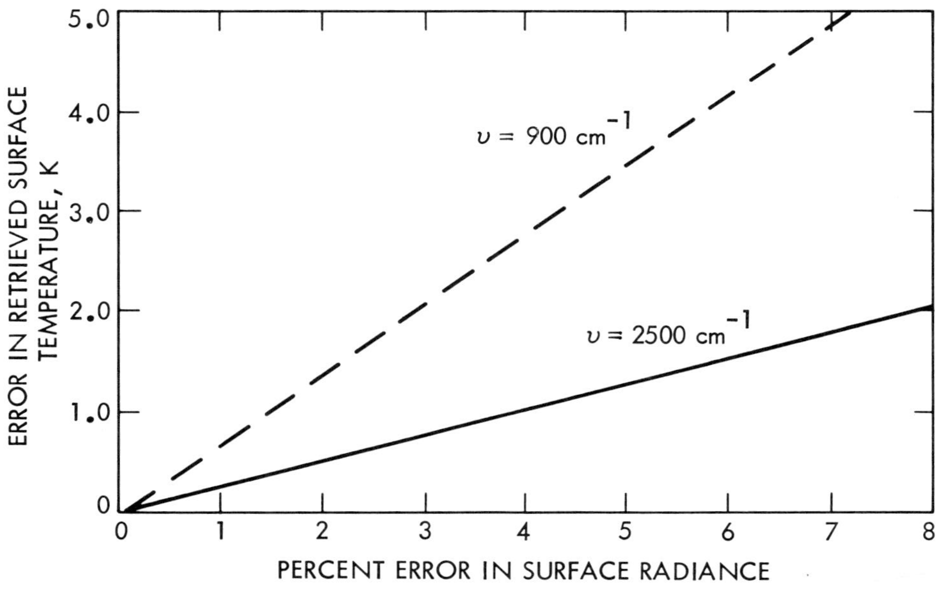

FIGURE 3. Effect of wavelength on the rate of amplification of radiance noise.

C. *Atmospheric Attenuations*

The energy emitted by the surface is attenuated by a factor $\tau(\nu,\theta,z_s)$ in passing through the atmosphere. In addition, the atmosphere emits energy equal to $I_a(\nu)$ as given in Eq. (13). If $\tau(\nu,\theta,z_s)$ is large, say $\tau(\nu,\theta,z_s) \approx 0.95$, then the atmospheric attenuation and emission will be small.

In the 3.7-μm regions the atmospheric transmittance for broad band observations with $\nu/\Delta\nu = 100$ can be large with $\tau(\nu,0,z_s) \approx 0.9$. Consequently, the corresponding atmospheric emission term $I_a(\nu)$ is small compared with the surface emission term with $\dfrac{I_a(\nu)}{B(\nu,T_s)} \approx 10\%$. The transmittance in the 3.7-μm region is generally independent of latitudinal locations because the corresponding atmospheric absorption is due mainly to CO_2, N_2, and N_2O, and weakly to H_2O and CH_4. (In certain very narrow parts of the 3.7-μm region, such as near 2686 ± 2.5 cm^{-1}, the atmosphere is more transparent with $\tau(\nu,0,z_s) \approx 0.98$.)

In the 11-μm region the transmittance is a strong function of the amount of water vapor in the atmosphere and thus τ tends to have a strong latitudinal and regional dependence. For relatively "dry" regions $\tau(\nu,0,z_s) \approx 0.9$ whereas for humid atmospheres $\tau(\nu,0,z_s)$ can become as small as 0.3. The atmospheric emission $I_a(\nu)$ can become large with $\dfrac{I_a(\nu)}{B(\nu,T_s)} = 60\%$. A method to account for the effects of water vapor has been proposed by Prabhakara (9).

D. *Reflection of Solar Radiation*

The reflected solar radiation $I_h(\nu)$ in the 3.7-μm region can be as large as the surface emission as shown in Fig. 4. This is due to the facts that (1) for the sun at $T_h = 5600$ k, the value of the corresponding Planck function $B(\nu,T_h)$ is larger for $\nu = 2500$ cm^{-1} than for $\nu = 900$ cm^{-1}, and (2) for the

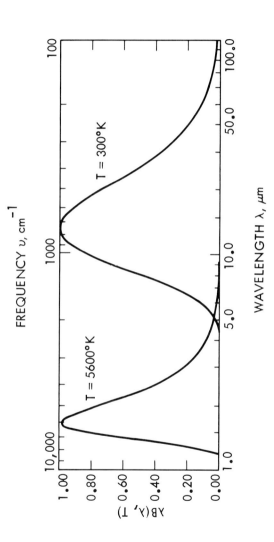

FIGURE 4. The Planck function at the equivalent temperatures of the Earth's surface and sun.

ocean surface as T_s = 300 k, the value of the surface Planck function at ν = 2500 is smaller than that of ν = 900. Hence, for observations in the 3.7-μm region $I_h(\nu)$ plays a very important role for daytime observations, but in the 11-μm region I_h can be safely neglected at all times.

E. The Clear Column Radiance $\overline{I}(\nu)$

The effects of scattering by clouds and hazes play an important role in the determination of clear column radiances, particularly in the short wavelength parts of the spectrum. For horizontally inhomogeneous clouds the method derived by Chahine leads to accurate determination of $\overline{I}(\nu)$ for both 3.7-μm and 11-μm observations (1-4). However, for horizontally uniform cloud and haze layers, the problem is difficult for both regions, but it is relatively less severe for the case of the 11-μm observations.

VI. EXPERIMENTAL VERIFICATIONS

To verify the advantages of determining the sea surface temperature from observations in the 3.7-μm or 11-μm regions, a multidetector grating spectrometer capable of measuring the outgoing radiance simultaneously in all sounding channels was constructed and flown on an aircraft. Details of the instrument design and calibration are described elsewhere by Aumann and Chahine (11). The multidetector approach is necessary here in order to ensure that all the sounding frequencies observe the same clouds at the same time.

Five separate flights of the multidetector instrument were carried out on the NASA-P3A aircraft during the month of July 1975. The flights were made over the Gulf of Mexico and its surroundings under different sun angles and types of clouds.

TABLE I. *The Set of Sounding Frequencies for the Aircraft-Mounted Sounder and Their Primary Objectives*

Channels	Frequencies	Objectives
1	2660 cm^{-1}	Surface temperature (day and
2	2601	night)
3	2517	
4	2298	
5	2281	
6	2260	Atmospheric temperature
7	2241	profile
8	2222	
9	2203	
10	2187	
11	2170	
12	1885	
13	1863	Water vapor profile
14	1843	
15	900	Surface temperature
16	773	
17	744	Clouds effects
18	726	

A list of the sounding frequencies is given in Table I with a description of their primary functions. The spectral resolving power $\nu/\Delta\nu$ is 100 for the first 15 detectors and 50 for the remaining ones. The experimental signal-to-noise ratio was in excess of 100:1 for the first 15 detectors and ranged between 100:1 and 40:1 for the remaining channels.

The data from *all* the channels were analyzed as described
by Chahine *et al.* to recover the clear column radiance, atmospheric temperature profile and the humidity profile (12).
A value for ε_s = 1 in the 11-μm region was established by
flying at very low altitudes over the sea surface and by comparing the observed brightness temperature with the measured
bucket temperature under the aircraft. This knowledge was used
to calculate the terms on the right-hand side of Eq. (17).

By comparing the results of T_s obtained from the 3.7-μm and
11-μm sounding channels, the 3.7-μm channels were found to be
more useful because of the possibility of eliminating the errors
introduced by variations in the surface emissivity and solar
reflections and because the surface temperatures derived from
the 3.7-μm region have the added advantage of being less
dependent on errors in measurement and on uncertainties in
the corresponding atmosphere attenuation. The results from this
experiment for one afternoon flight are given in Fig. 5. All
the sea surface temperatures given in Fig. 5 were derived from
the 3.7-μm channels.

The errors in recovering T_s from the 11-μm channel originated
mainly from the difficulty to account accurately for water vapor
atmospheric attenuation. This conclusion coincides with the
conclusions of Barnett *et al.* (13).

VII. DETERMINATION OF THE AIR-SEA TEMPERATURE DIFFERENCE

The possibility of determining the air-sea surface temperature difference was investigated by Chahine *et al.* (11). It was
shown that the difference between the sea surface skin temperature T_s and the air temperature at the bottom of the lowest air
slab $T_a(z)$ may be derived by using infrared remote sensing data.

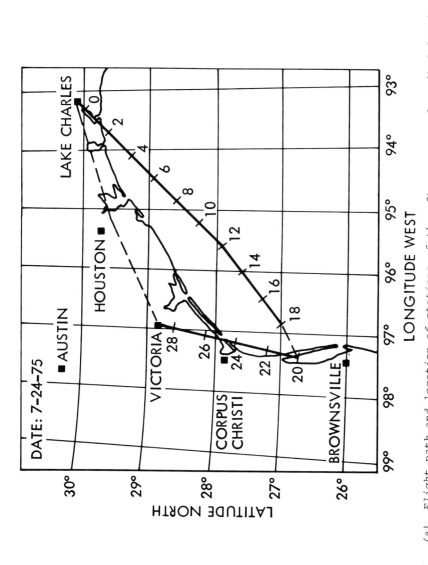

FIGURE 5. (a) Flight path and location of stations of the afternoon run described in Fig. 5b.

(b) Recovered surface, atmospheric and cloud parameters for an afternoon run.

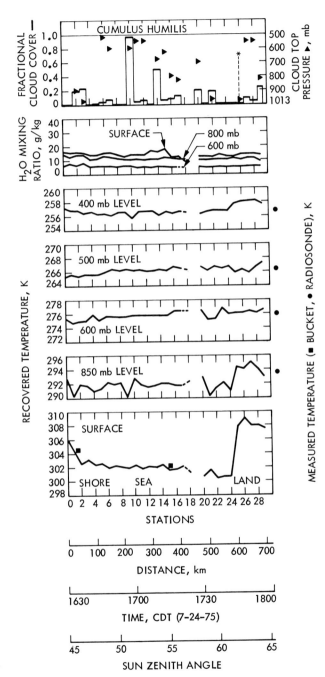

However, the experimental test illustrated in Fig. 6 was done
under very humid atmospheric conditions in the absence of any
atmospheric temperature inversions near the surface.

The determination of $[T_a(z_s) - T_s]$ was carried out
simultaneously with the determination of the vertical temperature
profiles using two sounding frequencies ν_1 and ν_2 which are
chosen such that

$$I(\nu_1) = I_s(\nu_1) + I_a(\nu_1) \qquad [I_s(\nu_1) \ll I_a(\nu_1)]$$

$$I(\nu_2) = I_s(\nu_2) + I_a(\nu_2) \qquad [I_s(\nu_2) \gg I_a(\nu_2)]$$

where I is the measured radiance, I_s is the contribution of the
surface and I_a is the contribution of the atmospheric emission.

The $I(\nu_1)$ is strongly dependent on the atmospheric temperature near the surface whereas the temperature derived from
$I(\nu_2)$ is strongly dependent on the sea surface skin temperature.
The derived temperature difference at the surface is therefore
a "weighted" difference because the vertical resolution of
current broad band temperature sounders is 2 km to 3 km near the
surface. Numerical and experimental verifications of this
approach under various atmospheric conditions are currently
being made.

SYMBOLS

a   constant defined in Eq. (4)
b   constant defined in Eq. (4)
B   Planck function
C   constant defined in Eqs. (10) and (16)
F   function defined in Eq. (22)
I   outgoing radiance function
T   temperature
z   local vertical axis

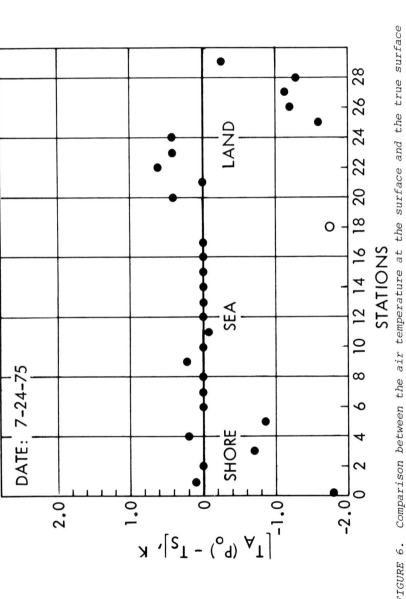

FIGURE 6. Comparison between the air temperature at the surface and the true surface temperature for run described in Fig. 5b.

α parameter defined in Eq. (21)

ϵ emissivity

ν frequency, cm^{-1}

θ zenith angle

ϕ azimuthal angle

ρ reflectivity

Ω solid angle

τ atmospheric transmittance

Subscripts:

  a   atmosphere

  d   downward flux

  h   sun

  i   incident radiation

  r   reflected radiation

  s   sea surface

REFERENCES

1.   Chahine, M. T., *J. Atmos. Sci.* 31, 233 (1974).

2.   Chahine, M. T., *J. Atmos. Sci.* 34, 744 (1977).

3.   Chahine, M. T., *J. Atmos. Sci.* 32, 1946 (1975).

4.   Chahine, M. T., *in* "Inversion Methods in Atmospheric
     Remote Sounding" (A. Deepak, ed.), p. 67. Academic
     Press, New York (1977).

5.   Shaw, J. H., *J. Atmos. Sci.* 27, 950 (1970).

6.   Wolfe, W. L., "Handbook of Military Infrared Technology."
     U.S. Government Printing Office, Washington, D.C. (1965).

7.   McCalister, E. D., *Applied Optics, 3,* 609 (1964).

8.   Kornfield, J., and Susskind, J., *Monthly Weather Review,*
     105, 1605 (1977).

9.   Prabhakara, C., Dalu, G., and Kunde, V. G., *J. Geophys.*
     Res. 79, 5039 (1974).

10. Chahine, M. T., *J. Opt. Soc. Am. 58,* 1634 (1968).

11. Aumann, H. H., and Chahine, M. T., *Applied Optics, 15,* 2091 (1976).

12. Chahine, M. T., Aumann, H. H., and Taylor, F. W., *J. Atmos. Sci. 34,* 758 (1977).

13. Barnett, T. B., Patzert, W. C., Webb, S. C., and Bean, B. R., *Bulletin American Meteorological Society, 60,* 197 (1979).

DISCUSSION

Susskind: You were doing these 3.7 μm retrievals during the day, so certainly you were getting some solar radiation and you were obviously doing a very good job of correcting for it, either with or without clouds. Do you have an estimate of how much solar radiation you had to make a correction for? Let us say you made no correction whatsoever, would it have been a 10 percent difference in your radiance that you accounted for very accurately or a 2 percent difference. How much effect were you correcting for under these conditions?

Chahine: It is very hard to tell because I had clouds. Roughly it was a factor of 2. The radiance was off by a factor of 2.

Susskind: So it was a 100 percent error that you were correcting to like 1 percent. Pretty good.

Fleming: I have a question about the last graph. Isn't it true that the relative accuracy or signal to noise ratio in the 3.7 μm region is quite a bit greater than in the 11 μm region? I mean the signal to noise ratio that is obtainable with present-day instrumentation.

Chahine: The present day detectors are more accurate in the 3.7 μm than in the 11 μm in terms of the signal to noise ratio.

Fleming: Even if the detectors are cooled?

Chahine: You need cooling for the 11 μm detectors; much less cooling is needed for the 3.7 μm detectors. You have to cool the 11 μm detectors way down to 100 K or less. But the 3.7 μm detectors, not that much.

Fleming: If your graph is read at the 3 percent error level, the values would be as shown; but should your graph be read at the same percentage-level on each curve?

Chahine: From an instrument point of view--measurement point of view--yes.

THE SPLIT WINDOW RETRIEVAL ALGORITHM FOR SEA
SURFACE TEMPERATURE DERIVED FROM
SATELLITE MEASUREMENTS

L. M. McMillin

National Environmental Satellite Service
National Oceanic and Atmospheric Administration
Washington, DC

Because remote measurements of the earth's surface from
satellite altitude are affected by the earth's atmosphere,
measurements of surface parameters must be adjusted. One of the
major factors which affects radiances in the atmospheric window
regions is absorption due to water vapor. Various techniques
for correcting measured earth radiances have been employed. One
of the more promising techniques is sometimes called the "split
window" technique, even though the basic technique has broader
applications than just a split window. A channel is being added
to the Advanced Very High Resolution Radiometer (AVHRR) specifi-
cally to allow use of this technique. Starting with NOAA D,
satellites in the TIROS-N series will provide two measurements in
the 10 to 12 μm region, one channel at 10.3 to 11.3 μm and one at
11.5 to 12.5 μm. Because the atmospheric transmittance is
greater at 10.3 to 11.3 μm than at 11.5 to 12.5 μm, the measure-
ment at 10.3 to 11.3 μm contains more signal from the earth's
surface and less signal from the earth's atmosphere. In its
simplest form, the retrieval method uses the difference between
these two measurements to estimate a second difference between
the measurement at 10.3 to 11.3 μm and a measurement at 10.3 to
11.3 μm that would be obtained if the instrument were located at
the earth's surface. This second difference is the correction
for atmospheric attenuation at 10.3 to 11.3 μm. If the area is
not too moist and the atmospheric transmittance is relatively
close to unity, the correction takes the form of a constant times
the difference in the two measurements. Basic prinicples of the
technique, its advantages, limitations, and modifications for
moist areas are discussed. The discussion also includes results
obtained from TIROS-N measurements at two separate windows at 4
and 11 μm to estimate the accuracy technique.

437

I.   INTRODUCTION

   One of the major difficulties in obtaining accurate measurements of sea surface temperatures from space is the need to
correct for atmospheric absorption. This correction is determined
by the atmospheric temperature and moisture profile as well as
the sea surface temperature. Perhaps the most direct estimate
of the correction can be determined from the difference between
two nearly transparent channels. Since the surface temperature
is equal to the measurement of a completely transparent channel,
the difference between two nearly transparent channels is proportional to the correction. Although this method has been
known since the beginning of the 1970s, satellite measurements
of the type required for this model have not been available.
With the launch of the TIROS-N series of satellites, measurements
to perform this correction will be available. In fact, later
satellites in the TIROS-N series will carry an instrument
specifically designed for this method. The coming availability
of data is generating new interest in this technique.

II. RADIATIVE TRANSFER

   Radiative transfer through a cloud-free atmosphere can be
written as

$$I(\nu) = B(\nu,T_s)\ \tau(\nu,P_o,\theta) + \int_{\tau(\nu,P_o,\theta)}^{1}$$

$$B\left[\nu,T(P)\right]\ d\tau(\nu,P,\theta) \tag{1}$$

where $I(\nu)$ is the radiance measured at the satellite, $B(\nu,T_s)$ is
the Planck radiance at wave number ν for the surface temperature
T_s, $\tau(\nu,P_o,\theta)$ is the transmittance at wave number ν, surface

pressure P_o and zenith angle θ. Equation (1) is valid for a
surface emissity of unity, a valid approximation in the infrared.
The first term in the equation is the radiation from the ground
that reaches the satellite. For humid atmospheres, this term is
often 50 percent or less in the 10 to 12 μm region. The second
term, the integral, gives the radiation emitted by the atmosphere. Because of a correlation between profiles of atmospheric
water vapor, profiles of atmospheric temperature, and the sea
surface temperature, differences between $I(\nu)$ and $B(\nu,T_s)$ never
reach 50 percent. However, they greatly exceed the 1 K value that
is desired for sea surface temperatures.

III.   ESTIMATES OF THE CORRECTION

At the satellite, measurements of $I(\nu)$ are available and a
value of $B(\nu,T_s)$ is desired. If $B[\nu,T(P)]$, which depends on the
atmospheric temperature profile, and $d\tau(\nu,P,\theta)$, which depends
on both the atmospheric temperature and water vapor profiles,
were known, Eq. (1) could be solved. However, over oceans,
knowledge of the temperature and moisture profiles is limited
and moisture, in particular, is variable from spot to spot. It
is desirable to produce an estimate based entirely on quantities
available from the satellite.

The method suggested empirically by Saunders, and Anding and
Kauth, and developed physically by McMillin is quite simple (1-3).
If the atmosphere between the satellite and the ground is
doubled, then the correction for the atmospheric term should be
doubled. Although it is physically impossible to obtain a
measurement unaffected by the atmosphere, it is possible to
double the atmosphere either by looking at a different angle or
a different spectral region. As usual, the actual application
is a compromise between conflicting constraints and the atmosphere may not be doubled. However, the basic concept is valid.

The first step is to use the mean value theorem on Eq. (1) to obtain

$$I(\nu,T) \simeq B(\nu,T_o) \tau(\nu,P_o,\theta) + \bar{B}(\nu,T_a)[1 - \tau(\nu,P_o,\theta)] \qquad (2)$$

where $\bar{B}(\nu,T_a)$ is the mean atmospheric radiance at wave number ν. Next, note that although it is possible to expand the Planck function in terms of temperature as

$$B(\nu,T) \simeq B(\nu,T_o) + [\partial B(\nu,T_o)/\partial T_o](T - T_o) \qquad (3)$$

where T_o is some reference temperature, it is considerably more accurate, especially when wave numbers are similar, to expand the Planck function in terms of a Planck function at the reference wave number as

$$B(\nu,T) \simeq B(\nu,T_o) + [\partial B(\nu,T_o)/\partial B(\nu_r,T_o)][B(\nu_r,T)$$

$$- B(\nu_r,T_o)] \qquad (4)$$

where ν_r is the reference wave number.

Setting $B(\nu,T_o)$ equal to $B(\nu,T_s)$ and substituting the result into Eq. (2) gives

$$B(\nu_r,T(I(\nu))) = B(\nu_r,T_s) \tau(\nu,P_o,\theta) + B[\nu_r,T(\bar{B}(\nu,T_a))]$$

$$[1 - \tau(\nu,P_o,\theta)] \qquad (5)$$

where $T(I(\nu))$ denotes the radiance temperature of $I(\nu)$. Now, select two spectral regions with different values of ν and let one be the reference ν_r. This gives

$$I(\nu_r) = B(\nu_r,T_s) \tau(\nu_r,P_o,\theta) + \bar{B}(\nu_r,T_a)$$

$$[1 - \tau(\nu_r,P_o,\theta)] \qquad (6)$$

Writing the explicit expression for $\bar{B}(\nu_r,T_a)$ gives

$$\bar{B}(\nu, T_a) = \int_{\tau(\nu,P_o,\theta)}^{1} B(\nu,T) \; d\tau(\nu,P,\theta) /$$

$$\int_{\tau(\nu,P_o,\theta)}^{1} d\tau(\nu,P,\theta) \qquad\qquad (7)$$

and for nearly transparent atmospheres, τ has the linear
approximation

$$\tau = e^{-ku} \approx 1 - ku \qquad\qquad (8)$$

where k is the absorption coefficient and u is the absorber
amount. Under these conditions $\tau(\nu_r,P,\theta)$ is linearly related
to $\tau(\nu,P,\theta)$ over the atmosphere and

$$\bar{B}(\nu,T_a) = \bar{B}(\nu_r,T_a) \qquad\qquad (9)$$

Eliminating $\bar{B}(\nu,T_a)$ from Eqs. (5) and (6) gives

$$B(\nu_r,T_s) = I(\nu_r) + \gamma[I(\nu_r) - B(\nu_r,T(I(\nu)))] \qquad\qquad (10)$$

where

$$\gamma = [1 - \tau(\nu_r,P_o,\theta)] / [\tau(\nu_r,P_o,\theta) - \tau(\nu,P_o,\theta)] \qquad\qquad (11)$$

Equation (10) gives the desired surface temperature as one of
the measurements plus a correction proportional to the difference
in the two measurements.

IV. PRACTICAL CONSIDERATIONS

    With Eq. (10) in hand, one of the decisions is the choice of
wave numbers. To keep transmittances (Eq. (8)) linear, it is
desirable to have both $\tau(\nu,P_o,\theta)$ and $\tau(\nu_r,P_o,\theta)$ near unity.
However, both $I(\nu_r)$ and $B(\nu_r,T(I(\nu)))$ in Eq. (10) are subject

to errors and the difference between these quantities is
multiplied by γ. Thus, it is desirable to keep γ small which
means making $\tau(\nu,P_o,\theta)$ small. In practice, γ values of 2 to 3
are typical in the 11 μm to 12 μm region.

It is also desirable to keep ν and ν_r as close as possible.
This reduces errors due to the approximation in Eq. (4) and
increases the probability that atmospheric absorption will be
similar in the two spectral regions. However, available
measurements may have a wide separation in wave number.

Finally, note that Eq. (10) contains selected terms of the
more general expansion

$$B(\nu_r,T_s) = I(\nu_r) + a_o + \sum_{i=1}^{n} a_i \, \Delta I^i \qquad (12)$$

where ΔI is the right-hand term in Eq. (10). Inclusion of a_o
and nonlinear terms can partially compensate for the various
approximations that are necessary.

V.   PREVIOUS RESULTS

Although use of the ratio method with actual data is
extremely limited, several experimenters have evaluated the
method using theoretical measurements. Although the method is
frequently referred to as the split window technique, the first
use of the ratio method used measurements at two angles rather
than two spectral regions. Saunders used the difference
between aircraft measurements at 0° and 60° as an estimate of
the atmospheric correction (1). In a series of papers, Anding
and Kauth, and Maul and Sidran tried the method with different
absorption models and found the results for specific wave
lengths to be model dependent (2,4,5). McMillin developed the
physical basis for the empirical results obtained by Anding

and Kauth (3,6). Prabhakara *et al.* applied the technique to
actual measurements from an interferometer (Iris) carried on
Nimbus 4 (7,8). All these studies concluded that the method,
when applied correctly, could provide sea surface temperatures
that approach the 1.0 K accuracy level.

VI. RESULTS WITH SIMULATED DATA

By using simultated data, satellite measurements were
calculated for a single atmosphere, but for a range of surface
temperatures. Then the ratio method was used to determine the
atmospheric correction (1). Corrected values were compared
with true values and the difference determined. Figure 1 shows
errors for several different values of γ. Notice the trade-off
between level of error and the range. Although γ of 1.799 is
the least accurate at a surface temperature of 300 K, it
maintains its accuracy over a wide range of surface temperatures.

Figure 2 demonstrates the capability of the method. It
compares the error in sea surface temperature when no correction
has been made to the error after correction with the ratio
method. The atmosphere is moderately moist with a transmittance
of 0.817 for the most transparent window channel. The reduction
in sensitivity of the error to the actual surface temperature
is exceptionally good.

Figure 3 shows the error for one value of γ for several
atmospheres taken from Wark *et al.* (9). The numbers refer to
the specific atmosphere. It is sufficient to know that they
cover a wide range of meteorological conditions. In each
atmosphere, the surface temperature was not allowed to range
over the whole scale shown in the figure, but was restricted
to ± 30 K from the 1000 mb temperature. This figure illustrates
that the accuracy of the technique is independent of the

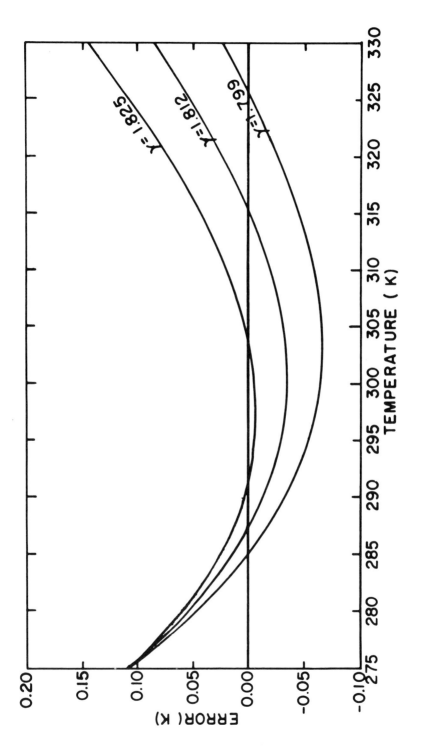

FIGURE 1. Errors in sea surface temperature derived by using the ratio method as a function of sea surface temperature for three values of Y. All values are for a single atmosphere.

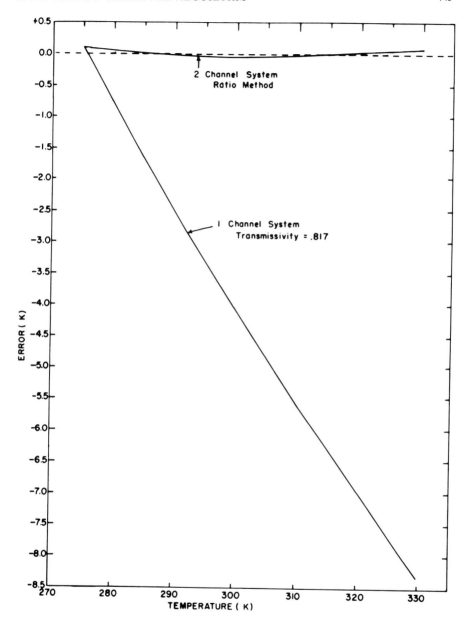

FIGURE 2. Errors in sea surface temperature for a fixed atmosphere as a function of surface temperature. For the single-channel system, the sea surface temperature is assumed to equal the radiance temperature of the satellite measurement. For the two-channel system, the ratio method is used.

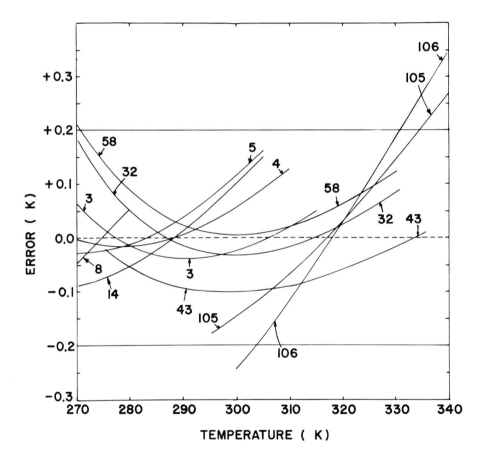

FIGURE 3.   Errors in sea surface temperature as a function of surface temperature for a variety of atmospheres.   Surface temperatures are limited to ± 30 K from the 1000 mb   temperature.

particular atmosphere. Atmospheres 105 and 106 appear to be
unique. These are desert atmospheres with vapor at high levels,
which makes the average atmospheric temperature $\bar{B}(\nu, T_a)$ cold,
even though the surface is warm. It should be noted that only
the range from 270 K to 300 K is of interest for sea surface
temperatures.

These three figures all show extremely low error levels,
because measurement errors have not been included in the
calculations. It is estimated that typical instrument errors
will raise the error level of the resulting estimate to about
1.0 K. McMillin shows the effect of additional parameters such
as surface emissivity, viewing angle, and aerosols on the
correction (3). The method worked with only slight increases
in error levels for all the methods that were tried.

VII. TIROS-N

Instruments carried on TIROS-N provide several opportunities
for using the method with real data. The High Resolution
Infrared Radiation Sounder (HIRS) provides measurements in
atmospheric windows centered at 900, 2515, and 2660 cm^{-1}.
Measurements for this instrument are available. The Advanced
Very High Resolution Radiometer (AVHRR) provides measurements
at 10.5 μm to 11.5 μm and 3.55 μm to 3.93 μm. Later instruments
in the TIROS-N series will provide measurements at 10.3 μm to
11.3 μm and 11.5 μm to 12.5 μm, specifically for use with the
ratio technique. Unfortunately, the 3.55 μm to 3.93 μm channel
on the first satellite of the series contains excessive noise.

VIII.   RESULTS FROM TIROS-N

    Measurements from the HIRS/2 have been provided by
M. Weinreb. These measurements were screened to eliminate
cloudy conditions and used to evaluate the method. Figure 4
shows a plot of the correction for channel 8 along the vertical
as a function of the difference between channels 8 and 18
(Fig. 4a) and channels 8 and 19 (Fig. 4b). The data were
limited to nighttime and only the center two retrievals were
used to eliminate angle effects. Sea surface temperatures
used as truth were taken from the National Meteorological
Center (NMC) sea surface temperature analysis. If the channel 8
radiance were used as a measure of the sea surface temperature
with no adjustment, the standard deviation of the error about
the mean would be 2.14 K. The error in the estimate resulting
from the ratio technique is 0.80 K. The improvement is also
obvious from the grouping of the points about the line in Fig. 4.
The results in the figure also show a curve resulting from a
tendency for the ratio of the correction (vertical axis) to the
difference (horizontal axis) to increase with the value of the
difference, a result that can be seen in Eq. (11), since the
denominator goes to zero faster than the numerator as transmittances approach zero. Inclusion of a second-order term
should lead to a slight improvement in accuracy.

IX.   CONCLUDING REMARKS

    The ratio technique of correcting measured radiances for
atmospheric absorption has been evaluated using both simulated
and real data. Although limited tests such as these never
reveal all the conditions that arise in a daily operation,
a root mean square difference of 0.8 K from an independent

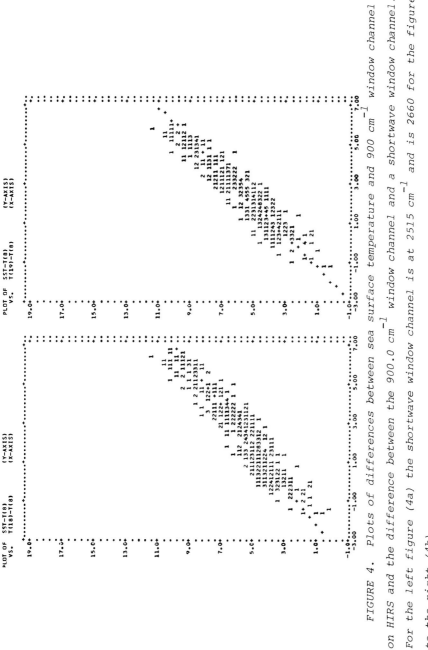

FIGURE 4. Plots of differences between sea surface temperature and 900 cm^{-1} window channel on HIRS and the difference between the 900.0 cm^{-1} window channel and a shortwave window channel. For the left figure (4a) the shortwave window channel is at 2515 cm^{-1} and is 2660 for the figure to the right (4b).

measure of sea surface temperature with its own inherent error
is an extremely accurate result. When the data become available
and the method is implemented, it should be possible to correct
clear atmospheres for atmospheric attenuation with an accuracy
of better than 1 K.

It should, however, be mentioned that this technique must
be used with some means of identifying clear areas. One feature
of the ratio technique is that it corrects for atmospheric
absorption above a cloud as well as above the ground. It could,
thus, be applied first and followed by a technique such as the
truncated normal technique by Crosby and Glasser to adjust for
clouds (10). This would lead to a complete package for solving
for sea surface temperatures on a global scale.

SYMBOLS

a_o, a_i	regression coefficients
$B(\nu,T_s)$	Planck radiance at wave number ν for a black body at temperature T_s
$\bar{B}(\nu,T_a)$	average atmospheric radiance with radiance temperature T_a at wave number ν
$I(\nu)$	radiance at wave number ν
k	absorption coefficient
$T(I(\nu))$	radiance temperature for radiance value $I(\nu)$ at wave number ν
u	absorber amount
γ	a constant
ΔI	difference in measured radiances after scaling to a common wave number by using the Planck equation
ν_r	reference wave number
$\tau(\nu,P_o,\theta)$	transmittance at wave number ν, pressure P_o, and zenith angle θ

ACKNOWLEDGMENT

I wish to express my thanks to Dr. Michael P. Weinreb for providing the graphs shown in Fig. 4.

REFERENCES

1. Saunders, P. M., *J. Geophys. Res.* 72(16), 4109-4117 (1967).

2. Anding, D., and Kauth, R., *Remote Sensing Environ. 1,* 217-220 (1970).

3. McMillin, L. M., A Method of Determining Surface Temperatures from Measurements of Spectral Radiance at Two Wavelengths. Ph.D. dissertation, Iowa State University, Iowa City, Iowa (1971).

4. Maul, G. A., and Sidran, M., *Remote Sensing Environ. 2,* 165-169 (1972).

5. Anding, D., and Kauth, R., *Remote Sensing Environ. 2,* 171-173 (1972).

6. McMillin, L. M., *J. Geophys. Res.* 80(36), 5113-5117 (1975).

7. Prabhakara, C., Connath, B. J., and Kunde, V. G., Estimation of Sea Surface Temperature from Remote Measurements in the 11-13 μm Window Region. NASA/GSFC X Doc. 651-72-358. U.S. Government Printing Office, Washington, DC (1972).

8. Prabhakara, C., Dalu, G., and Kunde, V. G., *J. Geophys. Res. 79(33),* 5039-5044 (1974).

9. Wark, D. Q., Yamamoto, G., and Lienesch, J. H., Infrared
    Flux and Surface Temperature Determinations from Tiros
    Radiometer Measurements. Tech. Rep. 10, U.S. Weather
    Bur., Washington, DC (1962).

10. Crosby, D. S., and Glasser, K. S., *J. Appl. Metero.*
    17(11), 1712-1715 (1978).

DISCUSSION

Susskind: I guess I am a little bothered about a couple of things you said. The degree of error you seem to be indicating is of the order of a few tenths of a degree. The work I have seen by Dr. Prabhakara was showing errors of 1 to 2 degrees, sometimes even a little more in the paper to be published. But we have done a little work on it. I think the problem is not as linear as you make it look. I am concerned about the effect of the water vapor continuum and the lines. Now each of the two channels has a different combination of effects due to lines and continuum. Essentially the emissions are coming from different parts of the atmosphere, the continuum emission is coming from much closer to the ground--the lines are coming from up higher. Absorption is proportional to the water density and one to the square of the water density. I do not think you can so quickly say that the mean atmospheric temperature in both channels is coming from more or less the same place because they are really coming from linear combinations of two different kinds of effects. Any comment on that?

McMillin: Yes. This has been a general treatment of the technique. In actual use, in fact, statistically, you get an advantage to adding a squared term. And with a squared term, it does quite well.

Susskind: Then, technically, you need another channel.

McMillin: No. One coefficient times the difference and another coefficient times the difference squared.

Walton: I notice that some people were looking at using the 3.7 μm and the 11 μm channels of the HIRS (High Resolution Infrared Sounder) and testing that in sort of a split window. Have you been able to do anything with that or do you have any comments on any results?

McMillin: I know some people who are going to look at it. I do not know anybody who has completed their evaluation.

Chahine: When you combine the 3.7 μm with the 11 μm, you have an emissivity difference, and you have a reflectivity difference between the two. You need 5 channels to solve your equations to get the true surface temperature.

McMillin: I think if you had two channels in the 3.7 μm region and two in the 11 μm region you could go a long way toward getting a total answer.

Lenoble: I have a comment. Dr. Deschamps[1] in our group did a
lot of simulation of this problem of atmospheric corrections in
remote sensing of the sea surface temperature. One point is that
he compared the use of two channels in the 11 μm band and the
use of only one channel with climatological corrections. He
found that an important point to be considered is the experimental
noise. If the noise is not low enough, the use of two channels
does not cause any improvement. It is only for low noise that we
get improvement. And the other point is that he tried also the
3.7 μm and his conclusion was that there is a great improvement
with the use of the 3.7 μm, at least for the night measurements.

Susskind: A quick comment which may or may not have been brought
out. Mous Chahine did not indicate magnitudes of effects. The
effect of water vapor on the 3.7 μm channels is not that large.
It may be a degree, even if you neglect it. While in the 11 μm
window, if you neglect it, you can make very large errors. So
there is not that much to account for. You may have other
problems with solar radiation but the water vapor effect at
3.7 μm is much smaller to begin with.

Chahine: From the results of Tim Barnett's experiment, we recall
that the largest error in the temperature retrieved from the
11 μm region are due to clouds and uncertainty in the amount of
water vapor.

McMillin: Those are both current techniques which are not
related to the procedure I am discussing.

Chahine: The derivation by Prabhakara and others indicated that
the value of the coefficient K used in your approximation is a
function of the water vapor distribution as well as the total
amount. When you try to put these approximations all together,
apparently they lead to an error of 3 to 4 degrees in the 11 μm
region.

McMillin: I did a study with some later data than this where I
did use the square term. The simulation used a good water vapor
model and included instrumental errors. Resulting accuracies
were within the range of a degree. Use of the squared term
resulted in a significant improvement in accuracy.

---

[1]*Deschamps, P. Y. and T. Phulpin, Atmospheric Correction of
Infrared Measurements of Sea Surface Temperature Using Channels
at 3.7, 11 and 12 μm (accepted for publication in "Boundary
Layer Meteorology").*

Kaplan: Where the square term does come in, it is a very narrow range in the lowest layers of the atmosphere, so there is not much of a temperature difference going from the bottom to the top like that. You can probably define it within one and two degrees, probably you can guess it.

McMillin: If you use this method, you would want to put in a squared term, that is a term not only proportional to the difference of those two channels but also proportional to the difference squared.

Kaplan: Another advantage of the 3.7 μm channel, besides the transmittance being larger and having less of a correction, of course, is that because you are on an exponential end of a black body curve, a 10 percent error in transmittance makes very little error in temperature.

ATMOSPHERIC CORRECTION OF NIMBUS-7 COASTAL
ZONE COLOR SCANNER IMAGERY[1]

Howard R. Gordon

University of Miami
Coral Gables, Florida

James L. Mueller

Laboratory for Atmospheric Sciences
National Aeronautics and Space Administration
Greenbelt, Maryland

Robert C. Wrigley
National Aeronautics and Space Administration
Moffett Field, California

The Coastal Zone Color Scanner (CZCS) on NIMBUS-7 is a scanning radiometer designed to view the ocean in six spectral bands (centered at 443, 520, 550, 670, 750, and 11,500 nm) for the purpose of estimating sea surface chlorophyll and temperature distributions. In the visible bands, the atmosphere obscures the imagery to the extent that at 443 nm, at most, only 20 percent of the observed radiance originates from beneath the sea surface. Retrieving this subsurface radiance from the imagery is complicated by the highly variable nature of the aerosol's contribution. In this paper, an algorithm for the removal of these atmospheric effects from CZCS imagery is described, a preliminary application of the algorithm to an image with very strong horizontal variations in the aerosol optical thickness is presented, and retrieval of the spatial distribution of the aerosol optical thickness is discussed.

[1]*This work received support from the National Aeronautics and Space Administration under contract NAS5-22963.*

I. INTRODUCTION

The Coastal Zone Color Scanner (CZCS) on NIMBUS-7 is a scanning radiometer designed to view the ocean in six coregistered spectral bands, which are listed in Table I along with their minimum and maximum saturation radiances. It views the ocean from approximately 955 km with a spatial resolution of 0.825 km and has an active scan of 78° centered on nadir producing an image area $1.3 \times 10^6 \text{ km}^2$. Since NIMBUS-7 is in a sun synchronous orbit crossing the equator at noon, sun glint would normally saturate the sensor near nadir. To compensate for this effect, the sensor can be tilted through an angle of ± 20° along the satellite track from nadir.

The purpose of the CZCS experiment is to measure the concentration of phytoplankton pigments and total seston in coastal regions and possibly relate these to sea surface temperature. To understand how such constituents of the water can be related to the signal available to the CZCS, it is useful to review the radiative transfer processes in the water as discussed by Austin, Gordon, and Morel and Prieur (1-3).

TABLE I. *Nominal Characteristics of the CZCS*

Band	Wavelength (nm)	Saturation radiance $(mw/cm^2 \ \mu m \ ster)$	
		Gain 0	Gain 3
1	433-453	11.52	5.32
2	510-530	7.97	3.88
3	540-560	6.45	3.11
4	660-680	2.82	1.29
5	700-800	24.00	24.00
6	10500-12500	--	--

If an irradiance at wave length λ E_d^λ enters the water, the upwelling irradiance E_u^λ just beneath the sea surface is given by

$$E_u^\lambda = R(\lambda) \; E_d^\lambda \tag{1}$$

where $R(\lambda)$ is called the irradiance reflection function. If E_u^λ is approximately diffuse, the upwelling radiance just beneath the sea surface is

$$L_w^\lambda = R(\lambda) \; E_d^\lambda / \pi \tag{2}$$

R is related to the optical properties of the water and its constituents through

$$R \approx \frac{1}{3} \, b_{b/a} \tag{3}$$

where a is the absorption coefficient of the medium and b_b is the backscattering coefficient. If there are n constituents in the water, a and b_b are given by

$$\left.\begin{array}{c} a = a_w + \displaystyle\sum_{i=1}^{n} a_i \\[2em] b_b = b_w/2 + \displaystyle\sum_{i=1}^{n} (b_b)_i \end{array}\right\} \tag{4}$$

where a_w and b_w are the absorption and scattering coefficients of pure water. The coefficients a_i and $(b_b)_i$ are linearly related to the concentrations C_i of the constituents through

$$\left.\begin{array}{c} a_i = a_i^o \, C_i \\[1.5em] (b_b)_i = (b_b)_i^o \, C_i \end{array}\right\} \tag{5}$$

Equations (1) through (5) show how variations in the constituent
concentrations can cause variations in the subsurface upwelling
radiance L_w^λ. These are the variations that were to be measured
with the CZCS. In particular, it should be noted that phytoplankton pigments show an absorption maximum near 440 nm, an
absorption minimum near 550 nm, and a backscattering coefficient
which depends only weakly on wavelength; these conditions suggest
that variations in the phytoplankton concentration could cause
strong variations in L_w^λ. This is indeed the case, as has been
well documented in the literature (3).

Unfortunately, L_w^λ is difficult to measure because scattering
by the atmosphere between the ocean and the sensor contributes
significantly to the radiance measured at the sensor. Hovis and
Leung have dramatically demonstrated this effect by noting the
difference in the radiance spectrum observed when the altitude
of the sensor is increased from 0.91 km to 14.9 km (4). In the
blue portion of the spectrum (near 400 nm), the radiance increased
about fivefold, whereas in the red (near 700 nm) the radiance
increased by a factor of about 2.5. It can be anticipated, then,
that as much as 80 percent of the radiance detected at satellite
altitudes could be due to atmospheric scattering. Removal of
these atmospheric effects is the subject of this paper.

II. ATMOSPHERIC CORRECTION ALGORITHM

In principle, removal of the radiance added by the atmosphere
would be simple if the optical properties of the aerosol were
known throughout an image. However, as the aerosol is highly
variable and hence unlike the Rayleigh scattering component, it
is not possible to predict *a priori* its effect on the imagery.
The method of correction used here follows that developed by
Gordon (5). Since the interest is centered on the retrieval

of L_w^λ from the imagery, an attempt is made to remove all surface effects (reflected skylight) from the imagery along with the atmospheric backscattered radiance.

If sun glint is ignored, in the single scattering approximation the normalized radiance at the top of the atmosphere due to photons which have not penetrated the sea surface (but may have reflected off the flat surface before or after interacting with the atmosphere) is given by

$$I_1^\lambda = I_R^\lambda + I_A^\lambda \tag{6}$$

where

$$I_x^\lambda = \frac{\omega_x^\lambda \tau_x^\lambda}{4\pi\mu} \left[P_x^\lambda(\theta_-) + \left[\rho(\mu_o) + \rho(\mu) \right] P_x^\lambda(\theta_+) \right] \tag{7}$$

$x = R$ or A, and

$$\cos \theta_\pm = \pm\mu \, \mu_o + \sqrt{(1 - \mu^2)(1 - \mu_o^2)} \, \cos(\phi - \phi_o)$$

(See "Symbols" for symbols not defined in text.) When the aerosol phase function is independent of wavelength

$$\left(P_A^{\lambda_1}(\theta) = P_A^{\lambda_2}(\theta) \right)$$

it is seen from Eq. (7) that

$$I_A^{\lambda_2} = \beta_{1,2} \, \varepsilon_{1,2} \, I_A^{\lambda_1} \tag{8}$$

where

$$\varepsilon_{1,2} = \tau_A^{\lambda_2} / \tau_A^{\lambda_1}$$

and

$$\beta_{1,2} = \omega_A^{\lambda_2} / \omega_A^{\lambda_1}$$

Considering the two wavelengths λ_1 and λ_2,

$$I_1^{\lambda_1} = I_R^{\lambda_1} + I_A^{\lambda_1}$$

$$I_1^{\lambda_2} = I_R^{\lambda_2} + I_A^{\lambda_2}$$

and using Eq. (8) yields

$$I_1^{\lambda_2} = I_R^{\lambda_2} + \beta_{1,2}\,\varepsilon_{1,2}\left(I_1^{\lambda_1} - I_R^{\lambda_1}\right) \tag{9}$$

Equation (9) is rigorously correct in the single scattering approximation; however, when the aerosol optical thickness exceeds about 0.1, Eq. (7) breaks down. An example of this effect is provided in Fig. 1, in which the results of a computation of I_1 at 750 nm as a function of scan angle θ and aerosol optical thickness τ_A using Eqs. (6) and (7) are compared with a calculation including all orders of multiple scattering for a geometry similar to the CZCS (sun behind the spacecraft with the scan normal to the spacecraft motion). The aerosol phase function used in the computations for Fig. 1 is a two-term Henyey-Greenstein (TTHG) function given by

$$P_A(\theta) = \frac{\alpha(1 - g_1^2)}{\left(1 + g_1^2 - 2g_1 \cos\theta\right)^{1.5}}$$

$$+ (1 - \alpha)\,\frac{(1 - g_2^2)}{\left(1 + g_2^2 - 2g_2 \cos\theta\right)^{1.5}}$$

with $\alpha = 0.9618$, $g_1 = 0.7130$, and $g_2 = -0.7596$, which was computed by Kattawar to provide an analytical expression to the phase function for a Haze L distribution (6). Clearly, the single scattering calculation underestimates the radiance in the geometry of Fig. 1. One would expect that when Eq. (7) is not

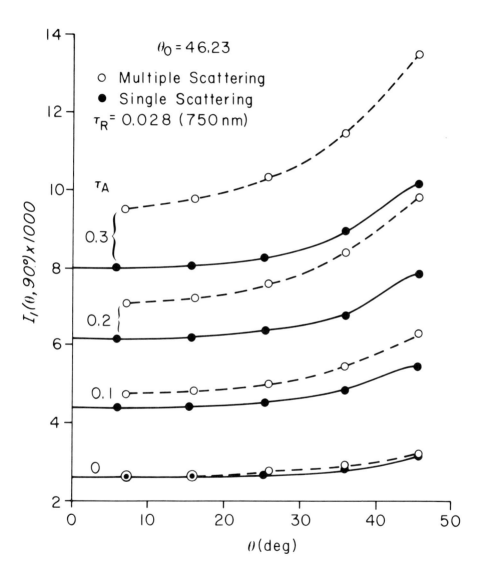

FIGURE 1. *Comparison of single and multiple scattering calculation of I, at 750 nm as a function of aerosol optical thickness and scan angle for a geometry similar to the CZCS.*

valid, it follows that Eq. (9) will also be a poor approximation. This, however, is not the case if all the radiances in Eq. (9) are computed by a code which includes all orders of multiple scattering rather than using Eq. (7). That is, if

$$I_R^{\lambda_1} \text{ and } I_R^{\lambda_2}$$

are computed exactly for a Rayleigh atmosphere bounded by an ocean which reflects according to Fresnel's law but absorbs all photons which penetrate the surface, and

$$I_1^{\lambda_1} \text{ and } I_1^{\lambda_2}$$

are similarly computed for a Rayleigh plus aerosol atmosphere, then Eq. (9) is still *usefully* accurate. This is shown in Table II, which gives the percent error in Eq. (9),

TABLE II. *Percent Error in Using Eq. (9).*

τ_A^{443}	$\varepsilon_{443,750}$	θ	$\theta_o = 0°$	$\theta_o = 21°$	$\theta_o = 46°$
0.2	1	7.4	8.85	3.00	-0.10
0.2	1	17.1	4.18	1.80	-0.28
0.2	1	26.8	1.92	0.63	-0.50
0.4	2	7.4	20.60	5.90	-1.63
0.4	2	17.1	10.50	3.00	-2.03
0.4	2	26.8	3.20	0.04	-2.41
0.6	3	7.4	33.40	9.20	-2.51
0.6	3	17.1	16.80	4.40	-2.94
0.6	3	26.8	4.90	-0.13	-3.36

$$\frac{(I_1^{\lambda_2})_{Eq.~(9)} - (I_1^{\lambda_2})_{exact}}{(I_1^{\lambda_2})_{exact}} ~ 100~\%$$

for λ_2 = 443 nm, λ_1 = 750 nm, $\beta_{1,2}$ = 1, τ_A^{750} = 0.2, and the sensor-sun geometry of Fig. 1. It is seen that the largest error occurs when the sun is near the zenith and the sensor views near the nadir. Since this is the region of strong sun glint, these errors are unimportant. Tilting the CZCS to avoid the sun glint results in excluding the cases in Table II which fall *above* the dashed lines; that is, when θ_o = 0, the sensor will be tilted 20° away from that nadir so the *minimum* value of θ is 20°. Taking this into consideration, it is seen that the maximum error in Eq. (9) will be less than about 5 percent and usually is considerably less than 5 percent.

Equation (9) forms the heart of the atmospheric correction algorithm. Consider a wavelength λ_0 for which diffuse reflectance of the ocean is negligible so that

$$L_w^{\lambda_0} = 0$$

and the total radiance at the sensor is

$$I_T^{\lambda_0} = I_R^{\lambda_0} + I_A^{\lambda_0}$$

For any other wavelength τ, the total radiance at the sensor is

$$I_T^{\lambda} = I_R^{\lambda} + I_A^{\lambda} + t^{\lambda}I_w^{\lambda}$$

where t^{λ} characterizes the transmittance of the radiance L_w^{λ} through the sea-air interface to the top of the atmosphere. Then, using Eq. (9) yields

$$t^{\lambda}I_w^{\lambda} = I_T^{\lambda} - I_R^{\lambda} - \beta_{\lambda,\lambda_0}~\varepsilon_{\lambda,\lambda_0}\left(I_T^{\lambda_0} - I_R^{\lambda_0}\right)$$

or converting from normalized radiance to actual radiance yields

$$t^\lambda L_w^\lambda = L_T^\lambda - L_R^\lambda - K_{\lambda,\lambda_0} \, \varepsilon_{\lambda,\lambda_0} \left(L_T^{\lambda_0} - L_R^{\lambda_0} \right) \tag{10}$$

where K_{λ,λ_0} is the known quantity

$$K_{\lambda,\lambda_0} = \frac{T_{O_3}^\lambda(\mu) \; T_{O_3}^\lambda(\mu_0) \; F_0^\lambda}{T_{O_3}^{\lambda_0}(\mu) \; T_{O_3}^{\lambda_0}(\mu_0) \; F_0^{\lambda_0}}$$

In Eq. (10) the only unknown is the combination

$$\delta_{\lambda,\lambda_0} \equiv \beta_{\lambda,\lambda_0} \, \varepsilon_{\lambda,\lambda_0}$$

since typical variations in the ozone concentration will have a negligible effect on K_{λ,λ_0} because of the smallness of $\tau_{O_3}^\lambda$. (In fact, K_{λ,λ_0} varies by less than 1 percent across a scan line.) It should be noted, however, that the extraterrestrial solar flux is required for computation of the Rayleigh radiances as well as K_{λ,λ_0} and that different investigators differ in their measurement of F_0^{443} by as much as 10 percent. Such errors as F_0^λ are immaterial as long as the F_0^λ used is that which the CZCS would measure if it viewed the sun. Measurements of L_T^λ, L_w^λ, and τ_A^λ on several very clear days could be used to adjust F_0^λ so that this compatibility with the CZCS is ensured.

The unknown in Eq. (10), $\delta_{\lambda,\lambda_0}$, can be measured in a simple manner from the ground only when the aerosol is non-absorbing (or when ω_A^λ happens to be nearly equal to ω_A^λ), in which case measurement of $\varepsilon_{\lambda,\lambda_0}$ gives $\delta_{\lambda,\lambda_0}$. When $\delta_{\lambda,\lambda_0}$ can be measured or estimated and $P_A(\theta)$ is independent of λ, Eq. (10) can be used to retrieve tL_w^λ from L_T^λ given a λ_0 such that $L_w^{\lambda_0} \approx 0$.

Thus far, it has been implicitly assumed that the atmosphere is homogeneous in the horizontal direction; that is,

$$\tau_A^\lambda \neq \tau_A^\lambda(x,y), \text{ etc.,}$$

where x and y represent the horizontal coordinates of a column of atmosphere. As shown below, this restriction is generally not satisfied. The algorithm given in Eq. (10), however, should still be valid even in an inhomogeneous atmosphere as long as $\delta_{\lambda,\lambda_0}$ and the aerosol phase function are independent of position over a large portion of the image. This is a much weaker restriction than the requirement that τ_A^λ be independent of position, and physically means that the aerosol size frequency distribution and composition must be independent of position and only its concentration is allowed to be position dominant.

Finally, if the aerosol phase function is weakly dependent on wavelength, it is to be expected that Eq. (10) will break down because Eq. (8) will no longer be as well satisfied as before. This effect is somewhat compensated, however, by the fact that when $P_A(\theta)$ increases slightly in the backward direction, it will decrease slightly in the forward direction and vice-versa. Equation (7), although not completely valid, shows that this behavior will tend to decrease the effect of such spectral variations in the phase function on Eq. (8) and hence on Eq. (10).

Summarizing the discussion above, it is expected that Eq. (10) can be used to recover $t^\lambda L_w^\lambda$ from L_T^λ even in an inhomogeneous atmosphere as long as (1) a suitable λ_0 can be found for which $L_w^{\lambda_0} \approx 0$, (2) $\delta_{\lambda,\lambda_0}$ is essentially constant over large portions of the image, and (3) a method of estimating $\delta_{\lambda,\lambda_0}$ can be devised.

III.   PRELIMINARY APPLICATION TO CZCS IMAGERY

The algorithm described has been applied to a CZCS image of
the Gulf of Mexico from orbit 130 on November 2, 1978. The
complete image is shown in Fig. 2. This is about a factor of 2
larger (along track) than the standard CZCS and was produced by
combining bands 1, 3, and 4 on an Image 100 Processing System at
Goddard Space Flight Center. Note that this is a very hazy day,
in particular, that there is very *strong horizontal structure*
associated with the haze. Clearly, if an algorithm designed to
remove its influence is based on a model that assumes a constant
aerosol concentration over even relatively small areas, it will
fail in this particular case.

As the first attempt at correcting an image, the algorithm
described has been applied to the subscene in the square off the
coast of Florida, which appears to be the clearest portion of the
image. Although most ideal spectrally, because of its low
radiometric sensitivity, Band 5 cannot be used as the correction
band and hence λ_0 was chosen to be 670 nm (Band 4).
Unfortunately, in the immediate vicinity of the coast and at the
mouths of rivers L_w^{670} is not likely to be small enough to be
considered negligible and the algorithm will yield meaningless
results in these areas. Since no atmospheric measurements were
available for this subscene, it was assumed that $\delta_{\lambda,670} = 1$;
that is, the scattering and absorption properties of the aerosol
are independent of wavelength. The results of this application
are given in Fig. 3, which shows in false color $t^\lambda L_w^\lambda$ for bands
1, 2, and 3, and $L_T^\lambda - L_R^\lambda$ for Band 4. It is seen that the
correction algorithm has removed most of the banded structure
evident in Band 4 from the other three bands; this effect
suggests that a considerable portion of this highly variable
atmosphere has been removed. Band 2, however, shows on the
western edge off Fort Myers some structure remaining which is

FIGURE 2. CZCS image of the Gulf of Mexico from orbit 130 (November 2, 1978) produced by combining Bands 1, 3, and 4 to form a color image.

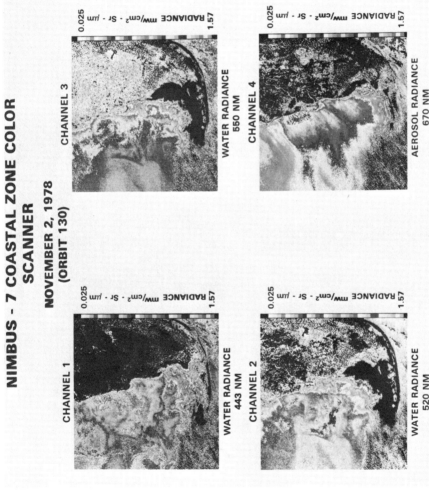

NIMBUS - 7 COASTAL ZONE COLOR SCANNER

NOVEMBER 2, 1978 (ORBIT 130)

CHANNEL 1
WATER RADIANCE 443 NM
RADIANCE mw/cm² - Sr - μm
0.025   1.57

CHANNEL 2
WATER RADIANCE 520 NM
RADIANCE mw/cm² - Sr - μm
0.025   1.57

CHANNEL 3
WATER RADIANCE 550 NM
RADIANCE mw/cm² - Sr - μm
0.025   1.57

CHANNEL 4
AEROSOL RADIANCE 670 NM
RADIANCE mw/cm² - Sr - μm
0.025   1.57

FIGURE 3. Corrected Bands 1, 2, 3 and aerosol radiance from Band 4 in the subscene indicated in Fig. 2.

likely still in the atmosphere, and suggests that $\delta_{520,670} > 1$.
The more important point to be considered here is that although
the radiance in Band 4 (and hence the aerosol optical thickness
τ_A^{670}) varies by a factor of 2.5 over the subscene, the apparent
success of the correction algorithm indicates that $\delta_{\lambda,\lambda_0}$ is, in
fact, nearly constant over the same subscene.

    To see whether this constancy of $\delta_{\lambda,\lambda_0}$ holds up over larger
areas, the algorithm has been applied to the cloud-free areas of
the entire Gulf of Mexico in this image. Figures 4, 5, and 6
show the uncorrected radiance in Bands 1, 3, and 4 with the
Mississippi outfall visible in the upper left corner of the
picture and Tampa visible at the middle of the right edge.
Careful examination reveals features in Bands 1 and 3 which are
not evident in Band 4. Over nearly all the open water areas of
the Gulf L_w^{670} should be approximately zero, so the conditions
for applying the algorithm should be satisfied. Figures 7 and
8 show, respectively, the corrected radiances in Bands 1 and 3.
Comparing Figs. 7 and 5 (the corrected Band 1 and the original
Band 4) shows no common features over the open ocean; thus, (1)
the assumed value of δ was close to correct, and (2) δ is
essentially constant over the entire Gulf. A similar comparison
of Figs. 8 and 5 shows that some of the haze structure remains
in the corrected Band 3; thus, δ for this band is greater than
1. When one examines the corrected and uncorrected Band 1
images, all the features seen in Fig. 7 are evident in Fig. 4;
however, it is not possible *a priori* to distinguish between the
features in Fig. 4 which are of atmospheric origin and those
which are of ocean origin.

    Finally, since several investigators have proposed or
presented algorithms for determining pigment concentrations from
the ratio of upwelling radiance in two spectral bands, the ratio
$t^{440}L_w^{440}/t^{550}L_w^{550}$ is presented in Fig. 9 for the Gulf from

FIGURE 4.   Uncorrected Band 1 from orbit 130.

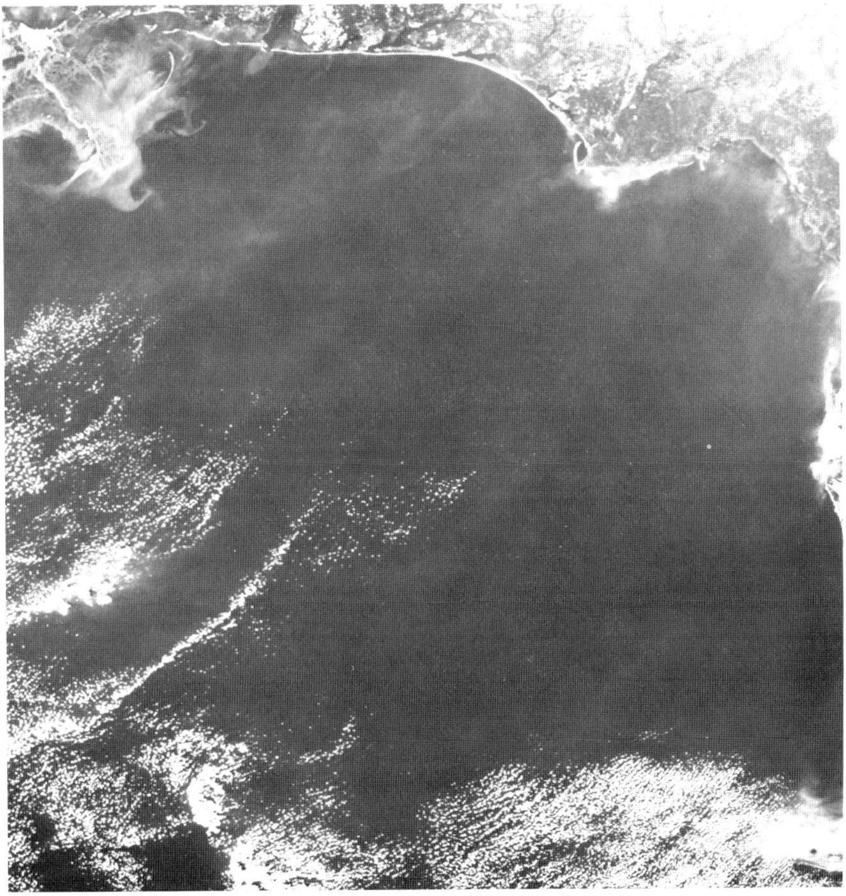

FIGURE 5.   Uncorrected Band 3 from orbit 130.

FIGURE 6. Uncorrected Band 4 from orbit 130.

FIGURE 7. Corrected Band 1 from orbit 130.

FIGURE 8.   Corrected Band 3 from orbit 130.

FIGURE 9. Corrected Band 1 divided by corrected Band 3 from orbit 130.

Figs. 7 and 8 (7-11). In this presentation, the light areas
represent low pigment concentrations, and the dark areas
represent high concentrations. The complex spatial structure
revealed in this corrected ratio is truly remarkable, considering
the very hazy atmosphere evident in the image before passing
through this correction algorithm.

IV. PROGNOSIS FOR AEROSOL MEASUREMENT WITH THE CZCS

The apparent success achieved in this preliminary application
of Eq. (10) to the imagery from orbit 130 strongly suggests that
the assumptions leading to Eq. (10) are reasonably valid.
Furthermore, success in this particular case, for which strong
horizontal inhomogeneities in the haze are evident, provides
convincing evidence that for this image

$$\delta_{\lambda,\lambda_0} \neq \delta_{\lambda,\lambda_0}(x,y).$$

When Eq. (10) is valid and $\delta_{\lambda,\lambda_0}$ is constant over the image, the
combination

$$L_T^\lambda - L_R^\lambda = L_A^\lambda$$

for any λ for which $L_w^\lambda \approx 0$, for example, λ_0, should be roughly
proportional to $\tau_A^{\lambda_0}$. Also a position independent $\delta_{\lambda,\lambda_0}$ implies
a position independent $P_A(\theta)$, so if the phase function can be
estimated at one point in the image, this estimate can be
extended to the entire image to retrieve $\tau_A^\lambda(x,y)$. As an example
of this procedure, it is assumed that $P_A(\theta)$ is the TTHG above,
and τ_A^{670} is to be determined just south of Mobile Bay where the
aerosol optical thickness at 670 nm was measured with a Volz
sun photometer with a 2.5° field of view. The measured aerosol
optical thickness was 0.20 and varied from 0.24 to 0.16 1 hour

before and after the overpass. Using Eq. (7) with $\omega_A^{670} = 1$
yields $\tau_A^{670} = 0.37$. Figure 1 suggests that Eq. (7) *underestimates*
the radiance or, vice-versa, using Eq. (7) to determine τ_A^{670}
overestimates this quantity. Taking this into consideration
yields a τ_A^{670} of about 0.29. Noting that the actual aerosol
phase function could differ substantially from the TTHG, it is
felt that the agreement between measured and derived τ_A^{670} is
satisfactory. At this time, the authors know of no way of
estimating $P_A(\theta)$ from the CZCS imagery.

V. CONCLUDING REMARKS

The atmospheric correction algorithm developed by Gordon (5)
has been reviewed and applied to CZCS imagery of the Gulf of
Mexico from orbit 130. The success of the algorithm in removing
most of this very inhomogeneous hazy atmosphere from an image
of area about half that of the entire Gulf suggests the
validity of the algorithm and the assumptions on which it was
based. In particular, in the context of Eq. (10), the haze in
this image appears to have a constant composition and size
distribution varying only in concentration. Although pigment
concentrations derived from the CZCS imagery have yet to be
compared with surface truth, it is clear from Fig. 9, especially,
that the sensor is producing imagery which shows small changes
in pigment concentrations. Finally, the results presented here
also indicate the potential of the CZCS for measuring the
aerosol optical thickness over very large spatial scales.

SYMBOLS

F_0^λ extraterrestrial solar irradiance at λ

I_x^λ normalized radiance, $L_x^\lambda/F_0^\lambda \, T_{0_3}(\mu) \, T_{0_3}(\mu_0)$, T, R, A, or G

L_A^λ contribution to L_T^λ by aerosol scattering at λ

L_G^λ contribution to L_T^λ by sun glint at λ

L_R^λ contribution to L_T^λ by Rayleigh scattering at λ

L_T^λ radiance measured by scanner at λ

L_w^λ Upwelling radiance just beneath sea surface at λ

$P_A(\theta)$ phase function for aerosol scattering through angle θ

$P_R(\theta)$ phase function for Rayleigh scattering through angle θ

$T_x(\mu)$ slant path transmittance through atmosphere,

                  $\exp[-\tau_x/\mu]$, $x = R, A, O$.

λ wavelength

μ cosine of angle θ between sensor and nadir

μ_0 cosine of solar zenith angle θ

$\rho(\mu)$ Fresnel reflectance for incident angle $\cos^{-1}(\mu)$

τ_A^λ optical thickness for aerosol scattering

τ_R^λ optical thickness for Rayleigh scattering

τ_O^λ optical thickness of ozone

$\phi-\phi_0$ relative azimuth of sun

ω_A^λ aerosol single scattering albedo

ω_R^λ Rayleigh single scattering albedo

REFERENCES

1. Austin, R. W., *in* "Optical Aspects of Oceanography" (N. G. Jerlov and E. Steeman Hielsen, eds.), pp. 317-344. Academic Press, London (1974).

2. Gordon, H. R., *Appl. Optics*, 15, 1974-1979 (1975).

3. Morel, A., and Prieur, L., *Limnology and Ocean*, 22, 709-722 (1977).

4.   Hovis, W. A., and Leung, K. C., *Opt. Eng. 16,* 153–166 (1977).

5.   Gordon, H. R., *Appl. Optics, 17,* 1631–1636 (1978).

6.   Kattawar, G. W., *J. Quant. Spectros. Radiat. Transfer, 15,* 839–849 (1975).

7.   Clarke, G. L., Ewing, G. C., and Lorenzen, C. J., *Science, 167,* 1119–1121 (1970).

8.   Ramsey, R. C., and White, P. G., Ocean Color Data Analysis Applied to MOCS and SIS data. Final Report. NOAA Contract No. N62306–72–C–0037.

9.   Arvesen, J. C., Millard, J. P., and Weaver, E. C., *Astronaut. Acta. 18,* 229–239 (1973).

10. Clarke, G. L., and Ewing, G. C., *in* "Optical Aspects of Oceanography" (N. G. Jerlov and E. Steeman Nielsen, eds.), pp. 389–414 (1974).

11. Gordon, H. R., and Clark, D. K., Atmospheric Effects in the Remote Sensing of Phytoplankton Pigments, *Boundary Layer Meterology* (1979).

DISCUSSION

Deepak: The photographs you showed were very interesting. Were any ground-truth measurements made about the particle size distribution?

Gordon: Not that I know of. Shortly after our experiment, there was a joint experiment organized in the area involving the Navy, LSU, and others. Someone was measuring aerosols at that time-- maybe a week or two after the images I have shown. There should be CZCS imagery available for that particular time period.

Deepak: I wonder if in some cases the dense aerosols or the dense fogs were close to the sea surface; or were they at higher altitudes?

Gordon: The horizontal visibility when that image was taken was very poor. I can say that for sure. A few days later it was a lot poorer. And I would be interested in looking at vertical distribution data, but we just do not have it here. The horizontal visibility was very poor which suggests, I think, that a lot of haze was near the surface. Also, it seemed that the wind was blowing from the northeast because the smoke plumes are all going toward the southwest. I guess this haze was generated in some fashion as the wind blew over the water.

Zardecki: Yesterday we had a paper by Deepak in which they neglected multiple scattering from aerosols while accounting for multiple Rayleigh scattering. What is your estimation of the aerosol contribution to multiple scattering because you accounted for both?

Gordon: Actually I really have no idea. I have never tried to separate the effects of single scattering and multiple scattering in that way. Although, in that one set of graphs that I showed, if you compute the radiance using a single scattering model with an aerosol optical thickness of about 0.3, you end up making an error of about 30% in the aerosol part. The error in the Rayleigh part probably is not more than about 3 or 4 percent so there seems to be a large contribution of multiple scattering due to aerosols. I think the reason is that they scatter often in the forward direction, and then occasionally scatter backwards. The single scattering calculation does not do that and hence enhanced backscattering is found when multiple scattering is included. The nice part about this algorithm is that the atmosphere does all the hard work for us. The atmosphere does the multiple scattering problem so we are not really faced with it. It can do it a lot better than we can anyway.

Deepak: I would like to clarify what Andrew Zardecki just mentioned. In our inversion technique for scattered radiance in which multiple scattering is included, we consider the effect of single scattering by aerosols and molecules and the multiple scattering by molecules alone—the effect of multiple scattering by aerosols per se being neglected. This makes the inversion problem tractable and at the same time improves the accuracy of the retrieved size distribution results in comparison to those obtained with the single scattering approximation. This has been demonstrated in recent papers in Applied Optics. The contribution due to multiple scattering by aerosols is, however, partially taken into account in our method by multiplying the molecular phase function by a quantity called the effective optical depth, which depends on the optical depth, sun zenith angle and the earth's albedo.

THEORY AND APPLICATION OF THE TRUNCATED
NORMAL DISTRIBUTION FOR REMOTELY
SENSED DATA[1]

D. S. Crosby

National Oceanic and Atmospheric Administration
National Environmental Satellite Service
Washington, DC

D. J. DePriest

Office of Naval Research
Arlington, Virginia

*A problem associated with operational procedures when infrared
radiometers are employed as sensor receivers is the presence of
clouds. Nevertheless, it is possible to estimate statistically
the surface temperature from satellite high resolution infrared
window radiation measurements. This procedure involves estimating the surface temperature from a large number of radiometer
measurements using the theory of the truncated normal distribution. In this paper, previously known results are presented and
a new result for determining the truncation point is developed.
Goodness-of-fit procedures are used as a screen to eliminate data
sets which are too contaminated.*

[1]*Part of this paper was written while Dr. Crosby was supported
by the Office of Naval Research Contract N00014-77C-0624.*

I.   INTRODUCTION

A fundamental problem in measuring the infrared radiation
emitted from the surface or lower atmosphere of the earth is
the presence of clouds. The usual effect of clouds in an
instrument field of view is to lower the measured radiation. In
order to be able to use many of the techniques of remote sensing,
it is necessary to have an estimate of the emitted radiance from
a clear or cloud-free field of view. The technique which is
developed in this paper uses the measured radiances from a large
number of neighboring small fields of view. If a sufficient
number of these small fields of view are cloud-free, then a
technique which uses the theory of the truncated normal distribution can be applied to the data to obtain an estimate of the
clear radiance (1).

Since the technique developed is a maximum likelihood
procedure, standard statistical procedures can be used to
determine whether the distribution in the tail is approximately
normal. This allows for a simple decision procedure to check
whether the model is satisfied (2,3). These decision procedures
use classical statistical goodness-of-fit tests. Finally, one
difficulty with the application of this procedure has been an
objective method of determining a truncation point. A
sequential procedure for determining this truncation point is
outlined.

II.   THE MODEL

In order to apply the technique of the truncated normal,
the following conditions must be satisfied:

1.   There are a number of neighboring measurements where the
radiances in the absence of the clouds would be the same.

2.   The presence of clouds lowers the radiance in the field
of view.

3.   Some of the fields of view are cloud-free.

4.   The difference between the measured radiances from two
cloud-free fields of view is due only to instrument noise, which
distribution is assumed to be normal.

It is assumed that the density of the measured radiances
from all the fields of view in a set takes the form

$$h(x) = \alpha \ f(x) + (1 - \alpha) \ g(x) \tag{1}$$

In Eq. (1), α is equal to the proportion of fields of view which
are cloud-free,

$$f(x) = (1/2\pi\sigma^2)^{1/2} \ \exp(- (x - \mu)^2)/2\sigma^2 \tag{2}$$

and $g(x)$ is some unknown density function.   It is further assumed
that there is some T such that

$$g(x) = 0 \qquad x > T \tag{3}$$

That is for the values of x above T, the density curve has the
form of a normal density curve with parameters μ and σ^2.   The
parameter μ is the true clear radiance; that is, μ is the
radiance that would be measured from a clear field of view if
there were no instrument noise.   The variance, σ^2, of the
instrument noise is assumed to be known.   The problem is to
obtain a good estimate of μ.

III.   THE ESTIMATION PROCEDURE

The statistical technique outlined in this section is well
known and is available in the literature.   See, for example,
Crosby and Glasser (1).

If the radiance data above some trunctation point T is
from a truncated normal distribution, that is, it has the form

given by Eq. (2), and if it has known variance σ^2, then the
maximum likelihood estimate for the mean μ is a monotonic
function of the arithmetic mean of the observations above that
truncation point.

Let $I_{\lambda i}$ ($i = 1, \ldots, n$) be the measured radiance values above
some truncation point T. Let

$$Z_i = (I_{\lambda i} - T)/\sigma \tag{4}$$

and let

$$\hat{v} = (T - \hat{\mu})/\sigma \tag{5}$$

Recall that T is a known truncation point, σ is the known
instrument noise, and $\hat{\mu}$ is the estimate of the clear radiance.
Let

$$\bar{Z} = \sum_{i=1}^{n} Z_i /n \tag{6}$$

then $\hat{v}$ is given by the solution to the equation

$$-\hat{v} + (\exp(-\hat{v}^2/2) / \int_{v}^{\infty} \exp(-x^2/2)\,dx) = \bar{Z} \tag{7}$$

Once $\hat{v}$ is known, $\hat{\mu}$ is found by the inverse of Eq. (5)

$$\hat{\mu} = T - \hat{v}\sigma \tag{8}$$

Since the function given in Eq. (8) is monotonic and smooth
so is its inverse. It is possible to find a good rational
polynomial approximation to the inverse function which gives
$\hat{v}$ as a function of $\bar{Z}$.

As an example, this procedure has been applied to the
10.5 μm to 12.5 μm infrared channel of the Scanning Radiometer
of the NOAA satellites. Table I gives the upper tail of one
such data set. The measurements are in counts and have been
transformed in order to keep the example as simple as possible.

TABLE I. *Data from the 10.5 μm to 12.5 μm Infrared Channel of the Scanning Radiometer*

Counts C	Frequency f_1	Estimated mean $\hat{\mu}$
73	2	
72	3	
71	8	
70	12	
69	22	
68	31	65.0

For the data in Table I, $T = 67.5$, $\sigma = 3$ and hence $\bar{Z} = 0.560$. Then $\hat{v} = 0.833$ and $\hat{\mu} = 67.5 - ((0.833)3) = 65.0$. The value for σ is found from independent data and $\hat{v}$ is found from $\bar{Z}$ by solving Eq. (7).

IV. GOODNESS-OF-FIT TESTS

The procedures outlined in the preceding section depend on the assumption that the upper tail of the distribution has the shape of a truncated normal distribution. If the shape is not normal, it may indicate that few, if any, of the fields of view are completely cloud-free. Hence, the shape of the data from the tail may be used to screen out situations of this type.

The test which is recommended is the chi-square goodness-of-fit test. For a description of other possible tests, see DePriest (2). The chi-square test is well known. See, for example, Cochran (4). However, for completeness, a short description of the procedure will be given here.

Given n observations grouped into K mutually exclusive
categories, let O_i and E_i (i = 1, 2, ..., k) denote, respectively,
the observed and expected frequencies for the K categories. The
statistics are

$$\chi^2 = \sum_{i=1}^{K} (O_i - E_i)^2/E_i \qquad (9)$$

Since the estimate of the parameter μ, $\hat{\mu}$, is a maximum likelihood
estimate, the asymptotic distribution of the statistics χ^2 will be
chi-square with K - 2 degrees of freedom. Since the radiance values
are already grouped, this natural grouping is used in the computation of the statistics. To illustrate the technique, two
examples are provided.

For the data in Tables II and III, it was assumed that
$\sigma = 3$. This value was determined from independent measurements.
The data are in counts. The estimated mean uses the data in the
tail above and including the corresponding count.

TABLE II. *Data from the Scanning Radiometer*

Counts C	Frequency f_i	Estimated Mean $\hat{\mu}$
73	2	
72	3	
71	8	
70	12	
69	22	
68	31	
67	42	
66	34	
65	37	
64	49	65.3

TABLE III.   Data from the Scanning Radiometer

Counts C	Frequency f_i	Estimated Mean $\hat{\mu}$
50	1	
49	9	
48	7	
47	9	
46	6	
45	8	
44	6	45.4

For the data in Table II, a truncation point of 63.5 was taken. This value gives an estimate of μ, $\hat{\mu}$ = 65.3. The number of groups was taken to be equal to 9. The data associated with 72 and 73 counts are grouped into a single class. Then the chi-square statistics will have seven degrees of freedom. It is found that χ^2 = 4.5. This value indicates a good fit. The model given is thus reasonable.

For the data in Table III, a truncation point of 43.5 was taken and $\hat{\mu}$ = 45.4. Grouping the counts 50 and 49 into one class, there are 6 classes. The chi-square statistics have four degrees of freedom. The value is found to be 12.3. The fit is not very good and the model of an upper tail which is normal is rejected.

V.   DETERMINING THE TRUNCATION POINT

Given a truncation point T, the procedure outlined in the earlier sections will estimate μ and test whether the tail is normal. However, one difficulty with the application of these techniques has been a determination of a good truncation point.

The following procedure is recommended. It is assumed that the
data are grouped. Then the count data from the instrument will
be in the format represented by Table IV. The minimum number
of classes used in the technique should cover approximately two
standard deviations. In the examples which have been presented,
σ was equal to 3, and a minimum of 7 classes was used. The
adjusted mean of the data of Table IV and including class j is

$$\bar{Z}_j = \sum_{i=1}^{j} (M_i - k_j)\, r_i \, / \, \sigma (\sum_{i=1}^{j} f_i) \qquad (10)$$

Then $\hat{v}_j$ is found by using Eq. (7), or an approximation to this
equation. If $\hat{v}_j > 0$, compute a $\bar{Z}_{j+1}$. Repeat this process
until $\hat{v}_m < 0$. The data above the point k_m is then checked by
using the goodness-of-fit criteria of the preceding section.
If the data do not fit the tail of a normal distribution, the
process is terminated and the retrieval discarded.

At this stage of the process, the truncation point is k_m.
The truncation point should be selected to include as much of
the data as possible. The reason for this is seen in Table V.

TABLE IV. *Count Data from Instrument*

Class	Class midpoint	Class frequency
$k_0 - k_1$	$m_1 = (k_1 + k_0)/2$	f_1
$k_1 - k_2$	$m_2 = (k_2 + k_1)\, 2$	f_2
$\cdot$	$\cdot$	$\cdot$
$\cdot$	$\cdot$	$\cdot$
$\cdot$	$\cdot$	$\cdot$
$k_{i-1} - k_i$	$m_i = (k_i + k_{i-1})/2$	f_i

TABLE V. Asymptotic Standard Deviation of $((\hat{\mu} - \mu)\sqrt{N})/\sigma$

$T*$ [a]	Asymptotic standard deviation
.5	3.474
0.0	2.346
-0.5	1.373
-1.5	1.178
-2.0	1.074
-3.0	1.007

[a] $T* = (T - \mu)/\sigma$.

Table V gives the asymptotic standard deviation of $((\hat{\mu} - \mu)\sqrt{N})/\sigma$ for various values of T. In this table N is equal to the total sample size if there were no truncation, σ is the known standard deviation, and $\hat{\mu}$ is the estimate of μ found by using the maximum likelihood technique described previously. It is seen from this table that there is a significant reduction in the standard error of estimate for $\hat{\mu}$ if the truncation point changes from μ to two standard deviations below μ.

A sequential procedure is used to determine wether the truncation point should be moved. Let $I_{\lambda 1}$, $I_{\lambda 2}$, $\ldots$, $I_{\lambda n1}$ be the radiance measurements above some T_1 and let $I_{\lambda 1}$, $I_{\lambda 2}$, $\ldots$, $I_{\lambda n1}$, $\ldots$, $I_{\lambda n2}$ be the radiance measurements above some T_2 $(T_2 < T_1)$. A decision must be made whether to move the truncation point to T_2 and hence increase the number of measurements used to estimate μ. The decision process is based upon the result that using the information from the values $I_{\lambda 1}$, $I_{\lambda 2}$, $\ldots$, $I_{\lambda n1}$, it is possible to estimate n_2. This estimate is called $\hat{n}_2$. It can also be shown that $(\hat{n}_2 - n_2)n_2^{-1/2}$ is asymptotical normal with mean 0 and

estimatable standard deviation. If $(\hat{n}_2 - n_2)/n_2^{1/2}$ is within
certain bounds, then the truncation point is extended to T_2.
If $(\hat{n}_2 - n_2)/n_2^{1/2}$ is outside these bounds, then T_1 is used
as the truncation point and $\hat{\mu}_1$, the estimate based on the
variables $I_{\lambda 1}$, $I_{\lambda 2}$, ..., $I_{\lambda n1}$ is used as the estimator for μ.
The process is repeated by using T_2 in place of T_1 and some
$T_3 (T_3 < T_2)$ in place of T_2. This procedure is repeated with a
sequence of T_1's until $(\hat{n}_i - n_i)/n_i^{1/2}$ is outside of the bounds
or until $T^* = ((T - \hat{\mu})/\sigma)$ is less than -2.

Let $\hat{\mu}_1$ be the estimator of μ based on the measurements
above T_1. Let

$$F_c(T_i, \hat{\mu}, \sigma^2) = (1/2\pi\sigma^2)^{1/2} \int_{T_i}^{\infty} \exp(-(x - \hat{\mu})/2\sigma^2)dx \qquad (11)$$

Then an estimator of n_2 is

$$\hat{n}_2 = [n_1/(F_c(T_1,\hat{\mu}_1,\sigma^2)/F_c(T_2,\hat{\mu}_1,\sigma^2))] \qquad (12)$$

where [y] denotes the greatest integer less than or equal to y.
Let n_2 be the actual number of measurements above T_2.
Assuming that $I_{\lambda 1}$, $I_{\lambda 2}$, ..., $I_{\lambda n2}$ come from a population which
is normal with mean μ and variance σ^2, the asymptotic variance
of $(\hat{n}_2 - n_2)/n^{1/2}$ is given by

$$s^2 = \frac{(F_c(T_2,\mu) \psi(T_1,\mu) - F_c(T_1,\mu) \psi(T_2,\mu))^2}{F_c(T_2,\mu)(F_c(T_1,\mu)) H(T_1,\mu)}$$

$$+ \frac{F_c(T_2,\mu) - F_c(T_1,\mu)}{F_c(T_1,\mu)} \qquad (13)$$

where

$$\psi(T,\mu) = (1/2\pi)^{-1/2} \exp-(T - \mu)^2/2 \qquad (14)$$

$$F_c(T,\mu) = (1/2\pi)^{-1/2} \int_T^\infty \exp{-(x - \mu)^2/2}dx \qquad (15)$$

$$H(T_1,\mu) = F_c^2(T_1,\mu) + (T_1 - \mu) F_c(T_1,\mu) \psi(T_1,\mu)$$
$$- \psi(T_1,\mu)^2$$

Table VI gives the asymptotic standard deviations for some values of T_1 and T_2 when $\mu = 0$.

This sequential procedure has been applied to data from the 10.5 μm to 12.5 μm channel of the Scanning Radiometer. Table VII gives the upper tail of such a data set. The data are in counts.

TABLE VI. The Asymptotic Standard Deviations of
$(n_2 - n_2)/n_2^{1/2}$

T_1	T_2	S
1.0	0.67	1.064
0.5	0.17	0.833
0.0	-0.33	0.632
-0.5	-0.83	0.460
-1.0	-1.33	0.316
-1.5	-1.83	0.202
-2.0	-2.33	0.119
-2.5	-2.83	0.064

TABLE VII. Data from the Scanning Radiometer

Counts	Frequency	Estimated Mean	$(n_2 - n_2)/n_2^{1/2}$	S
88	1			
87	3			
86	11			
85	18	78.3		
84	31	79.0		
83	47	79.3		
82	67	79.4		
81	68	79.7		
80	71	79.9*	0.57	0.58
79	75	79.8	-0.10	0.48
78	60	79.9	0.13	0.37
77	57	79.8	-0.52	0.29
76	45	79.7	-0.40	0.23
75	44	79.4	-0.89	0.18
74	28	79.3		
73	22	79.2		
72	16	79.1		
71	26	78.8		
70	21	78.5		
69	9			
68	21			
67	20			
66	13			

*$\hat{v}$ is less than zero.

For this example, $\sigma = 3$. The estimate of μ, $\hat{\mu}$ uses the data in the tail above and including the corresponding counts. For example, the estimate corresponding to 80 uses a truncation point of 79.5. This is the first estimate where $\hat{v}$ is less than zero. The sequential procedure is then applied. For the other entries corresponding to 80, $T_1 = 79.5$, $T_2 = 78.5$, $n_2 = 317$, $\hat{n}_2 = 327$, and $n_1 = 246$. Note that $(\hat{n}_2 - n_2)n_2^{1/2}$ is not large compared with S and hence the truncation point would be moved to T_2 or 79.5. Continuing this process, if a conservative rule were used, the process would terminate at $T = 75.5$ which would yield an estimate of $\hat{\mu} = 79.7$. By any reasonable rule the process would terminate at $T = 74.5$ which gives a $\hat{\mu} = 79.4$. Note for this last entry the value of the statistic $(\hat{n}_2 - n_2)/n_2^{1/2}$ is greater than four standard deviations from zero.

VI. CONCLUDING REMARKS

In this paper a procedure which uses standard and new statistical techniques to estimate clear radiances for a high spacial resolution, a multiple field of view infrared instrument has been developed. The technique does not completely solve the cloud problem. For example, it will not distinguish between a low level uniform cloud layer and the surface. Also, certain types of inversions in the temperature profile will cause difficulty. However, the technique is computationally simple and can easily be used with other types of procedures. This technique has been applied to real data and seems to work very well.

SYMBOLS

I_λ measured radiance

T   truncation point

T*   normalized truncation point

$\hat{v}$ estimated adjustment parameter

[y]   greatest integer less than or equal to y

Z   normalized measured radiance

α proportion of cloud-free fields of view

μ true clear radiance

$\hat{\mu}$ estimated clear radiance

σ^2 known variance of instrument noise

REFERENCES

1. Crosby, D., and Glasser, K., *J. Applied Meteorology,*
    11, 1712-1715 (1978).

2. DePriest, D., Testing Goodness-of-Fit for the Tail of the
    Normal Distribution. Ph.D. Thesis, American University,
    Washington, DC (1976).

3. Crosby, D., and DePriest, D., *in* "Proceedings of the Fifth
    Biennial International Codata Conference" (B. Dreyfus, ed.),
    p. 31, Pergamon Press, New York (1977).

4. Cochran, W. G., *Am. Math. Statist. 23,* 315-345 (1952).

DISCUSSION

Staelin: Your method is rather robust in the sense that it makes
almost no assumptions about the character of the perturbing noise.
What can you say about the possibility of improving the technique
by modeling the statistical character of that perturbing signal?
In other words, might you improve you performance if you could
do so even if the technique became less practical?

Crosby: I would suppose so but again you would lose the great
simplicity of this particular approach. Of course, if one adds
lots of things together almost always you get something which
is approximately normal. I would be reluctant to abandon that
particular assumption but of course one could. There exists a
whole series of papers on estimating parameters involving just
the tail of various distributions. Almost always they assume
known distributions such as the normal.

Susskind: Could we just take a look at that distribution—the
good one—again? I must be missing something. It did not look
that normal to me. In a sense it almost looks like a bimodal
distribution. You have 136.5 as your mean which is a lot on the
warm side. It almost looks like there is another gaussian at
132 or something like that.

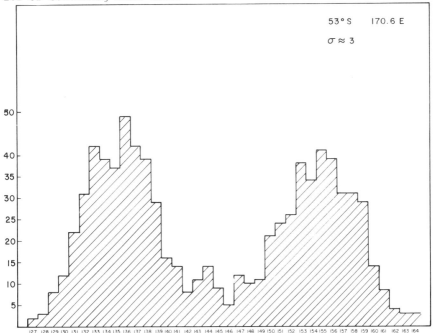

Figure D-1.

Crosby: But all I can say is that I fit it with a normal and it fits well. The measure of the deviation from normal is the square of the deviation divided by the expected frequency of the cell. In the cell we are looking at, the expected frequency is about 50 and the deviation is about 10. So the measure of the deviation is about 2. That is marginal, but it is not a large deviation.

Susskind: Does this represent a side scan of an instrument looking at many spots?

Crosby: With the ones I usually picked out, they were usually 1,024 measurements. Those are pretty close to the nadir. You have blocks of 1,024 spread out and I usually pick the ones that are looking straight down to avoid the difficulties with the atmospheric effects.

Susskind: That is one of the things. I was wondering that as you are changing the zenith angle, you will have differences in the atmospheric effects. But it is possible that you are really looking at two slightly different temperatures differing by a degree.

Crosby: I would expect of course that the surface is not perfectly homogeneous. You are talking about in this case a fairly large area—(8 x 32)km by (8 x 32)km.

Susskind: It looked like two Gaussians put together. I was wondering if you had ever tried anything like that.

Crosby: I think you would have trouble separating out the two that were that close together. There are techniques which are supposed to do that.

Chahine: My question is that very often you have a bias--a uniform background--above the surface and is there no way for you to get rid of this uniformity?

Crosby: I would not say that this is the complete answer to the problem of clouds. You certainly could have a uniform cloud deck just above the surface and it is going to look just like the surface does. What I am suggesting is that it can be used as a preprocessor for other techniques. If you want to associate high resolution results with the larger resolution results, one may want to do something like this. It is sometimes very difficult to have to put the two together on an operational basis, as we have discovered.

Chahine: Another question I have here is that if you have to use a field-of-view of a 1000 by 1000 km, the assumption of a uniform ocean surface temperature is physically unacceptable.

Crosby: That is correct. These just happen to be the ones that they had. You will see that the technique is being applied at the present time to squares about 50 km on a side.

APPLICATION OF THE TRUNCATED NORMAL DISTRIBUTION
TECHNIQUE TO THE DERIVATION OF
SEA SURFACE TEMPERATURES [1]

Henry E. Fleming

National Environmental Satellite Service
National Oceanic and Atmospheric Administration
Washington, D. C.

The truncated normal distribution technique is applied to the derivation of sea surface temperatures from high-resolution satellite measurements in the 11-μm window region. The purpose of this technique is to remove cloud contamination from the measurements. This procedure produces sea surface temperature maps on a mesoscale level so that a reasonably small-scaled temperature structure, such as that found in the Gulf Stream, can be determined. Because not all the cloud contamination is removed by the truncated normal distribution technique, the resulting temperature fields are modified by an error detection and correction technique. Results from both simulated and real data are presented. No attempt has been made to account for atmospheric attenuation by water vapor; therefore, the results are accurate in only a relative sense.

I. INTRODUCTION

Maps of sea surface temperature contours have been available
for years; however, the temperature structure of these maps is
generally on the macroscale level. The objective of this paper
is to present techniques for producing sea surface temperatures

[1]*This research was done at the Naval Post Graduate School and was supported by the Naval Air Systems Command.*

on a mesoscale level so that a fine-scaled structure, such as that of the Gulf Stream, can be delineated. Furthermore, most users of such maps require accuracies of 1° C, or better, on a daily, global basis.

At the present time, the only satellite measurements that can meet this accuracy requirement are the high resolution infrared (IR) measurements made from polar orbiting satellites by scanning radiometers that have a spatial resolution of the order of 1 to 3 km in the nadir direction. Global coverage is possible twice a day from a single satellite, but if visible cloud image data are used in conjunction with the infrared measurements, then global coverage is limited to once a day. Fortunately, this coverage is adequate for most purposes because the thermal structure of the ocean changes very little over the period of a day.

By far the most difficult problem in deriving sea surface temperatures (SSTs) from infrared measurements is the contamination of the data by clouds. Therefore, the numerical method employed is twofold. First, the truncated normal distribution (TND) technique, described in the preceding paper in this volume (also see Ref. 1), is used to remove cloud contamination from the measurements. Then the resulting sea surface temperature fields are modified by an error detection and correction (EDC) technique, which is used to remove residual cloud contamination and to correct data that have been filled in for missing measurements. Each of these techniques are applied to both real and numerically simulated data.

The four names in this introductory section to which acronyms have been assigned occur so frequently in the paper that only their acronyms are used throughout most of the text. These acronyms are defined here for easy reference:

EDC--error detection and correction (technique)

IR--infrared

SST--sea surface temperature

TND--truncated normal distribution (technique).

II.   THE TRUNCATED NORMAL DISTRIBUTION TECHNIQUE

A. *Assembling the Data Arrays*

Even though the preceding paper in this volume describes the
TND technique in detail, a very brief summary of the technique is
given here to provide a smooth transition between papers.   The
individual high resolution IR measurements from which the SSTs
are derived are referred to as *pixels.* Because sea surface
temperatures are not required on the scale or density of the
pixel size, even for mesoscale work, the pixel data are assembled
into arrays for processing. After the data comprising an array
are processed, a single SST results which is representative of the
entire array.

To clarify matters, one should consider Fig. 1 in which an
array of pixels is represented by the fine grid. For the applications of this paper, the pixel array will be square and will
vary in size from a maximum of 1024 (32 x 32) entries to a
minimum of 64 (8 x 8) entries. The processed arrays produce
individual SSTs on the scale of the larger grid in Fig. 1 and the
temperature values are located in the center of each square of the
larger grid, as indicated by the round, solid dots.

It should be pointed out that a "raw" pixel value is a voltage
count value which is readily converted to a temperature value by
means of a nonlinear calibration curve that is available for each
instrument. Usually the SST data processing is done with voltage
counts because they are integer values. Then conversion of voltage
to SST becomes the final step in the processing procedure. Also,
photographic images are made from the arrays of pixel values by
converting the voltage counts to a gray scale whose shading depends
upon the voltage intensity.

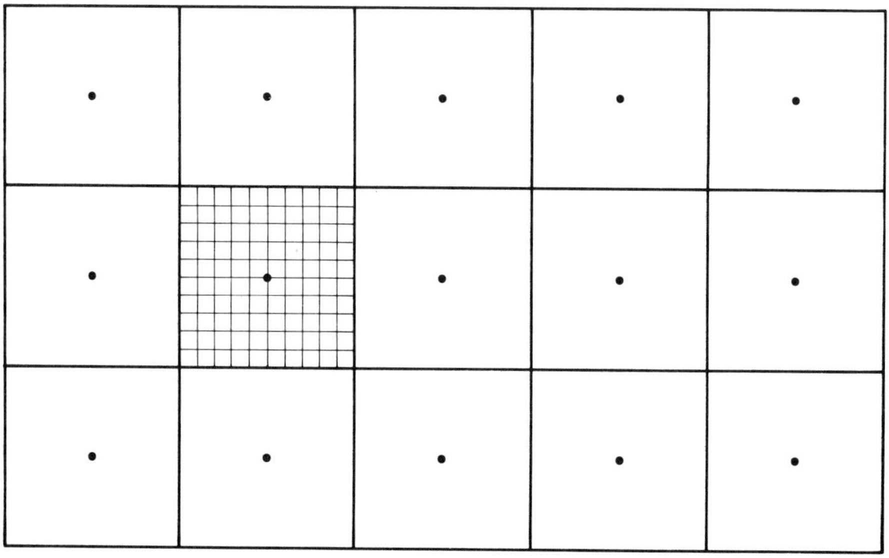

FIGURE 1. The fine grid illustrates a typical data array;
the course grid is the SST grid; and the dots locate each SST.

B. *Deriving the Sea Surface Temperatures*

Suppose for the sake of illustration that one wishes to derive SSTs from arrays of pixels of dimension 32 x 32. The first step in processing these 1024 data is to plot them as a frequency histogram. Figure 2 shows such a histogram with class intervals of 1° C. Notice that the histogram has two modes and that the data come from an area that is 75 percent cloud covered. Data associated with the cold mode come from the cloudy portion of the pixel array, whereas the data associated with the warm mode arise from the cloud-free portion of the array. The warm mode does not fall into a single class interval (a spike) because of instrumental noise.

Since the data used for Fig. 2 were numerically simulated, the true density curve is known and is plotted as the continuous solid curve in the figure. It is a normal density function with a standard deviation of 4° C and a mean of 285 K. If all 1024 pixels were clear, the histogram would approximate the entire continuous density curve. Consequently, the sea surface temperature being sought is just the mean (peak) value of the density curve, which is 285 K in this case.

Figure 2 succinctly illustrates the problem of extracting the sea surface temperature from a histogram of pixel values. The required SST is determined by the mean of the underlying density curve, given the following assumptions:

1. The underlying probability density function is normal (Gaussian).

2. The cloud-free (warm) mode of the histogram is sufficiently populated that the normal density function can be approximated from the warmest part (the wing) of the histogram.

3. The standard deviation of the normal density function is known. (The value used usually is the root mean square (rms) noise level of the instrument.)
The determination of the mean of the density function (i.e., the

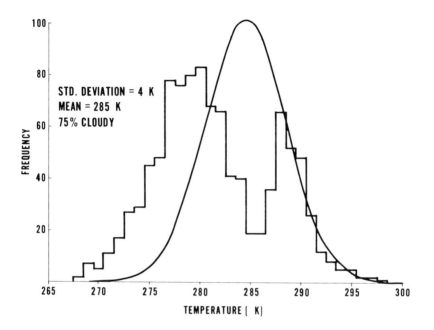

FIGURE 2. A typical SST histogram of 1024 elements with the
theoretical cloud-free density curve superimposed.

determination of the SST) under these three assumptions is precisely the classical truncated normal distribution problem in
statistics. The solution to this problem, and hence the solution
to the SST problem, is given in the preceding paper in this volume,
and so will not be repeated here.

III. NUMERICAL SIMULATION STUDY

A. Details of the Simulation

Numerical simulation is an important tool for testing
numerical techniques because all aspects of the experiment can
be controlled and because the true answer is known. Therefore,
it is only by simulation that one can answer the fundamental
question: What spatial resolution is required to achieve a given
SST accuracy?

The numerical simulation was initiated by simulating the values
of the individual pixels with a random number generator, subject
to the following constraints:

1. The pixel values were generated in degrees Kelvin to
simplify the calculations.

2. The entire sea surface was made isothermal at 285 K, i.e.,
the mean values of all the density curves were fixed at 285 K.

3. The values of the pixels were distributed normally with a
standard deviation of 4° C.

4. Each pixel was assumed to have a resolution of 1.5 n mi
on a side.

5. Whenever it was required that y percent of the sky was
cloudy, y percent of the pixels were made colder than their given
value by means of a second random number generator with a uniform
distribution. In other words, if a given pixel was cloud contaminated, its temperature was reduced by the amount $2\alpha X$, where
α is the known standard deviation and X is the random value

produced by the random number generator with a uniform density
curve of unit amplitude.

6. Enough pixels were generated to simulate a square sea
surface area 384 n mi on a side, that is, 65,536 individual pixel
values.

The histograms resulting from these random numbers were similar
in general appearance to that shown in Fig. 2.

The truncated normal distribution technique for removing
cloud contamination was applied to these simulated pixel data for
different cloud amounts and array sizes. Specifically, the TND
technique was applied to the following combinations:

1. Cloud amounts of 0, 25, 50, and 75 percent.

2. Square pixel arrays of sizes 32 x 32, 16 x 16, and 8 x 8,
which yielded resolutions of 48, 24, and 12 n mi, respectively,
for each derived SST.

Results from these various combinations are summarized in the next
section.

B. Root-Mean-Square and Maximum Errors

Since the true SST value for the simulated data was known to
be 285 K, the rms and maximum errors between the calculated and
true temperatures could easily be determined for each of the combinations cited in the previous section. The results are listed
in Table 1. The rms and maximum error values under discussion
are those listed in the columns labeled "before." The columns
labeled "after" are discussed subsequently.

The cogent points to be made about Table I are that (1) both
the rms and maximum errors increase with increasing resolution, and
(2) the requirement of 1° C accuracy is not met for cloud cover
of 75 percent or more and for the 12-n mi resolution cases.
Fortunately, these results are preliminary; more can be done to
increase the accuracies of the "before" entries of Table I by

TABLE I. Root-Mean-Square and Maximum Errors for Various
Combinations of Spatial Resolution and Percent Cloud Cover before
and after Application of the EDC Technique for the 285 K
Isothermal Surface

	RMS Errors (^{o}C)					
	48-Mi. Resolution		24-Mi. Resolution		12-Mi. Resolution	
Cloud Cover	Before	After	Before	After	Before	After
Clear	0.33	0.29	0.62	0.38	1.16	0.37
25%	.36	.32	.63	.34	1.15	.37
50%	.39	.32	.68	.41	1.16	.42
75%	1.46	1.47	1.32	1.29	1.34	1.11
	Maximum Errors (^{o}C)					
Clear	0.85	0.84	1.93	1.02	4.00	3.01
25%	.84	.84	2.70	-.99	3.74	-1.78
50%	1.44	1.12	2.13	1.73	4.56	3.26
75%	1.97	1.97	2.56	2.01	3.86	2.01

applying the error detection and correction technique, mentioned
previously, to the field of SSTs produced by the TND technique.

C. The Error Detection and Correction Technique

In the simulation study, the number of SST values comprising
the complete field varied with resolution size, since resolution
size was determined by the number of pixels used in processing a
single SST value. However, the overall size of the field
remained invariant at 384 n mi on a side. Thus, the 48-mile
resolution study yielded a field of 64 (8 x 8) SST values,
whereas the 12-mile resolution study yielded a field of 1,024
(32 x 32) values.

Since the individual values that make up a SST field are produced independent of one another, the associated errors tend to be
uncorrelated and result from cloud contamination that cannot be
distinguished by the TND technique. These kinds of errors lend
themselves to detection and correction by an objective technique
reported by Fleming and Hill (2). The technique is too involved
to describe here, but the idea behind the method is so simple that
it will be explained.

The basic concept behind the EDC technique is the fact that
when finite differences are applied to equally spaced data, the
errors are magnified in the higher differences, while the higher
differences of the accurate data approach zero, provided that the
accurate data have polynomial smoothness. Table II illustrates
this point. For simplicity, suppose that the correct data lie on
the cubic polynomial

$$y = x^3 + x^2 + x + 1 \tag{1}$$

and that the data are given at the points x = -2, -1, 0, 1, and
2. Then the corresponding values of y and its higher differences
up to the fourth one are given in Table II. Clearly, the fourth
differences of the values of y that lie on the cubic polynomial
are all zero.

TABLE II. *Values of the Polynomial of Equation (1) and
Higher Differences up to the Fourth Corresponding to the Values
of the Independent Variable X Indicated.*

X	Y	ΔY	$\Delta^2 Y$	$\Delta^3 Y$	$\Delta^4 Y$
-2	-5		-10		0
		5		6	
-1	0		-4		0
		1		6	
0	1		2		0
		3		6	
1	4		8		0
		11		6	
2	15		14		0

On the other hand, suppose an error ε is introduced into the third value of y. Table III shows how this error is propagated throughout the difference table. Note that while the fourth differences in Table II have gone to zero, the fourth differences of Table III have picked up the original error and amplified it in three out of five entries. If the fourth difference column in Table III is examined for the largest value, it is found in the third entry just opposite the original error ε in the y-column and is six times ε. Consequently, the error detection procedure for an isolated error is to search the fourth difference column for the largest entry. That entry identifies the value of y that carries the isolated error ε. Furthermore, if that largest entry of the Δ^4y-column is divided by 6 and then subtracted from the corresponding value of y, the error in y will be removed, that is, corrected. This is the essential idea behind the error detection and correction technique.

Detection and correction of clusters of errors is more compli-cated and is discussed in Fleming and Hill (2). Additional dif-ficulties arise at the edges and corners of two-dimensional arrays because the higher differences cannot be calculated there. The remedy for this is to use a least-squares approach, details of which can be found in Lanczos, pages 317-320 (3). (Complete details for the multidimensional error detection and correction

TABLE III. *Propagation of an Error ε Throughout a Fourth Difference Table*

Y	ΔY	$\Delta^2 Y$	$\Delta^3 Y$	$\Delta^4 Y$
0		0		ε
	0		ε	
0		ε		-4ε
	ε		-3ε	
ε		-2ε		6ε
	$-\varepsilon$		3ε	
0		ε		-4ε
	0		$-\varepsilon$	
0		0		ε

technique used in the examples to follow will be given in a
forthcoming paper by Fleming and Hill.)

D.   Errors after Application of EDC Procedure

The rms and maximum errors between the calculated and true
SSTs for the simulated data set described in the section, "Details
of the Simulation," were completely different before and after the
EDC technique was applied to the data. The errors before the
technique was applied were discussed in the section, "Root
Mean Square and Maximum Errors," and are listed in Table I in the
columns labeled "before." On the other hand, the errors that
remained after the EDC technique was applied are listed in Table I
in the columns labeled "after." Except for the 75 percent cloud-
cover entries, the rms errors in the "after" columns are comparable, being of the order of 0.3° to 0.4° C. This is well
within the accuracy requirement of 1° C. Clearly, 75 percent
cloud cover is too much to be handled adequately by any of the
techniques.

The conclusions to be drawn from Table I are that when the EDC
technique was applied to the SST fields, the rms errors were
reduced in all cases, except the one at 75 percent cloud cover,
and the higher the resolution, the more the rms errors were
reduced. In fact, the most dramatic improvements occurred in the
12-mile resolution cases where the improvements were as large as
68 percent. The improvements were not as dramatic for the
maximum errors, but, nevertheless, they are impressive. The
previous conclusion that the errors increase with increasing
resolution is no longer valid when the EDC technique is combined
with the TND technique. Instead, one now may conclude that for
an isothermal sea state, the rms errors are virtually the same for
the high 12-mile resolution fields as for the lower 48-mile
resolution field, and all but the cases of 75 percent cloud cover
are well within the 1° C accuracy requirement.

There is, however, a caveat in these conclusions. An
isothermal sea surface is an unrealistic assumption in any case,
but especially for mesoscale analyses. For example, in the Gulf
Stream it is quite possible to have temperature changes as large
as 1° in 2 n mi, or even larger.

E. *Simulated Variable SST Field*

   In order to make the simulation study discussed more realistic,
the contoured surface of Fig. 3 was used in place of the isothermal surface. Although the shape of this surface may not be
typical of the thermal structure of the oceans, it nevertheless
is useful for a simulation study in that it has constantly
changing structure, with gradients as large as $1/3^\circ$ C per n mi,
and individual features that are somewhat typical. The contour
lines in Fig. 3 are plotted every 3°; the lowest temperature
of a contour line is 2° C, whereas the highest one is 29° C.
This surface tends toward the other extreme of excessive temperature change, but it and the isothermal surface establish bounds
between which an actual situation can be interpolated.

   The same random numbers that were used in the simulation study
discussed previously are incorporated into the temperature field
of Fig. 3, but this time the random numbers are perturbed
about the values of the contoured field instead of about the constant value of 285 K. In other words, each pixel value is the
sum of the temperature of the contoured field at the pixel
location and a perturbed temperature provided by the random
number generator. Clouds were handled in the same manner as
described in the section, "Details of the Simulation."

FIGURE 3.   Simulated SST field.

F. Errors Arising from Use of the Variable SST Field

As one might expect, the accuracies from the simulation study using the variable thermal sea surface of Fig. 3 are not nearly as good as those for the isothermal case. Both the rms and maximum errors are listed in Table IV. As before, the entries of the columns labeled "before" and "after" are, respectively, before and after the EDC technique was applied. A comparison of Table IV with Table I immediately shows a dramatic reduction in accuracy. The major source of error is now the temperature gradient existing across each array of pixels, which distorts the histogram. Furthermore, the EDC technique does not provide very much improvement except for the 12-mi resolution cases. In fact, it is only the 12-mi resolution case in Table IV, exclusive of the 75 percent cloudy case, that yields an accuracy better than the 1° C requirement.

There is an obvious relationship between SST accuracy and sea surface thermal structure. From Table I, one deduces for ocean areas of near-constant thermal state that the greatest accuracy is achieved when using the largest pixel array that is practical (i.e., the lowest resolution). On the other hand, from Table IV one concludes that for a variable thermal sea state, one achieves the greatest accuracy with the smallest array (i.e., highest resolution) that is practical. In short, one concludes that for mesoscale applications a resolution size of 12 to 16 n mi is required, based on a typical pixel size of 1.5 n mi. Of course, if the area of each individual pixel is reduced, then the SST retrieval resolution size also can be reduced.

TABLE IV. Rms and Maximum Errors for Various Combinations
of Spatial Resolution and Percent Cloud Cover before and after
Application of the EDC Technique for the Variable SST Field

	RMS Errors (^{o} C)					
	48-Mi Resolution		24-Mi Resolution		12-Mi Resolution	
Cloud Cover	Before	After	Before	After	Before	After
Clear	8.37	8.73	1.68	1.62	1.32	.87
25%	8.57	8.86	1.95	1.78	1.29	.85
50%	11.54	11.42	2.33	1.96	1.26	.86
75%	12.77	12.77	6.03	5.25	2.13	1.77
	Maximum Errors (^{o} C)					
	48-Mi Resolution		24-Mi Resolution		12-Mi Resolution	
Cloud Cover	Before	After	Before	After	Before	After
Clear	22.00	22.00	6.76	4.65	5.07	3.04
25%	18.18	18.18	10.04	10.04	7.24	3.48
50%	22.36	22.36	9.64	6.75	5.50	3.45
75%	-20.51	-20.51	18.09	17.18	-10.46	10.42

IV.   REAL DATA

A. Preliminaries

Real infrared data in the 10.5 to 12.6 μm spectral interval
were acquired from the Visible and Infrared Spin Scan Radiometer
(VISSR) aboard the Geostationary Operational Environmental
Satellite (GOES) stationed over the equator at 75° W longitude
and known as GOES-East. The infrared measurements were received
at the Naval Environmental Prediction Research Facility (NEPRF)
through a terminal and interactive display device called the
"SPADS" (the SATDAT Processing and Display System).

The processing of these data was similar to that described for the simulated data. First, the data were grouped into 8 x 8 arrays from which histograms of 64 pixels each were constructed. Each pixel has a resolution of about 2 n mi at the subsatellite point, so the SST grid resolution is about 16 n mi. Next, the TND and EDC techniques were applied to produce SST fields, and, finally, contour maps of the SST fields were machine plotted.

Because the research reported in this paper is limited to producing SST maps having relative accuracies, only the correctness of the mesoscale temperature structure is of interest here. Therefore, the following limitations must be considered when interpreting the results of this paper.

1. The data are not corrected for atmospheric attenuation, which is due mainly to water vapor absorption.

2. The data are not corrected for varying zenith angles; however, this is a minor problem for a geosynchronous orbit 36,000 km above the earth, provided the measurements are not taken too close to the limb of the earth.

3. The data are digitized to 8-bit words, that is, the dynamic range of the signal is from 0 to 255 voltage counts. Thus, both the class intervals and the standard deviation were taken to be one voltage count.

4. The voltage counts are converted to temperatures through a calibration curve that is correct in a relative sense, but not in absolute terms.

Studies of two distinct cases are made. Both studies are of the Gulf Stream covering an area roughly 63° to 73° W longitude and 31° to 42° N latitude and are for March 23, 1979 and May 9, 1979. These cases were selected because of their interesting mesoscale structures and the availability of the data. (It is unfortunate that the Advanced Very High Resolution Radiometer (AVHRR) data from the TIROS-N satellite were not available for this study because they have 10-bit word accuracy and a resolution of about 1.1 km in the nadir direction.)

The objective of these real-data studies is to determine how well the derived SST field reproduces the actual thermal structure of the sea surface. Ground truth for the derived SST field is twofold: (1) picture images of the satellite infrared measurements and (2) the Fleet Numerical Weather Central (FNWC) fine mesh Gulf Stream SST analyses, which are contour maps. In all the infrared images presented, the light shadings represent cold temperatures and the dark shadings represent warm temperatures.

B. A Case Study

The first case studied is for March 23, 1979, and the visible image of the area under study is shown in Fig. 4. Note the large cloud-free area to the east of the U.S. coast from Virginia to Maine which presents no difficulty in obtaining SSTs. There are two heavy concentrations of clouds south of the clear region over which it is virtually impossible to obtain SSTs, but the area of scattered clouds between the two heavy concentrations does lend itself to accurate SST derivations.

The infrared image of the same area, as shown in Fig. 4, is shown in Fig. 5. Close examination of the picture reveals small boxes representing the individual pixel values of the infrared measurements which are the cause of the infrared image appearing more blurred than the higher resolution visible picture of Fig. 4. Recall that the darker areas represent warm areas and the lighter areas, therefore, represent cold areas. The salient features of Fig. 5 are:

1. The land mass in the upper left-hand corner is the east coast of the U.S. and Canada running from Virginia to Nova Scotia.

2. The white area running east of the land mass to almost the right-hand side of the picture is the cold continental shelf water. Contrary to its appearance, this area is not cloud covered, which can be verified by the visible picture of Fig. 4.

FIGURE 4. Visible picture for March 23, 1979, off the
ortheast coast of the U.S.

FIGURE 5. Infrared picture for March 23, 1979, covering
the same area as that shown in Fig. 4.

Furthermore, there is a very steep temperature gradient between
the white area and the dark area south of the white area.

3.   The elliptical dark spot in the upper center of the
picture is a pool of warm water over 100 n mi across.

4.   The Gulf Stream is partially hidden by clouds, but it
appears as a dark wavy band below the elliptical spot with an
east-west orientation.

5.   The large dark region below the Gulf Stream is warm water,
a little colder than the Gulf Stream, but more thermally
homogeneous than the other areas described.

Because the main concern in this paper is to be able to
delineate mesoscale features, the features described by items (2),
(3), and (4) are of greatest interest.   Therefore, sea surface
temperatures are derived only for the area shown in Fig. 6,
which is an infrared picture of a subsection of Fig. 5.   This
limited area is one-fourth the area shown in Fig. 5 and is
obtained by zooming the data in Fig. 5.   Zooming is accomplished
by repeating the data twice in each direction.   The particular
area zoomed can be identified by comparing the features in
Fig. 6 with those in Fig. 5.   The details in Fig. 5 described
earlier are even more evident in Fig. 6.

The numerical processing procedure for obtaining SSTs is
begun by grouping the numerical infrared pixel data, used to
construct the picture in Fig. 6, into 8 x 8 arrays.   Next,
histograms are formed from the sets of 64 pixel values and the
TND technique is applied to the histograms to derive SST values.
The resulting SST field is corrected by the EDC technique and,
finally, the field is displayed as a contour map.

The contour map of the processed SSTs is shown in Fig. 7.
The contour lines are drawn for each 1° C interval and the range
of temperatures runs from a minimum of 6° C for the contours at
the top of Fig. 7 to a maximum of 22° C for the small circles in
the center and the left-center of the picture.   In order to orient

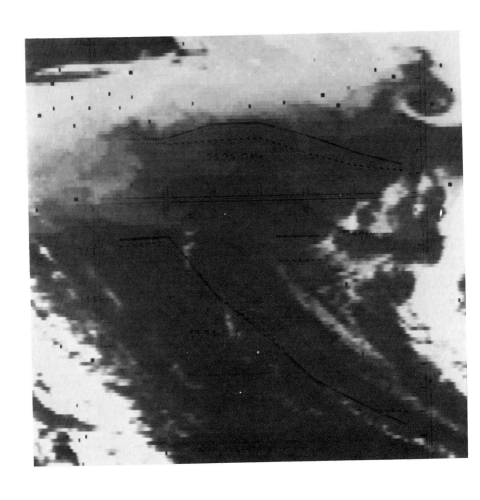

FIGURE 6.   Zoomed infrared picture of Fig. 5 covering one-
fourth of the area of that picture.

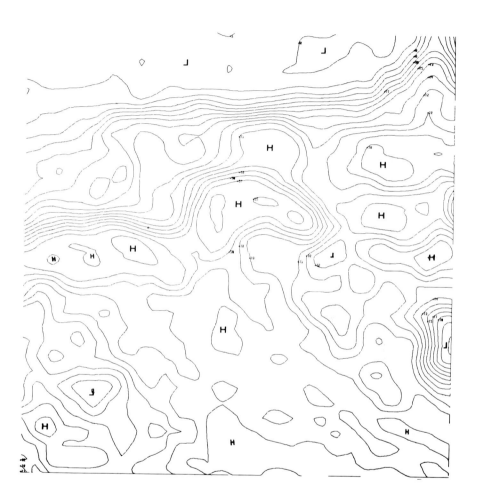

FIGURE 7.   Contour map of the processed SSTs corresponding
to Fig. 6.

the contour map in the same direction as Fig. 6, it had to be
reversed before printing; consequently, the letters indicating
the high (H) and low (L) temperature areas are reversed. Also,
the contour value labels, which are barely discernible, are
upside down due to a programming error. It also is important to
remark that the contour map of Fig. 7 is based on a 32 x 32-point
SST grid, as are all the subsequent contour maps shown in this
paper.

The features described in items (2) to (5) are very evident
in Fig. 7, namely, item (2) the cold shelf water, starting at
6° C at the top of the map, and the steep gradient between the
shelf water and the Gulf Stream, item (3) the elliptical warm pool
whose center contour is 17° C, item (4) the Gulf Stream is represented by the series of high-temperature spots at 22° C across
the middle of the map, and item (5) the large, rather homogeneous
area in the lower half-center of the map with temperatures varying
between 18° and 19° C.

Finally, the low-temperature areas at the lower left and lower
right-hand sides of the map follow the cold dense cloud structure.
The lowest temperature on the left is 13° C, and the center of the
bull's eye on the right is at 8° C. Because the histogram of the
infrared data from a solid cloud overcast appears no different
than that of a clear area, except that all the temperatures are
lower, the TND technique produces temperatures that follow the
cloud patterns in these areas. Also, if these overcast areas are
sufficiently large, the EDC technique will not change the values
produced by the TND method because it has no indication that such
large scale features are not real. A partial solution to this
large-scale cloud problem is presented in a subsequent section.

The most dramatic method of presenting the results shown in
Fig. 7 is to superimpose its contours on the picture of Fig. 6.
The effect of doing this is shown in Fig. 8. All the major
features of the contours match the major features of the picture,
that is, the contours are an excellent reproduction of the

FIGURE 8. Superposition of Figs. 6 and 7.

shadings in the picture. But, another important feature is also
evident. In the broken-cloud areas, the contours do not follow
the cloud-top temperatures; rather, they follow the warmer sea
surface temperatures in a continuous manner. Two mechanisms are
at work. First, in partially cloudy areas, the warm pixels are
segregated from the cold ones in the histogram, and since the
TND technique uses only the warm wing to deduce the SST, the cold
pixels are ignored. Second, the EDC technique corrects isolated
cold spots and small groupings of cold spots created by residual
cloud contamination. Thus, the EDC method builds the continuity
of the larger field into the contours.

Clearly, if the infrared picture of Fig. 6 is used as the
ground truth, then the results of Fig. 7 are in excellent agreement with the truth, as is illustrated by Fig. 8. On the other
hand, if the Fleet Numerical Weather Central (FNWC) fine mesh
Gulf Stream SST analysis for March 22, 1979 is used as the ground
truth, the results of Fig. 7 are not very good. The SST contours
produced by FNWC, which are also in 1° C intervals, are shown in
Fig. 9. The discrepancy in time between Figs. 7 and 9 is insignificant because the thermal structure of the oceans is very
conservative. (The letter L in Fig. 9 is reversed, but the contours are oriented correctly.)

Agreement between Figs. 7 and 9 exists only in the broadest
terms. In particular, the gradient directions (but not the
amplitudes) generally agree, the Gulf Stream flow is in the same
general direction, and the thermally homogeneous (relatively)
mass of water below the Gulf Stream is present in both figures;
but, beyond these general features all similarity ends. For
example, missing in Fig. 9 are the warm elliptical pool of water,
the very steep gradient from coastal waters toward the Gulf Stream,
and most of the mesoscale features. To understand the reasons
for these discrepancies, one must understand how the FNWC-SST
analysis is constructed.

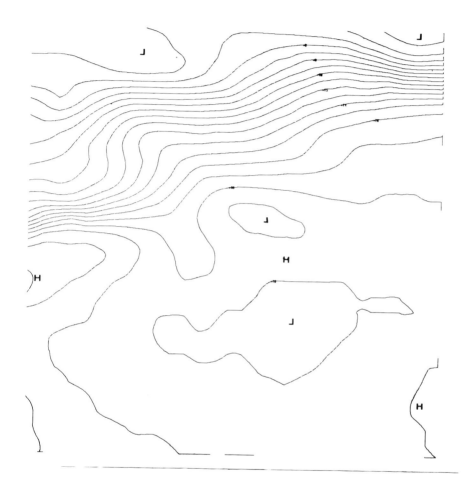

FIGURE 9. The FNWC fine mesh Gulf Stream SST analysis for March 22, 1979.

C. *The FNWC-SST Analysis*

This FNWC-SST analysis is a composite of the following data:

1.   U.S. Navy and NATO ship reports, and surface weather observations that are sent to the World Meteorological Organization by nonmilitary ships.

2.   Bathythermograph measurements of which, typically, there are 200 per day for the Northern Hemisphere.

3.   Airborne Remote Temperature Sensed Temperatures (ARTST), which are obtained from aircraft flying about 500 feet above sea level.

4.   Anchored buoys in the Northern Hemisphere and drifting buoys in the Southern Hemisphere.

5.   SSTs derived from satellite infrared measurements. Numerical weights ranging from 0.1 to 1.0 are assigned to each individual report in items (1) through (5) as follows: items (2) and (4) are given weights of 1.0, item (1) has a weight of 0.5, and items (3) and (5) have a weight of only 0.1.

The FNWC fine mesh Gulf Stream SST analysis is computed on a grid mesh of 40 km, which is one-eighth the mesh size of the standard FNWC grid for the Northern Hemisphere polar stereographic projection SST maps. An analysis is one every 12 hours and is begun by interpolating the data from items (1) to (5) to their nearest grid point, using the gradient values of the previous 12-hour map. The originally assigned numerical weight is changed in direct proportion to the distance each data point must be moved by the interpolation process. This produces what is called the "assembled data field" in which all raw data have been assembled to the nearest grid point.

The new 12-hour SST analysis is constructed in a series of data-blending steps by a process known as the Fields by Information Blending (FIB) technique. The first step is to blend the "first guess field" with climatology. The first guess field is the previous 12-hour analysis whose assigned weights at each grid point are decayed proportionately to the lapsed time, while the climatological field is a blend of the climatological SST fields for the previous, present, and past months whose resulting grid-point values are given a weight of 0.001. When the climatological and first guess fields are blended into a single field, a corresponding numerical weight field also is calculated by blending the corresponding individual weight fields.

The second step of the FIB method is to blend the field from the first step with the assembled data field. This is a point-by-point blending at each grid point during which a new weight value also is calculated. If a particular grid point has no assembled data assigned to it, that grid point will carry only the blended first guess and climatology value, along with the associated weight.

Horizontal blending is performed next to produce horizontal consistency between adjacent grid points. This process is followed by one in which each of the original raw-data values is compared with the newly blended grid-point value. The weight of each original raw-data value is either increased, left unchanged, decreased, or removed altogether, depending upon the amount of agreement between its value and the blended grid-point value. This produces a "refined" set of data (and associated weights) which is horizontally blended again for consistency to produce the final analyzed SST field for the given 12-hour period.

It now should be clear why the FNWC-SST contours in Fig. 9 are so different from the satellite-derived SST contours of Fig. 7. Only if a large cluster of satellite-derived SSTs occur near a grid point can they have a sufficient weight to strongly influence the FNWC analysis. Furthermore, discrepancies between satellite-derived SSTs and measurements of types (1), (2), and (4) can easily occur because the satellite's infrared scanning radiometer measures only the skin temperature whereas the other measurements are representative of deeper layers of water. Also, the

combination of high numerical weights assigned to measurements
of types (2) and (4) and the sparsity of these measurements can
cause a strong smoothing of the resulting contours.

V.   THE CLOUD PROBLEM

A.   *Combined Use of Infrared and Visible Data*

In the section, "A Case Study," difficulties were encountered
in obtaining the SST field because of the interference
(contamination) by dense cloud cover. This problem cannot be
completely eliminated, but techniques can be employed to extract
the maximum amount of SST information in these situations. An
example of what can be done is provided by Fig. 10, which is a
zoomed infrared image of very nearly the same ocean area as that
shown in Fig. 6, except it is about a month and a half later,
namely, May 9, 1979.

The clusters of small, solid white boxes in Fig. 10 represent
blank areas from which clouds have been removed by using visible
data corresponding in place and time with the infrared data.
Before discussing this, it should be pointed out that blocking
out cloudy areas using infrared data alone (that is, deleting all
infrared pixels whose values are below some threshold value as
being cloud contaminated) is not a reasonable approach. The
difficulties with the scheme of classifying infrared data by a
threshold value can be illustrated by Fig. 6 where the continental
shelf water is just as cold as the cloud tops. Hence, if the
infrared-only cloud detection scheme were applied to this situation,
most of the shelf-area pixels would be deleted along with the
cloud-covered pixels.

A more reasonable scheme for detecting and deleting cloud
contaminated pixels is to use the simultaneous visible data.
Because cloud tops have a much higher brightness value in the
visible region of the spectrum than does the surface of the ocean,

FIGURE 10. Zoomed infrared picture for May 9, 1979, off the northeast coast of the U.S. with cloud-contaminated pixels blocked out.

the two surfaces readily can be distinguished from one another.
Hence, the cloudy pixels can be blocked out of the infrared image
by deleting those pixels that correspond to areas in the visible
images that are brighter than some preassigned brightness value.
This is precisely what was done to the image in Fig. 10, which
was produced by Roland Nagle of the Naval Environmental Prediction
Facility (NEPRF) on the SPADS terminal.

The description of the cloud-free portion of the infrared
image in Fig. 10 is very similar to that given for Fig. 5 in the
section, "A Case Study." Its features include cold continental
shelf water with a steeply increasing temperature gradient toward
the Gulf Stream; two large warm eddies north of the Gulf Stream;
the Gulf Stream flowing easterly, then toward the northeast, next
turning southeasterly, and finally flowing eastwardly again; and
a large, warm thermally (relatively) homogeneous mass below, and
slightly colder than, the Gulf Stream. In addition, Fig. 10 displays a dominant tongue of warm water below the Gulf Stream
flowing eastwardly, which appears to have resulted from a branching
of the Gulf Stream.

After the TND and EDC techniques were applied to the infrared
data of Fig. 10, a contour map was constructed of the resulting
SST field, which is shown in Fig. 11. All the features described
in the preceding paragraph also appear in Fig. 11, including the
protrusion of warm water in the lower part of Fig. 10. The contour lines are drawn for each 1° C interval and they range from a
minimum of 10° C at the top of the figure to a maximum of 20° C
along the Gulf Stream and along its secondary branch. In Fig. 11,
as in Fig. 7, the contour maps had to be reversed before printing
to orient the contours in the same direction as in Fig. 10;
consequently, the letters indicating the warm (W) and cold (C)
temperatures are reversed. Also, the barely discernible contour
labels are upside down because of a programming error.

The most notable feature in Fig. 11 is the relative continuity of the SST field in the sense that low temperature

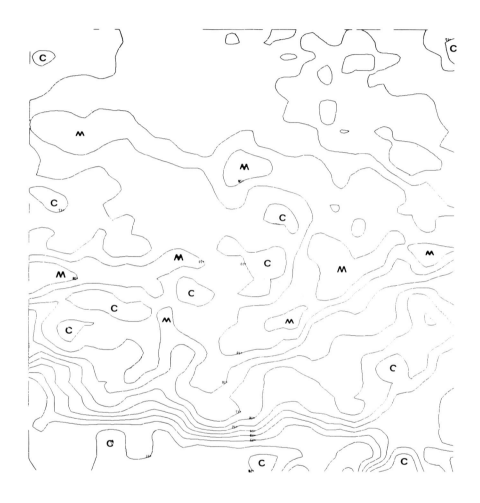

FIGURE 11. Contour map of the processed SSTs corresponding to *Fig. 10.*

bull's-eyes do not appear in the vicinity of clouds as they do
in Fig. 7. The reason for this is that cloud-contaminated data
have been removed and the contour lines have been interpolated
through the areas of missing data. There are two basic ways to
draw contours for missing data: (1) draw the contours on the
basis of the available data, or (2) first complete the data grid
by some means and then draw the contours in the usual manner.

The first of these two approaches is basically a least-
squares approach in which the interpolation through the regions
of missing data is determined by the basis functions used and the
relative density and position of the available data. This approach
is not very satisfactory because the interpolation process is so
arbitrary and can result in major distortions of the overall SST
field. The second approach can be just as arbitrary as the first
one, depending upon how the missing data are filled in, and can
lead to the same objections as in the first approach. A novel
and objective method for completing the data grid when some of
the data are missing, which is relatively free of the objections
raised previously, is described in the next section, and is the
method used to produce Fig. 11.

B.   *Treatment of Missing Data*

Once a grid point in the SST field is cloud contaminated,
knowledge of the underlying SST is lost forever. However, the
SSTs of the cloud-free grid points surrounding the cloud-
contaminated grid points do give a strong indication of the
SST structure in the cloudy regions. Consequently, an interpolation between the cloud-free grid points around the boundary
of the cloudy grid points yields a reasonable guess of the SST
field under the cloud. Of course, the smaller the cloudy area,
the more accurate the interpolation becomes. Details of the
scheme used to fill in the missing values at cloud grid points

are given in this section. Once the SST data grid is completely
filled in, the EDC technique is applied to the data field to make
the filled-in data consistent with the rest of the SST field.

The fill-in process involves two arrays of data: (1) the SST
field itself, and (2) a corresponding array of weights assigned
to each grid point. Initially, the array of weights carries unit
value in each position. The data fill-in procedure is begun with
the position in the upper left-hand corner of the SST data array.
If it has a SST value, move on to the next position to the right
of it. On the other hand, if the first SST value is missing,
assign it the best guessed value available, such as the previous
day's value or climatology, and increase its corresponding weight
value by one.

Now move to the second position in the first row. If the
SST is missing, assign it the value of the position immediately
to the left of it (that is, the first position) and assign the
weight associated with the left-hand position plus one as the
weight of the missing SST position. In other words, the missing
position will have weight 3 or 2, depending upon whether the first
position in the row had a missing SST. On the other hand, if the
second position already has a valid SST value, it is left alone
and the corresponding weight is one. Continue this process across
the first row; assign each position with a missing SST value the
value of its neighbor to the left and give the missing point its
neighbor's weight plus one. If any position has a valid SST value,
change nothing--just skip over it.

Suppose for example that the first row had a valid SST value
in the ith position, but the next n values were missing. The
fill-in process would put the value of the ith position into each
of the n positions with missing SST values and the associated
weights of these positions would be, respectively, from left to
right 2, 3, ..., n + 1. That is, the further removed the filled
in point is from its uncorrected antecedent, the larger its
weight will be.

Now move to the second row. From the second row onward, the filled-in value can come from either the point immediately to the left of the point in question, or from immediately above it, or from both. The value actually assigned is the one with the lowest weight, or it is the average of the two points if their weights are the same. The new weight associated with the point in question is the weight of the replacement point (or the average weight, if the filled-in value is an average) increased by one. The procedure is illustrated in Fig. 12. The left-hand array represents a SST field with known values T_1, T_2, ..., T_6 and missing values indicated by the letter b for "blank," and the array on the right-hand side represents the associated current weights. After the blank entries in Fig. 12 are filled in, using the scheme just described, the resulting data and weight arrays are shown in Fig. 13. Notice that the filled-in SST values in each case are those of closest proximity to the initial values and the weights are directly related to the distance an individual value had to be moved. Furthermore, it is important to note that the sequential process by which the missing data are filled in guarantees that there always will be neighboring data available

SSTs				WEIGHTS			
T_1	T_2	T_3	T_4	1	1	1	1
T_5	b	b	b	1	1	1	1
T_6	b	b	b	1	1	1	1

FIGURE 12. Initial array of SSTs and missing values and the initial array of weights.

SSTs			
T_1	T_2	T_3	T_4
T_5	$\dfrac{T_2+T_5}{2}$	T_3	T_4
T_6	T_6	$\dfrac{T_3+T_6}{2}$	T_4

WEIGHTS			
1	1	1	1
1	2	2	2
1	2	3	3

FIGURE 13. Filled-in array of SSTs and the corresponding array of weights.

with which to complete the array. This, however, is not necessarily the case if data below or to the right of the missing point had been used to fill in the missing values.

After the missing-data points have been filled in and weights have been assigned to each point, one can apply the EDC technique. However, a problem remains in that the scheme described does not take advantage of the SST information below and to the right of the missing data points. It was pointed out that these values could not be used initially because it is possible that some of them too could be missing-data points.

One way of rectifying this problem is to apply the sequential fill-in scheme all over again, but this time the process is started in the lower right-hand corner and only the points below and to the right of a given point are used to fill in data. Of course, this second sweep of the data must also be applied to the original array of data as depicted in Fig. 12, not to the filled-in array of Fig. 13. After both sweeps of the data have been completed, the two completed SST arrays are averaged pointwise to produce the final SST field. Note that by this process, points that had missing data can be filled in with the average of

two very different numbers, but any point that originally had a
value always will retain that value because the average is of that
number with itself.

During the second sweep of the original data, a second array
of weights is also constructed as before, and the resulting two
arrays of weights are averaged pointwise just as the two arrays
of SST data are averaged pointwise. Clearly, the weights
associated with a point that originally had a value will retain
their value of unity, and so their average also will be one. All
other average weights will be greater than unity and can be the
average of two quite different values.

The double sweep procedure can be extended to a quadruple
sweep by initiating the fill-in process at each of the four
corners of the original SST data array and generalizing the
procedure just described in an obvious manner. In the work
reported here, only the double sweep was used along with the
simple averaging of the data values and of the associated weight
values.

Finally, each SST in the field can be assigned a quality, or
reliability, indicator whose value is just the reciprocal (or the
square of the reciprocal, if one wishes) of the corresponding
weight value. Obviously, those SST values that were present
originally will have a quality indicator of unit value and all
SST values that had to be filled in will have a quality, or
reliability, of less than unity. In fact, for the points that
have a reliability of less than one, the lower the value the more
removed the position of the filled-in SST value (or values) is
from the location of the data that were its antecedents.

C. Application of the EDC Technique

It is clear that the fill-in procedure for missing data described in the preceding section can result in SST fields having a
tendency toward flatness in the filled-in regions, as can be seen
in Fig. 13. However, if the EDC procedure then is applied to the
SST field, it can be made more realistic because any filled-in
value that is inconsistent with the original SST field can be
detected and corrected by the procedure. This is what was done
to produce Fig. 11.

An indication of how well the fill-in and EDC procedures work
is given by superimposing the contours of Fig. 11 on the infrared
image of Fig. 10. This is shown in Fig. 14. The contours match
the various shadings in the picture very well. Thus, as in
Fig. 8, when the infrared image is used as ground truth, the contours are in good agreement with the truth. Equally important,
however, is the fact that the contour lines appear to be blind
to the regions of missing data in that there is no correspondence
between the missing-data regions and the shape of the contour
lines through those regions. In every case, the contour lines
are consistent with the surrounding given pixel shadings. In
fact, in many instances there is a surprising amount of structure
in the contour lines in areas of missing data. It appears that
the fill-in and EDC procedures work very well.

VI. SUMMARY AND CONCLUSIONS

A viable approach to obtaining SSTs on a mesoscale level from
high resolution 11-μm window data has been described. The main
difficulty in using infrared data is contamination from clouds.
This problem essentially has been solved by (1) using visible
data to detect and eliminate cloud-contaminated pixels;
(2) applying the TND technique to obtain the mesoscale SSTs;
(3) filling in missing data by neighboring SST values and

FIGURE 14.   Superposition of Figs. 10 and 11.

weighting the results; and (4) applying the EDC technique to eliminate residual cloud contamination and to make the entire SST field consistent. Of course, this approach has limitations in that it becomes increasingly unreliable with increasing cloud cover as do all approaches.

Several conclusions can be drawn from the results of this paper. First, results of the simulation study suggest that in ocean areas for which the SST field has steep gradients, such as the Gulf Stream, the SSTs should be derived from data arrays that represent areas no larger than about 12 to 16 n mi across to achieve accuracies of the order of 1° C. Second, the approach of this paper provides SST data at every point on the predetermined grid and provides a quality, or reliability, index of the SST data at each grid point. Other conclusions are the following:

1.   The FNWC operationally produced SST maps disagree sub-stantially with the results of this paper for reasons mentioned in the section "The FNWC-SST Analysis."

2.   Visible data should be used to eliminate cloud-contaminated infrared data. The fact that these data are available only during daylight should cause no problem because usually one analysis per day is adequate.

3.   The TND technique is a viable approach to obtaining mesoscale SSTs.

4.   Filling in missing data with neighboring values is an objective and satisfactory approach.

5.   The EDC technique is a useful method for correcting erroneous and filled-in SST data, without disturbing the valid data.

ACKNOWLEDGMENT

    The author would like to thank the officers and staff of the
Naval Environmental Prediction Research Facility for the use of
their equipment and data, and for their helpful assistance and
encouragement in this research. The special help from and discussions with the Head of the SATDAT Processing and Display
Branch, Roland Nagle, are gratefully acknowledged.

REFERENCES

1. Crosby, D. S. and Glasser, K. S., *J. Appl. Meteor.* *17*, 1712
    (1978).
2. Fleming, H. E., and Hill, M., "Third Conference on Atmospheric
    Radiation," Davis, California, June 28-30, 1978. (Preprint
    volume available from Am. Meteor. Soc., Boston, MA 02108.)
3. Lanczos, C., "Applied Analysis," p. 320. Prentice-Hall,
    Englewood Cliffs, New Jersey (1957).

DISCUSSION

McMillin: On your case where you had a resolution 48 n mi and that was more accurate, that was simply because you had more points to deal with in the sample, right?

Fleming: Yes, that was an array size of 32 by 32 or 1,024 data points, while the 12 mile case was an 8 x 8 array, or 64 data points.

McMillin: So it is points, not resolutions, that is the key to the difference between the two results?

Fleming: Yes. It was the large number of data points that caused the reduction in the rms value for the isothermal case study.

Westwater: Were there some ship data to compare at least in some of the locations so that you had some measure of ground truth? How did the derived temperatures compare with those data?

Fleming: I didn't attempt to do that because I have not corrected the data for zenith angle or water vapor.

Westwater: But when you talk about people not believing satellite results, though, there has to be some measure of ground truth with which to compare.

Fleming: The classical measures of ground truth are the ship temperatures. Just as radiosondes, whether they are accurate or not, are accepted by the meteorological community as being the ground truth for atmospheric temperature sounding, so are the ship data accepted as ground truth for sea surface temperature determinations. Even though these ground truth data sources are also subject to error, people are so familiar with them that they treat them as absolute truth.

Chahine: I agree with you. The oceanographers usually ask for sea surface temperatures with an accuracy of one-tenth of a degree because their thermometers can read down to one-tenth of a degree. My question is if you have a uniform background biased linearly, could you filter it out with your truncated distribution function?

Fleming: It would filter out only to the degree that it would filter an arbitrarily shaped gradient. However, if it were known in advance that such a bias existed, it could be removed from the data before processing it.

DERIVING SEA SURFACE TEMPERATURES
FROM TIROS-N DATA

C. Walton

National Oceanic and Atmospheric Administration
National Environmental Satellite Service
Washington, DC

Operational procedures for deriving sea surface temperatures on a global basis from the TIROS-N data are described. The statistical maximum likelihood technique is applied to a histogram of AVHRR 11-μm channel data to provide the fundamental surface temperature measurement at a resolution of 50 kilometers. The strong correlation relating the brightness temperature difference of two of the HIRS tropospheric sounder channels to the amount of cloud contamination in the sounder's field of view provides the means for cloud detection. A HIRS water vapor channel is used to correct the fundamental measurement for variations in atmospheric moisture.

I.   INTRODUCTION

Sea surface temperature (SST) is one of the primary quantitative products derived from the National Oceanic and Atmospheric
Administration (NOAA) series of polar orbiting satellites which
are operated by the National Environmental Satellite Service
(NESS). The current model of global operational sea surface
temperature computation (GOSSTCOMP) employs coincident data from
the Advanced Very High Resolution Radiometer (AVHRR) and the High
Resolution Infrared Sounder (HIRS) which are included in the
recently launched TIROS-N satellite system. A frequency histogram is produced from an 11 x 11 array of raw count values
obtained from the 11-micron channel of the AVHRR providing
approximately a 50-km square target. An analysis of the warm
end of the histogram using the Maximum Likelihood Technique (MLT)
provides the fundamental measurement of sea surface temperature
(1). The MLT procedure estimates the mean of a normal histogram
using only data from the warm tail and its use represents a significant change in the GOSSTCOMP model which was implemented
with the operational use of TIROS-N data. The primary sources
of error in the fundamental measurement of sea surface temperatures with the 11-micron AVHRR data are cloud contamination and
the atmospheric water vapor absorption of surface radiance. The
single pixel spacially coincident measurement from the 20-channel
HIRS instrument provides the means for cloud detection within the
AVHRR retrieval area and for estimation of the effect of atmospheric absorption by water vapor.

This paper describes the original development of discriminant
functions for cloud detection using sounder data from the Vertical Temperature Profile Radiometer (VTPR) of the improved TIROS
satellite system (ITOS). Although these procedures were developed well before the scanning radiometer (SR) became inoperative
on the last of the ITOS series, discriminant functions have only

been used operationally in the nominal GOSSTCOMP model since the advent of TIROS-N. Another paper titled "Deriving Sea Surface Temperatures from Satellite Infrared Sounder Data--A Single Field-of-View Approach" is currently being submitted for publication in one of the JGR journals. It describes in detail the application of sounder data to the current GOSSTCOMP model considering both the problems of cloud detection and of atmospheric absorption by water vapor.

II. DATA BASE

A data base consisting of 83 orbits of coincident SST retrievals and VTPR spot measurements was collected over the period from 9/22/76 to 9/27/76. Included in the data base for each retrieval are the following parameters: (1) latitude of SST retrieval; (2) longitude of SST retrieval; (3) SST retrieval brightness temperature; (4) zenith number of the SST retrieval; (5) zenith angle of the SST retrieval; (6 to 7) packed GOSSTCOMP field information which is coincident with the SST retrieval spatially and temporally including satellite temperature, climatology temperature, Navy temperature, gradient information, and a confidence value of the satellite temperature; (8) the year, month and day of the retrieval; (9 to 16) the coincident radiances of the VTPR channels 1 to 8 which are obtained from the VTPR disk archive I file; (17) the GOSSTCOMP SST field temperature coincident with the VTPR channel measurements which is also obtained from the VTPR disk archive I file.

The SST retrievals are obtained from an analysis of the warm side of the histogram composed of 1024 samples of SR-IR data (2). The coincident VTPR radiances are obtained from the single VTPR spot which is closest to the center of the SST retrieval area (3). Because the VTPR resolution is half that of the SST retrieval (60 km. compared with 120 km at nadir), the VTPR measurements

are nearly always included within the SST retrieval area.   This
data base has been stored on magnetic tape.

III.   PARAMETRIC CLASSIFICATION

   The so-called parametric classifier is virtually identical
to the classification procedure first described in the original
NESS memo and was previously used operationally to isolate the
Class 1 set of SST observations (4).   As is described in the
subsequent sections, in the development of the so-called discriminant classifier, the parametric classification is used as a
first guess estimate of which VTPR measurements are actually
clear.

   A complete description of the parametric classifier is given
elsewhere and will not be repeated here (5).   Its outstanding
virtue is that it yields an accurate classification of the VTPR
spot measurements into either "clear" or "not clear" categories
which is independent of any sea surface field temperature estimate.   Its primary failings or difficulties are that it requires
the use of coincident SR SST retrievals and VTPR measurements,
data which are difficult to gather, and the test which compares
the VTPR window channel measurement with the measurement in the
lowest of the CO_2 channels is much more stringent in the polar
regions than in the equatorial regions.   Another failing is that
there are several parameters in the model which must be obtained
by trial and error, some of which are difficult to evaluate such
as the calibration bias between the SR-IR and VTPR instruments.
The development of the discriminant classifier is a significant
achievement primarly because it eliminates these objections to
the parametric classifier.

IV. DISCRIMINANT CLASSIFIER

A. Justification and General Description

The development of the discriminant classifier is inspired
by the desire to isolate the clear VTPR spot measurements without
need for gathering other satellite data. Such a procedure could
be very useful in VTPR sounding processing, specifically in
determining clear column radiances from which the temperature
and moisture profiles are produced. As will be seen, this effort
has been successful and results in discriminant classifiers
which under most conditions yield a more accurate classification
of the VTPR measurements than does the parametric classifier.
An application of the procedure to the production of sea surface
temperatures with high resolution SR-IR data is also described.

The general approach to this problem is to attempt to find
a relationship between two or more channels of VTPR data in a
clear scene which is lost in the presence of clouds. It was
thought that a statistical multiple linear regression against
the appropriate features (i.e., combinations of VTPR channel
measurements, and other parameters), if restricted to clear
scenes, might yield the correct relation. A multiple linear
regression procedure has been adopted for this purpose which
attempts to satisfy the following relationship for all clear
VTPR measurements:

$$\sum_i b_i T_{i,n} = c \tag{1}$$

The regression procedure specifies values of the regression
coefficients which minimize in a least square sense the
quantity

$$\sum_n (c - \sum_i b_i T_{i,n})^2 \tag{2}$$

Here C is an arbitrary nonzero constant, b_i is the regression
coefficient of the ith feature, and $T_{i,n}$ is the value of the
ith feature obtained from the nth clear measurement of the VTPR.
The regression is nonstandard in that it does not include a
constant coefficient term and the dependent parameter is a constant. One of the problems with this procedure is that although
the regression will minimize Eq. (2) for any set of features,
this does not guarantee that the functional relationship is
useful for cloud detection since the same relationship may also
apply with a cloudy set of measurements. Accordingly, the
choice of the features which are supplied to the regression
analysis is very important. In the language of statistical
classification theory, the expression to the left in Eq. (1) is
a two-category discriminant function which, following the
specification of the coefficients, may be evaluated for any VTPR
measurement--clear or cloudy. The user determines a threshold
value so that if the discriminant function is greater than this
value, the measurement is classified "clear" whereas if the
function is less than the threshold value, the measurement is
classified "not clear."

B. *Regression Development and Feature Selection*

As indicated previously the regression procedure requires a
set of clear VTPR measurements. For this purpose the parametric
classifier produces a first guess estimate of which of the SST
retrievals and coincident VTPR measurements within the data base
are clear. Of a total of 48,911 VTPR measurements and coincident
and coincident SST retrievals, 24,344 are classified "clear"
with the parametric classifier and were therefore included in
the statistical regression.

The choice of possible features from the eight VTPR
channels to be included is quite large, many of which would be
nearly useless for the function of cloud detection. The

features which have been found in this study to be most useful
are (1) a feature involving the difference or a combination of
differences between various VTPR channel brightness temperature;
(2) a feature linearly proportional to an estimate of sea
surface temperature, and (3) a feature involving the zenith
angle of the measurement. The first feature is perhaps not
surprising since it is also incorporated in the parametric
classifier as a threshold test in which the difference between
the brightness temperatures of the window channel and the lowest
of the CO_2 channels must be greater than the threshold value for
the measurement to be classified "clear." The physical explanation of why this test is useful for cloud detection has been
explained in terms of the radiative transfer equation (5). The
test is unique in that it represents the only means which
currently exists to distinguish between a uniform cloud top
measurement and a sea surface measurement without the use of
visible data and/or surface temperature information. As was
mentioned previously, this threshold test is more restrictive
in the polar regions than in the equatorial region because the
standard atmospheres in these regions are very different.
Physically, the threshold value has a surface temperature
dependence which is accounted for in the discriminant analysis
with the second feature. Correspondingly, the third feature is
included to account for the zenith angular dependence of the
threshold value. Obviously, other features could be incorporated.
For instance, a field estimate of sea surface temperature has
been used in this study because only data over the ocean has been
included in the data base. For general application over land
and sea, one would want to replace this feature with a surface
air temperature estimate or perhaps a 1000 millibar air
temperature estimate which can be obtained from the global
National Meteorological Center (NMC) analysis. Additional

features which might be useful include estimates of surface emissivity and relative humidity.

Before describing the results obtained with various discriminant functions it is perhaps worthwhile to consider the difference in purpose between the parametric and discriminant classifiers. Both represent an attempt at cloud detection without the use of visible data. However, the parametric classifier is designed to be independent of all field analyses of any physical parameters and attains its objective by combining low resolution sounding data (VTPR) with higher resolution IR window channel data (SR). In contrast, the discriminant classifier is designed to work with data from a single instrument (VTPR or SR) and attains its objective by including analyzed field estimates of the important physical parameters such as surface temperature in its development. Thus, each classifier has its own goals and application and neither classifier supersedes the other.

V.   DESCRIPTION OF RESULTS

A.   *Regression Analysis*

Several discriminant functions have been analyzed using different combinations of the VTPR channels in the difference feature in order to find which combination of the channels is optimal for cloud detection. In all cases the discriminant function is given by

$$D = b_1 \, \Delta T_{i,j} + b_2 \, T_0 + b_3 \, ZEN \tag{3}$$

Here, $\Delta T_{i,j}$ is the positive brightness temperature difference between channels i and j (Channel 1, 2, 3, ... = window, water vapor, lowest CO_2 channel, ...), T_0 is the sea surface temperature in Kelvin degrees minus 290°K and ZEN is the zenith angle of the measurement in radians. The regression procedure minimizes in a least squares sense the quantity

$$\frac{1}{N} \Sigma_n \ (1 - D_n)^2 \tag{4}$$

for the clear set of measurements. The results obtained with various discriminant functions are shown in Table I.

In this table, the residuals are simply an evaluation of expression (4) for both the clear and cloudy sets of VTPR measurements as classified with the parametric classifier. The ratio column is the residual of the cloudy set divided by that of the clear set. In general, the discriminant functions with the highest ratios are the most useful for cloud detection. As can be seen the discriminant functions produced from the window channel and the higher tropospheric CO_2 channels produce the highest ratios. Apparently, the functions which do not include the window channel cannot discriminate against low-level cloudiness. The functions which use the water vapor channel are obviously overly sensitive to the moisture content of the atmosphere. This same condition, but to a lesser degree, also applies to the discriminant functions which use the lowest of the CO_2 channels, channel 3. Finally, the functions which use the highest CO_2 stratospheric channels are probably too sensitive to the atmospheric temperature profile above the troposphere to be useful for cloud detection. Considering the regression coefficients, it should be noted that the coefficient associated with the sea surface field temperature is less than that associated with the difference in the VTPR channel measurements in all cases indicating that the influence of the sea surface temperature in the discriminant analysis is relatively minor. The last discriminant function, which uses the difference between the window channel measurement and the

TABLE I. *Statistical Description of Various Discriminant Functions*

Function number	Channels i,j	Regression coefficients			Residuals		
		$\Delta T_{i,j}$	T_0	ZEN	Clear	Cloudy	Ratio
1	1,2	0.60262E-01	-0.20525E-01	0.35563+00	0.03460	0.08864	2.6
2	1,3	0.50657E-01	-0.92241E-02	-0.11525E+00	0.00265	0.05651	21.0
3	1,4	0.29659E-01	-0.77430E-02	-0.94899E-01	0.00148	0.04363	29.0
4	1,5	0.19920E-01	-0.83037E-02	-0.61858E-01	0.00113	0.03524	31.0
5	1,6	0.15586E-01	-0.14561E-01	+0.95686E-02	0.00132	0.02484	19.0
6	3,4	0.70664E-01	-0.51740E-02	-0.36307E-01	0.00261	0.03431	13.0
7	3,5	0.32584E-01	-0.74028E-02	-0.94679E-02	0.00183	0.02880	16.0
8	3,6	0.22266E-01	-0.16390E-01	+0.89581E-01	0.00317	0.02292	7.2
9	4,5	0.60230E-01	-0.91472E-02	+0.22483E-01	0.00202	0.02578	13.0
10	4,6	0.32274E-01	-0.21156E-01	+0.16376E+00	0.00503	0.02475	4.9
11	5,6	0.67075E-01	-0.32625E-01	+0.39789E+00	0.01610	0.04574	2.8
12	1,0	0.34939E-02	+0.79367E-03	+0.36035E-02	0.00004	0.00142	35.0

coincident sea surface field temperature in the regression, has particular application in the production of sea surface temperature observations in that it does not require the use of sounder data. The special properties of this function are described in a later section.

B. Discriminant Analysis

In the previous section various discriminant functions were compared with regard to their relative ability to isolate the clear VTPR field of view (FOV) measurements. In this section the behavior of the discriminant functions is described as a function of degree of cloud contamination and geographical location. A scattergram or three dimensional graphical display is utilized for this purpose. Figure 1 is an example in which the vertical ordinate represents the value of the discriminant function number 3, whereas the horizonal abscissa represents the difference between the GOSSTCOMP SST estimate and the VTPR window channel brightness temperature measurement corrected for atmospheric absorption with the water vapor channel data using procedures described elsewhere (5). The Z axis indicates the frequency of occurrence of a given (X,Y) value with a letter code (A represents 1 through 10, B represents 11 through 20, etc.). A large positive value along the X axis is indicative of significant cloud contamination. As can be seen in Fig. 1 there is a high correlation between the amount of cloud contamination and the value of the associated discriminant function. All the VTPR measurements from the data archive are included in this graph. The statistics which follow the graph provide a quantitative estimate of the indicated correlation. The statistics are compiled for various ranges of the Y-axis parameter. In Fig. 1 the first line of statistics is compiled for lines 60 through 80 which corresponds to values of the discriminant function X100 of 100 through 120. The count value

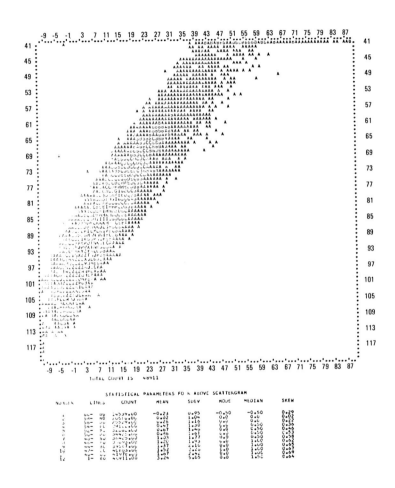

FIGURE 1. *Discriminant function value times 100 derived from the difference between channels 1 and 4 vs. temperature difference between window channel measurement corrected for the atmosphere and the sea surface temperature in 0.5°C intervals.*

indicates the number of measurements for which the associated
discriminant function has values within the appropriate range.
The statistics which follow pertain to the X-axis parameter.
Thus, from line 1 it is seen that the overall mean of the VTPR
window channel measurements corrected for the atmosphere is
0.23°C warm against the GOSSTCOMP sea surface temperature
estimate and has a standard deviation of 0.95°C. The subsequent
lines of statistics show that as the allowable range of the
discriminant function is increased, the overall mean temperature
of the measurements drops while the standard deviation increases,
and reflects an increasing presence of cloud contamination.
This data is very useful for establishing a threshold value for
the discriminant functions. Figure 2 is similar to Fig. 1 but
applies to the discriminant function number 2. It is apparent
that the value of the discriminant function is useful not only
for isolating the clear measurements, but also for indicating
the severity of contamination in the cloudy set. For the
measurements with little or no cloud contamination, i.e., small
abscissas, it is apparent from Figs. 1 and 2 that the values of
discriminant function number 3 contains less scatter than that
of function number 2 which illustrates the superiority of the
former function as a cloud classifier.

The sea surface temperature field estimate is included as a
feature to make the values of the discriminant functions
independent of geographical location or temperature zone. The
ability of the COSSTCOMB SST to achieve this purpose is indicated
with Figs. 3 and 4. In these figures the ordinate represents
the GOSSTCOMP SST estimate in 0.5°C intervals whereas the
abscissa represents the value of the discriminant function.
Only the "clear" measurements are included in these graphs as
determined with the parametric classification. Represented in
Fig. 3 is the discriminant function number 3 derived from the
difference between the window channel and the middle tropospheric

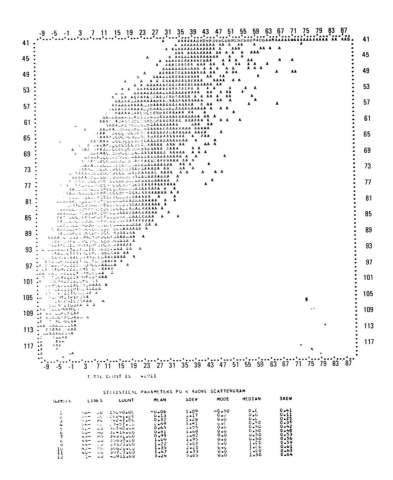

FIGURE 2. *Discriminant function values times 100 derived from the difference between channels 1 and 3 vs. temperature difference between window channel measurement corrected for the atmosphere and the sea surface temperature in 0.5°C intervals.*

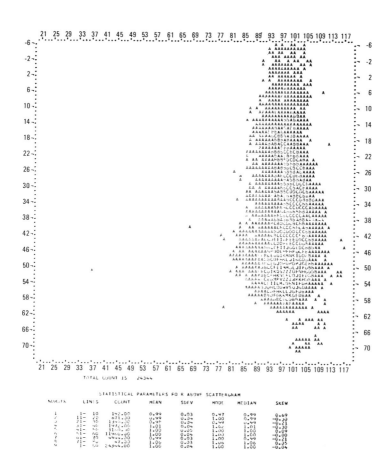

FIGURE 3. Sea surface temperature on 0.5°C intervals vs. discriminant function derived from the difference between channels 1 and 4, including as a feature the surface temperature.

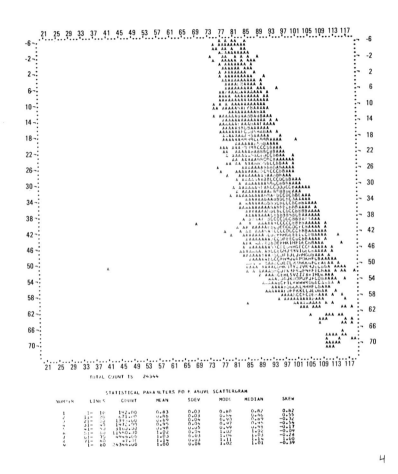

FIGURE 4. *Sea surface temperature in 0.5°C intervals vs. discriminant function derived from the difference between channels 1 and 4, not including as a feature the surface temperature.*

CO_2 channel 4 and including the GOSSTCOMP SST estimate as a feature. As is indicated with the associated statistics, this function is independent of surface temperature as desired. The behavior of the discriminant function which does not include the sea surface temperature estimate as a feature is shown in Fig. 4. A strong temperature dependence is indicated. The ability to describe a function whose value is indicative of cloud contamination while being independent of surface temperature is significant as this implies that the function can be applied to all areas of the globe without need for regional adjustments. Therein lies the power of the discriminant technique.

C. Temperature Analysis

In the previous two sections the relative accuracy of various discriminant functions as well as their behavior in the presence of cloud contamination have been described. In this section the cloud detection ability of the more promising discriminant functions is considered in an absolute sense. Again scattergrams are employed to describe the results. The ordinate represents the sea surface temperature in 0.5°C intervals and the abscissa represents the temperature difference between the window channel measurement corrected for the atmosphere and the GOSSTCOMP sea surface temperature estimate. Figure 5 shows the resulting scattergram when both the clear and cloudy window channel measurements are included. Correspondingly, Figs. 6 through 9 represent the scattergrams resulting when only the "clear" measurements are included, using first the parametric classifier and then the discriminant classifiers numbers 2, 3, and 4, respectively. A comparison of the various figures and their associated statistics indicates that the classifiers eliminate nearly all the cloud-contaminated measurements in each temperature zone, although the discriminant functions

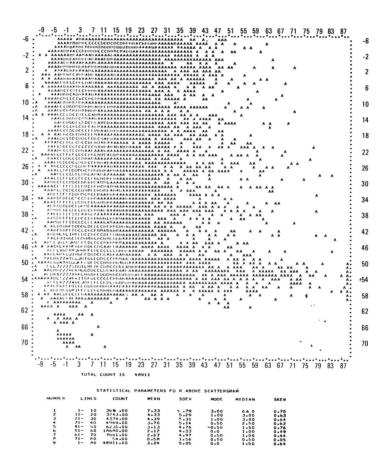

FIGURE 5. *Sea surface temperature in 0.5°C intervals vs. temperature difference between window channel measurement corrected for the atmosphere and the sea surface temperature in 0.5°C intervals using the discriminant function derived from the difference between channels 1 and 3 for cloud detection.*

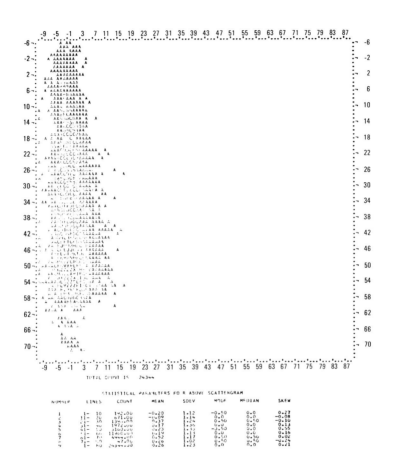

FIGURE 6.   *Sea surface temperature in 0.5°C intervals vs. temperature difference between window channel measurement corrected for the atmosphere and the sea surface temperature in 0.5°C intervals using the parametric classifier for cloud detection.*

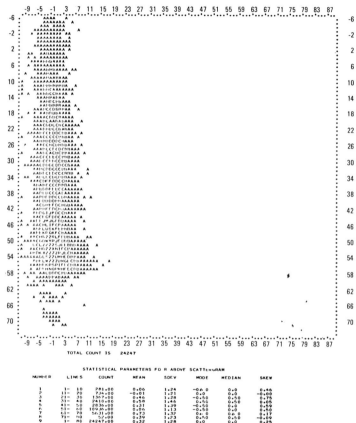

FIGURE 7.   Sea surface temperature in 0.5°C intervals vs.
temperature difference between window channel measurement corrected for the atmosphere and the sea surface temperature in
0.5°C intervals using the discriminant function derived from the
difference between channels 1 and 3 for cloud detection.

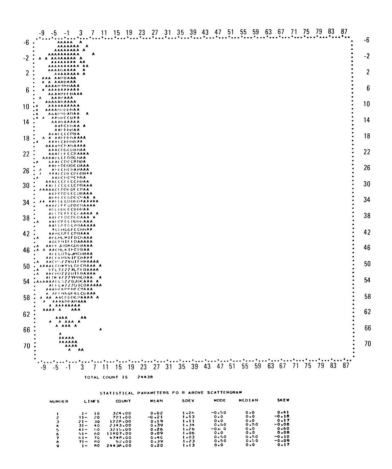

FIGURE 8. Sea surface temperature in 0.5°C intervals vs. temperature difference between window channel measurement corrected for the atmosphere and the sea surface temperature in 0.5°C intervals using the discriminant function derived from the difference between channels 1 and 4 for cloud detection.

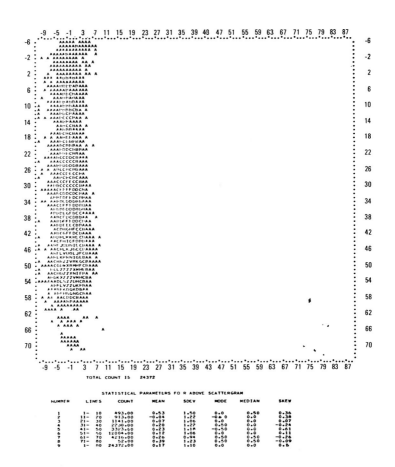

FIGURE 9.   Sea surface temperature in 0.5°C intervals vs.
temperature difference between window channel measurement
corrected for the atmosphere and the sea surface temperature in
0.5°C intervals using the discriminant function derived from the
difference between channels 1 and 5 for cloud detection.

number 3 and 4 appear to be most successful. One of the
interesting parameters which can be obtained from a comparison
of these figures is an estimate of the percentage of measurements
which are classified as clear in the various surface temperature
zones. As may be seen, the percentage of "clear" measurements
decreases from over 50 percent in the equatorial regions to
less than 25 percent in the polar regions. This pattern
generally agrees with the mean climatological global distribution
of cloud cover (6). The choice of the threshold value for the
discriminant functions in this comparison has been arbitrarily
chosen so that the number of "clear" measurements nearly equals
the number occurring with the parametric classifier. In
practice the user must decide the value which is optimal for
his application. The choice represents a compromise between the
desire to eliminate as much cloud contamination as possible and
not at the same time misclassifying a significant number of
clear measurements as "cloudy."

It is important to realize that the standard deviation
statistic associated with these figures includes errors in the
GOSSTCOMP analysis and in the correction for atmospheric
attenuation in addition to the errors resulting from residual
cloud contamination. Recent studies indicate that $1^{\circ}C$ is a
reasonable estimate of the global average root-mean-square error
resulting from residual cloud contamination in the "clear" set of
measurements as determined with the parametric classifier. The
global root-mean-square error of the best of the discriminant
function is somewhat less than this value.

D. *Cloud Detection Without Sounder Data*

All discriminant functions described previously have
included as a feature the difference between two of the sounder
channels. The last function included in Table I, however,
replaces this feature with the difference between the GOSSTCOMP

SST field estimate and the window channel brightness temperature.
This last discriminant function does not require sounder data at
all and, therefore, has particular application in the production
of sea surface temperature with high resolution infrared data.
The ratio of residuals given in Table I indicates that this
discriminant function may be a very accurate cloud detector if
the GOSSTCOMP SST field temperature is correct. The effect
of errors in the GOSSTCOMP field is described in the next
section. As with all the discriminant functions, the user must
establish a threshold value. In this case, the threshold value
establishes a maximum expected temperature difference between
the surface temperature and window channel measurement resulting
from atmospheric attenuation. This maximum expected atmospheric
attenuation feature is a linear function of surface temperature
and the zenith angle of the measurement. Measurements for which
the temperature difference is greater than this maximum expected
value are classified as cloudy with this discriminant function.
By contrast the previous operational procedures for cloud
detection without sounder data are first to correct the
retrieval for atmospheric attenuation and then require that the
corrected retrieval agree closely with the GOSSTCOMP surface
temperature estimate. Advantages associated with the discriminant procedure are (1) it separates the cloud contamination
problem from that of correcting for the atmospheric absorption
and (2) it is less sensitive to errors in the GOSSTCOMP surface
temperature estimate than is the operational procedure.

   A temperature scattergram for this last discriminant function
is shown in Fig. 10. A comparison with the previous scattergrams
indicates that this discriminant function is the most accurate
classifier in nearly all temperature regions if the surface
temperature estimate is correct. As with the previous temperature scattergrams, the threshold value of the discriminant
function has been chosen so that the number of "clear"

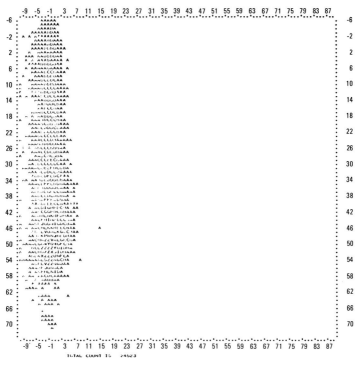

FIGURE 10. Sea surface temperature in 0.5°C intervals vs. temperature difference between window channel measurement corrected for the atmosphere and the sea surface temperature in 0.5°C intervals using the discriminant function for cloud detection derived from the difference between channel 1 and GOSSTCOMP SST.

measurements approximately equals that of the parametric
classifier. The scattergram shown in Fig. 11 is similar to the
previous one with the VTPR window channel measurement being
replaced with the coincident SR retrieval brightness temperature
in the discriminant analysis while the threshold value is held
constant. It is seen from the number of "clear" SR retrievals,
each of which are computed from a set of 1024 SR samples, is
approximately 45 percent greater than the number of "clear"
VTPR window channel measurements. On the other hand, the
standard deviation of the "clear" retrievals is significantly
greater and the mean temperature is warmer because of errors
in the correction for atmospheric attenuation resulting from
the use of cloud-contaminated VTPR water vapor information.
These results illustrate both the advantages and problems
associated with the use of high resolution retrieval schemes.

E. Effect of Errors in Surface Temperature Estimate

   The possibility of misclassification resulting from errors
in the GOSSTCOMP surface temperature analysis is considered in
this section. Errors are simulated by replacing the GOSSTCOMP
temperature with the corresponding National Center for
Atmospheric Research climatology value in the discriminant
analysis. The resulting "clear" set of measurements is then
compared against the GOSSTCOMP temperatures to determine the
veracity of the classification. Temperature scattergrams are
again employed to illustrate the results. The results achieved
with discriminant function number 3 are shown in Fig. 12. This
scattergram should be compared with that of Fig. 8 to evaluate
the effect of the use of climatology. The same threshold value
is applied in both cases. Although there is a reduction of
10 percent in the number of "clear" measurements, the simulated
errors have little or no effect upon the quality of these
measurements as indicated by the statistics generated. Such is

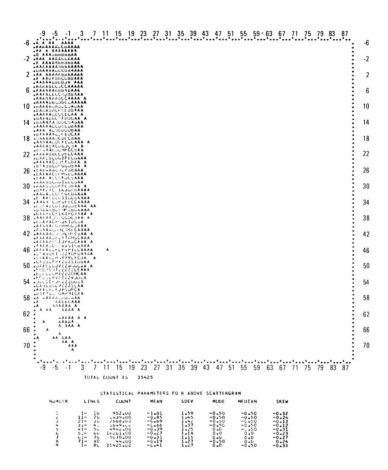

FIGURE 11. Sea surface temperature in 0.5°C intervals vs. temperature difference between SST brightness temperature corrected for the atmosphere and the sea surface temperature in 0.5°C intervals with the discriminant classifier using the difference between GOSSTCOMP SST and retrieval brightness temperature.

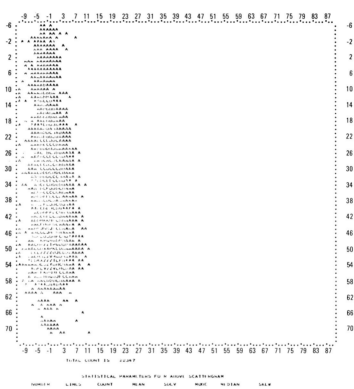

FIGURE 12.   Sea surface temperature in 0.5°C intervals vs. temperature difference between window channel measurement corrected for the atmosphere and the sea surface temperature in 0.5°C intervals using the discriminant function for cloud detection derived from the difference between channels 1 and 4, but utilizing climatology for the surface features.

not the case with the discriminant function number 12. The
effect of the simulated errors is indicated in Fig. 13 which
should be compared with Fig. 10. Not only is there a 20 percent
reduction in the number of "clear" measurements, but the
associated statistics indicate a significant increase in the
amount of cloud contamination within the "clear" set of measurements. These results indicate that the discriminant functions
which employ the difference between two of the sounder channels
as a feature are relatively unaffected by errors in the surface
temperature estimate. In fact, a comparison of Fig. 12 with
Fig. 6 shows that even with the simulated surface temperature
errors included, the discriminant function number 3 is somewhat
more accurate than the parametric classifier. On the other
hand, the discriminant function number 12 has a relatively large
dependence upon surface temperature and, consequently, should
be considered as a backup cloud detection procedure when no
sounder data are available.

VI.   CONCLUDING REMARKS

    This paper describes a technique for cloud detection within
a single field of view of the sounder using multiple channel
data. When applied in the GOSSTCOMP model, the procedure
represents a conservative approach to the cloud contamination
problem. That is, if the HIRS data at a resolution of 25 km is
clear or nearly so, then one can be sure that an analysis of
the warm side of a histogram of coincident 4 km AVHRR data will
provide an estimate of sea surface temperature which is also
cloud free. Alternatively, with a significant number of AVHRR
histograms, it will be possible to derive a clear estimate of
surface temperature although the sounder data is cloud contaminated. Observations derived from these data are assigned a lower

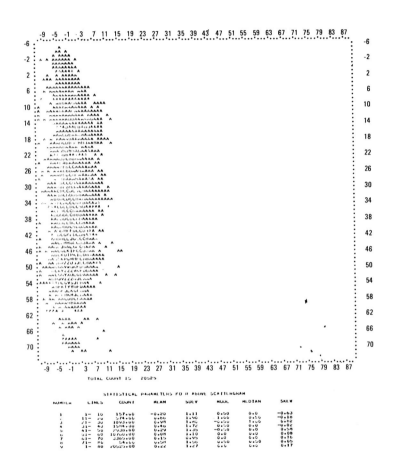

FIGURE 13. Sea surface temperature in 0.5°C intervals vs. temperature difference between window channel measurement corrected for the atmosphere and the sea surface temperature in 0.5°C intervals using the discriminant function for cloud detection derived from the difference between channel 1 and climatology.

reliability in the GOSSTCOMP model than when the HIRS data is clear, because the sounder data can no longer be used to estimate the effects of absorption by water vapor.

As mentioned previously, discriminant functions have been applied in the nominal GOSSTCOMP model only since the advent of TIROS-N. It is important to determine what improvements in quality, if any, have resulted from the use of these functions for cloud detection. A recently published paper indicated that the satellite-derived sea surface temperatures were signifi-cantly cold relative to AXBT measurements in tropical regions during the period of their study November 1977 to January 1978 (7). At that time, the VTPR sounder data was used for cloud detection but only through the use of the parametric classifier which provides a much more stringent cloud test in polar regions than in tropical regions. In the current GOSSTCOMP model, it has been found that by increasing the threshold value of the discriminant functions in the tropical regions relative to what is applied over the remainder of the globe, this mean negative bias as determined through a comparison with reports from Navy ships has been significantly reduced or eliminated. Thus, the application of the discriminant procedures in the current GOSSTCOMP model has significantly improved the overall quality of the satellite-derived sea surface temperature product.

REFERENCES

1. Crosby, D. S., *in* "Fourth Conference on Probability and Statistics in Atmospheric Sciences," pp. 163-168. American Meteorological Society, Boston, Massachusetts (1975).

2. Brower, R. L., Gohrband, H. S., Pichel, W. G., Signore, P. L.,
    and Walton, C. C., Satellite Derived Sea-Surface
    Temperatures from NOAA Spacecraft. NOAA Technical
    Memorandum NESS 78. U.S. Government Printing Press,
    Washington, DC (1976).

3. McMillan, L., Wark, D. Q., Siomkagld, J. M., Abel, P. G.,
    Werbowetzki, A., Lauritson, L. A., Pritchard, J. A.,
    Crosby, D. S., Wools, H. M., Luebbe, R. C., Weinreb, M. P.,
    Fleming, H. E., Bittner, F. E., and Hayden, C. M.,
    Satellite Infrared Soudings from NOAA Spacecraft. NOAA
    Technical Memorandum NESS 65. U.S. Government Printing
    Press, Washington, DC (1973).

4. Walton, C. C. Analysis of Coincident Data Base of SST
    Retrievals and VTPR Spot Measurements Yielding Procedures
    for Cloud Detection and for Correcting Atmospheric
    Attenuation in the Window Channel. NESS Memo, OA/S14/CW,
    Washington, DC (November 20, 1975).

5. Walton, C. C., Brower, R. L., and Signore, T. L., *in*
    "Symposium on Meteorological Observations from Space: Their
    Contribution to the First GARP Global Experiment," pp.
    155-159. COSPAR, Philadelphia, Pennsylvania (1976).

6. Miller, D. D., and Feddes, R. G., Global Atlas of
    Relative Cloud Cover. U.S. Department of Commerce and
    United States Air Force, U.S. Government Printing Office,
    Washington, DC (1971).

7. Barnett, T. B., Patzert, W. C., Webb, S. C., and Bean, B. R.,
    Bulletin of the American Meteorological Society, 60(3),
    197-205 (1979).

DISCUSSION

Staelin: As I understand it, you were estimating the sea surface temperature using only two channels at a time. If you are not using a large number of channels, why not?

Walton: For the atmospheric attenuation problem? I have done it with many channels; and you really do not improve the results using many as opposed to two that correlate to the answer you want. And in some cases, results you get are actually not as good because the statistics can fool you because of the effect of the residual cloud contamination; usually just a different function of the appropriate two channels gives the best result.

Susskind: I am sure there have been some comparisons made of your derived sea surface temperatures with a ground truth ship. Can you give a general idea about the quality of the soundings?

Walton: Yes. Well, operationally we get comparisons every day in 18° latitude bands and compare ships with sea surface temperature observations; and the search areas are 100 km to get match-ups, and I refer to the problem of the cold bias in the tropics. Our comparisons of satellite measurements with ships show now the means are very close to zero, the standard deviations are usually quite low. As a matter of fact, there seem to be more fluctuations, more variations in the ship reports when you compare them against themselves than there are in the satellite measurements against themselves, and you reach the point where-- if you get a large deviation in one region--you do not know if it is the satellite measurement or a bad ship report.   In the tropics we typically get standard deviations of 1° in comparisons of ships with satellites.

Susskind: Do you get any disasters that are way off.

Walton: Occasionally we will get one. Usually it is a bad ship report. Sometimes ships report in Fahrenheit instead of Centigrade or are mislocated. A recent problem we have had this spring is atmospheric warming and satellite observations in the $36^{\circ}N$ to $54^{\circ}N$ latitude region now are too warm, like maybe one or two degrees warmer when compared with ships. The satellite measurements however are extremely consistent internally.

INTERPRETATION OF SOLAR EXTINCTION DATA
FOR STRATOSPHERIC AEROSOLS[1]

T. J. Pepin

Department of Physics and Astronomy
University of Wyoming
Laramie, Wyoming

This paper discusses the inversion problem for aerosols using the solar extinction method. A series of numerical experiments is described in which solar extinction measurement systems are modeled. A numerical model of a solar extinction measurement system has been coupled with model atmospheres that exhibit fine scale structures to produce numerically generated data signals. These signals were then inverted to study the effect that measurement errors and desired vertical resolution produce in the inverted results. Knowledge of the trade off between vertical resolution and the accuracy of inversion aid in the interpretation of the inverted results.

[1] *This research was accomplished as a part of the author's Nimbus Experiment Team and SAGE Experiment Team activities with funds from the National Aeronautics and Space Administration under contract with the Langley Research Center in Hampton, Virginia.*

I.   INTRODUCTION

The author's group at the University of Wyoming has
developed a system of algorithms to model solar extinction
experiments and test inversion methods for solar extinction
data. This system includes a number of model atmospheres that
have been derived from observational data as well as standard
models. Included are models for aerosols, ozone, NO_2, and
temperature. Various vertical resolutions as well as latitude
differences are included in the model sets which have been
described by Pepin and Cerni (1).

Using the atmospheric models and the algorithms systems'
description of a remote sensing experiment configuration,
signals can be generated that model an experiment's responses
to various atmospheric conditions.

Error generators are incorporated which allow for the study
of error sources that are present in solar extinction experiments. With the model, signals for a range of atmospheric
conditions can be constructed that contain trackable errors.
The synthesized signals can then be inverted using an inversion
method that one would like to study. The inverted signal
result can be compared with the model, allowing for insight into
the system's sensitivity to both measurement errors and atmospheric conditions.

Studies of this type have been used in designing the ASTP-
SAM experiment (2,3) and the SAM II and SAGE experiments (4).
They have also been used in setting up the data reduction
systems for these experiments.

This paper concerns itself with several numerical experiments
that have made use of the Wyoming algorithms to study the
effect that different vertical resolutions have on inverted
results for aerosols.

II. ALTITUDE RESOLUTION STUDY

Figures 1 to 4 show the results of the numerical experiments
on inversions that have been done to study the effect of
different vertical resolutions on the inverted results for
total extinction due to aerosols and molecules.

For these experiments, a highly structured model derived
from an early 1975 post-volcanic dustsonde observation was
used to first construct radiometric signals as would be measured
by a 10-bit radiometer with ± 1/2 bit digitization error (1).
The observing system picked for this study was a model of the
SAM II optical system on Nimbus VII (4). Randomized errors of
± 1/3 percent of signal were added to model the radiometric
errors for the observing system. A wavelength of 1.0 μm was
used.

Figures 2 to 4 show the results for several cases in which
these errors were included and they are labeled "all errors."
Figure 1 shows the results when the model was run when digitization and randomized errors were not included. It has been
labeled the case of "no errors."

The synthesized signals were inverted using the method of
Twomey and Phillips of constrained linear inversion in which the
second derivative of the solution vector is minimized. For
the examples shown, the constraint parameter was zero and hence
the solutions qualify as the least-squares type (5,6). For
these experiments the path matrix and kernel matrix were
rectangular ones. The number of rows was determined by the
total number of singular data points (1500 to 2000). The
number of columns was determined by the vertical resolution
sought. Three different vertical resolutions were sought:
1/2 km, 1 km, and a variable resolution scheme consisting of
1 km resolution to 26 km altitude, 2 km from 26 to 30 km
altitude, 3 km from 30 to 36 km, and 4 km resolution above.

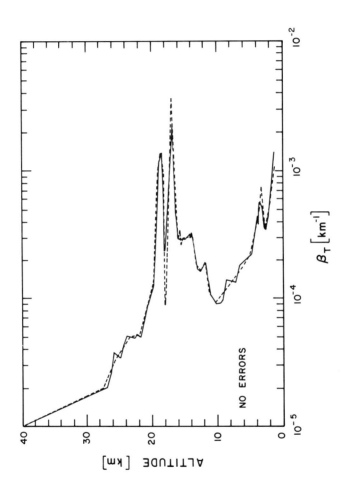

FIGURE 1. Comparison between model and inverted results for the total extinction due to molecules and aerosols at a wavelength of 1.0 µm. For this inversion, vertical resolution of 0.5 km was used. "No errors" were added to the model's signals.

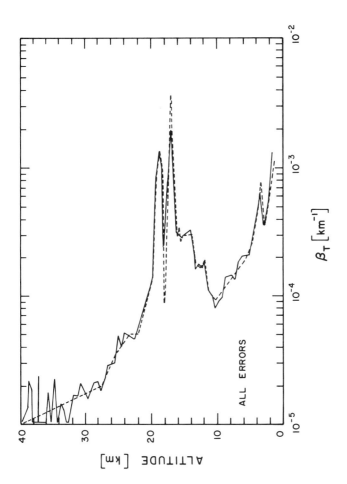

FIGURE 2. *Comparison between model and inverted results for the total extinction due to molecules and aerosols at 1.0 μm wavelength. This inversion used 0.5 km vertical resolution and had "all errors" as described in the text.*

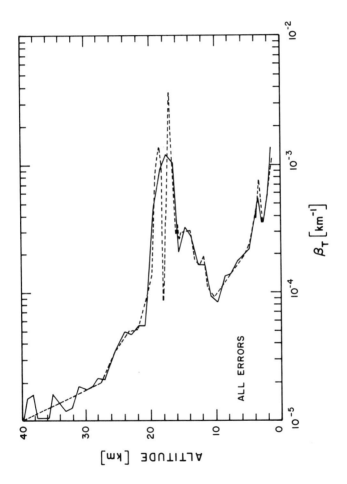

FIGURE 3. Comparison between model and inverted results for the total extinction due to molecules and aerosols at 1.0 μm wavelength. This inversion used 1.0 km vertical resolution and had "all errors" as described in the text.

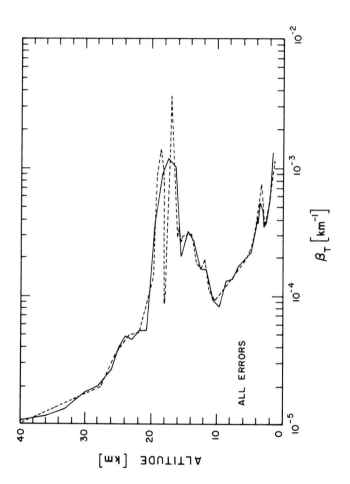

FIGURE 4. *Comparison between model and inverted results for total extinction due to molecules and aerosols at 1.0 µm wavelength. For this inversion, the variable vertical resolution scheme identified in the text was used along with "all errors."*

The input data profiles, shown with dashed symbols, are compared
with the inverted results in Figs. 1 to 4. In Fig. 1, the
"no error" case at 1/2 km resolution, the quadrature error
of the inversion system is illustrated.

The results of these experiments illustrate the effect,
which is a fundamental behavior of the inversion process, that
resolution can be traded for accuracy. They also illustrate
the vertical resolutions that can be achieved with the SAM II
and SAGE experiments, and the desirability when structured
atmospheres are present, of using variable resolution in the
inversion process for these experiments.

ACKNOWLEDGMENTS

The author would like to acknowledge the help of Dr. T. Cerni
and Mrs. F. Simon in performing the calculations required for
this study.

REFERENCES

1. Pepin, T. J., and Cerni, T. A., Definition of Model
    Atmospheres for Application in Preliminary Studies of the
    Stratospheric Aerosol Experiment (SAM II Aboard the Nimbus G
    Spacecraft). University of Wyoming publication APP-11,
    Laramie, Wyoming (1977).

2. Pepin, T. J., McCormick, M. P., Chu, W. P., Simon, F.,
    Swissler, T. J., Adams, R. R., Fuller, W. H., Jr., and
    Crumbly, K. H., Stratospheric Aerosol Measurement in
    "ASTP Summary Science Report," Vol. 1, NASA SP 412 (1977).
    (NTIS, Springfield, Virginia.)

3. Pepin, T. J., in "Inversion Methods in Atmospheric Remote
    Sounding" (A. Deepak, ed.), pp. 529-554. Academic Press,
    New York (1977).

4. McCormick, M. P., Hamill, P., Pepin, T. J., Chu, W. P.,
    Swissler, T. J., and McMaster, L. R., *Bul. of AMS, 60(9),*
    1038-1046 (1979).

5. Phillips, D. L., *J. Assoc. Comp. March 9,* 84 (1962).

6. Twomey, S., *J. Assoc. Comp. March 10,* 67 (1963).

DISCUSSION

Remsberg: Ted, how does the uncertainty in the density profile
affect the retrieval of the ozone in SAGE? Is this a small
error for your ozone retrieval?

Pepin: At the altitude of the peak of the ozone distribution
there is approximately a order of magnitude difference in
extinction between the background and the peak density. If there
is a factor of an order of magnitude and you know the background
you are trying to subtract off, namely, the molecules in the
atmosphere to let us say 10 or 20 percent, you are approaching
the 1 or 2 percent or less error from that source. Those type
of errors were included in simulations Bill Chu showed.

Remsberg: Can you believe the wiggles in the ozone profile as
being due to ozone and not density?

Pepin: The answer is yes. It depends which wiggles you are
talking about. I am talking about the wiggles near the peak.

Chu: I would like to make a comment on the SAGE inversion and
climate density. We retrieve density from SAGE. We do not use
a climatology, so the fluctuation in density is correct in the
ozone channel.

Hamill: Could you tell me what the altitude discrimination in
your one-channel measurments in PAM is in the 18- to 25-km range?

Pepin: Near the peak of the aerosol and near the peak of the
ozone the simulations we have done indicate that we can approach
2- to 3-km vertical resolution with the full sun instrument with
the orbital configuration that PAM has. Above that, of course,
the resolution degrades and will go upward of 5 km resolution
toward the top of the profile. Now, one of the things I did not
show in great detail is where the sensitivity to ozone is seen
in the PAM data. It turns out that when you get to 50 km of
altitude, that is, when the lower limit of the sun is at 50 km
altitude, the attenuation in the ozone channel is larger than
the blue channel. This says that at 50 km already the main
absorber we are looking at in the atmosphere is ozone. Because
of this I believe we will be able to invert the ozone with this
measurement to somewhere on the order of 50 km.

Russell: Ted, in the profiles you showed and the inversion
studies you were doing, did you have a finite field of view
included in those studies?

Pepin: Yes they included the field of view that was present in
the SAM or SAGE type instrument--the half arc minute field of view.

A FAST AND ACCURATE RADIANCE ALGORITHM
FOR APPLICATION TO INVERSION
OF LIMB MEASUREMENTS

Larry L. Gordley

Systems and Applied Sciences Corporation
Hampton, Virginia

James M. Russell III

NASA, Langley Research Center
Hampton, Virginia

The time-consuming nature of limb relaxation type inversion
algorithms is due primarily to the numerous integrations over
an absorption band to obtain forward radiance values with which
to compare measured values. A new method has been devised for
the quick and accurate (<0.5 percent error) calculation of broad
band ($\approx 100 \text{ cm}^{-1}$) limb radiance which is based on a precalculated
data base consisting of homogeneous path emissivity as a function
of mass path data for a wide range of temperatures and pressures.
Splicing together interpolated information from these broad-band
emissivity curves in a unique fashion has been shown to give
accurate estimates of limb radiance with no reliance upon a
priori statistical knowledge. The method has been applied in a
simulated inversion study to the inference of O_3, H_2O, NO_2, and
HNO_3. A 50-km deep, 1-km resolution, constituent inversion
employing this method requires under 1 second of computational
time.

I.   INTRODUCTION

The limb emission experiment for remote sensing of the
atmosphere has been analyzed extensively. (See, for example,
Refs. 1 to 7.) The early studies were aimed at determination
of the temperature profile using measured emission from CO_2.
Later studies assumed knowledge of the temperature profile and
analyzed the feasibility of remotely sensing concentrations of
variable gases such as H_2O and O_3. McKee *et al.* and House and
Ohring were the first to publish results of limb inversions
(for temperature and H_2O) using data obtained in the Project
Scanner rocket flights (2,3). The first satellite experiment
employing the limb sounding approach was the Limb Radiance
Inversion Radiometer (LRIR) experiment flown on the Nimbus 6
satellite in June 1976. This experiment was designed to measure
upper atmospheric temperature, ozone, and water vapor. A
follow-on experiment, the Limb Infrared Monitor of the
Stratosphere (LIMS) experiment, was flown on the Nimbus 7
satellite in the last quarter of 1978 to measure even more
tenuous gases in the upper atmosphere including O_3, H_2O, NO_2,
and HNO_3. Also on the Nimbus 7 satellite, is the Stratosphere
and Mesosphere Sounder (SAMS) experiment which is a pressure-
modulated radiometer limb sounder for measuring H_2O, CO, NO, N_2O,
and CH_4.
The maximum and most efficient use of data from these
experiments calls for fast, efficient algorithms. This
represents a formidable challenge since nonlinear iterative
types of approaches are required. This paper describes an
inversion technique for this purpose which does not require
a priori statistical information and provides both speed
advantage and high accuracy. The technique is conceptually
analogous to that proposed for transmission calculations by
Weinreb and Neuendorffer (8).

II. PHYSICS OF LIMB RADIANCE INVERSION

The limb experiment geometry provides a number of advantages for remote sensing applications (Fig. 1). The horizon path contains up to 60 times more emitter than a corresponding nadir view and provides greater sensitivity for measurement of tenuous species. The combination of the spherical geometry and the exponential decrease of gas density with height provides data heavily weighted around the tangent point and gives high vertical resolution. Most of the emitted energy comes from a layer only 2- to 3-kilometers above the tangent point (Fig. 2). Also, the background viewed by the instrument is the cold blackness of space, which simplifies data interpretation. Since atmospheric emission is being observed, measurements can be made day or night and in any view direction. The limb geometry has the disadvantage that horizontal resolution is poor, most energy being emitted over $\tilde{\ }$ 300-km horizontal distance. This is not considered a serious limitation, however, since the most suitable region for study by limb sounding is the upper atmosphere which is not expected to have rapid horizontal variations.

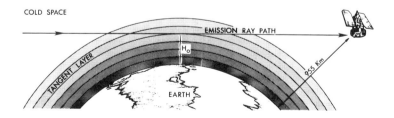

FIGURE 1. Limb experiment geometry.

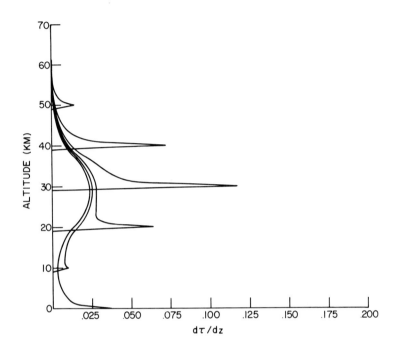

FIGURE 2. Ozone limb weighing functions.

The radiance measured by a limb-viewing instrument can be
represented by

$$N(\Delta v, H_0) = \int_{\Delta v} \int_{\tau_b}^{1} \phi(v) \ B[v, T(H)] \ d\tau[v, q(H), T(H), P(H)] dv \tag{1}$$

where $N(\Delta v, H_0)$ is the radiance measured over a bandwidth Δv,
$\phi(v)$ is the instrument response function, $B[v, T(H)]$ is the Planck
intensity at wave number v and temperature $T(H)$,
$\tau[v, q(H), T(H), P(H)]$ is transmittance, $P(H)$ is pressure, H is
altitude, H_0 is tangent height or the point of closest approach
of a ray path to the earth (refer to Fig. 1), and τ_b is the

transmittance from the instrument to a point at the outside edge
of the atmosphere for a given gas. A profile of N is measured
as a function of H_0 for each spectral band by causing a mirror to
scan the view direction vertically across the horizon or earth
limb. It has been assumed that the temperature solution is
available by a method like that of Gille and House which yields
temperature as a function of pressure (5). Therefore, the
height variable in Eq. (1) should be replaced with pressure (P).
The emphasis in this paper is on the constituent inversion
problem.

Virtually all inversion algorithms which operate on limb
radiance data require iteration due to the nonlinear nature of
the problem. A simple form for a relaxation equation can be
determined by using a Taylor's series expansion which relates
the measured radiance N_m to calculated radiance N_c in the
following way:

$$N_m = N_c + \frac{\partial N_c}{\partial q} \Delta q$$

$$\Delta q = (N_m - N_c) \times \left(\frac{\partial N_c}{\partial q} \right)^{-1} \qquad (2)$$

where the quantity Δq represents the amount the initial mixing
ratio guess (q_0) needs to be changed in order to make ($N_m - N_c$)
approach zero. This relaxation equation can be applied at each
tangent altitude and the tangent point mixing ratio can be
inferred, for example, in an onion peeling fashion (6). This
can be a very time-consuming process since, as Eq. (1) shows,
each N_c value requires integration over many layers and many
spectral intervals. As an example, for a 10-km tangent height
and a spectral band pass of 200 cm^{-1}, as many as 120 layers and
40 spectral intervals or 4800 integrations and exponentiations
are required. Thus, iterative solutions are very time consuming
even on modern, fast computers. This is an important

consideration when faced with the task of reducing large
quantities of data from a long-duration satellite experiment.

The authors have developed a new radiance calculation
technique which is fast, efficient, accurate, and does not suffer
from the computational difficulties just described. The method
is based on the use of an emissivity plotted against mass path
data base (curves of growth) calculated for a large number of
homogeneous paths covering a wide range of temperatures and
pressures. When the method is employed in inversion algorithms,
significant time reduction for a solution results. Details of
the technique are described in the next section.

III. EMISSIVITY GROWTH APPROXIMATION TECHNIQUE

The emissivity growth approximation (EGA) technique
developed allows the $d\tau$ calculation in Eq. (1) to be done
incrementally over a limb ray path without the time-consuming
integration over wave number. All the information normally
obtained by integration can be obtained from a precalculated
data base sufficient to cover any atmospheric condition.

The data base consists of emission growth curves for paths
of constant T and P. The integration through the atmosphere is
done by breaking the ray path into homogeneous segments. For
limb viewing, these segments are altitude-layer intersect paths.
A growth curve $\bar{E}_i(U)$ is determined for each layer i, according
to the formulation

$$\bar{E}_i(U) = \frac{\int_{\Delta\nu} \phi(\nu) B(T_i,\nu) E(T_i,P_i,U,\nu) d\nu}{\bar{B}_i} \tag{3}$$

where, as before, ϕ and B are the instrument filter and Planck
function, respectively. U is mass path, and E is the monochromatic emission. $\bar{B}_i$ is defined as

$$\bar{B}_i = \int_{\Delta\nu} \phi(\nu) \ B(T_i,\nu) \ d\nu \tag{4}$$

The radiance for a tangent ray is computed by using the relation

$$\varepsilon^i = \bar{E}_i (U_i + U_i^a) \tag{5}$$

where U_i is the tangent ray mass path through layer i. The quantity U_i^a is a pseudo mass path determined by the procedure graphically depicted in Fig. 3. The radiance coming from layers 1 through i can be found by using the recursive formula

$$R_i = R_{i-1} + \bar{B}_i \ (\varepsilon^i - \varepsilon^{i-1}) \tag{6}$$

The total radiance R_T can be calculated by carrying this procedure through the tangent layer to cold space. This summing procedure is depicted in Fig. 4. It should be noted that if atmospheric spherical symmetry is assumed

Layer 1 = Layer N

Layer 2 = Layer N - 1

Layer N/2 = Layer N/2 + 1

where tangent layer path has been split into two layer paths. Equation (6) can be viewed as building an emissivity curve of

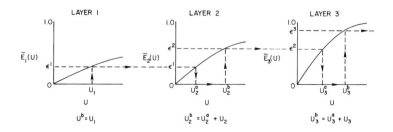

FIGURE 3. Graphical interpretation of emissivity calculation.

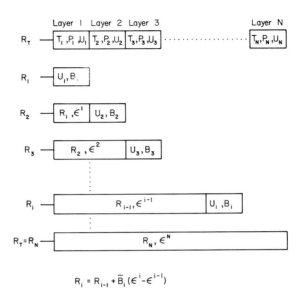

$$R_i = R_{i-1} + \bar{B}_i(\epsilon^i - \epsilon^{i-1})$$

FIGURE 4. EGA multilayer emission radiance.

growth for the entire path by taking a segment from the curve
of growth of each individual layer path. The method used for
calculating the emissivity curves for each cell will be described
in a later section.

IV. ACCURACY DISCUSSION

Although it will not be done here, it can be shown that if
any of three conditions exists, i.e., optically thin paths,
nearly homogeneous paths, or very dense spectra (approaching a
continuum), then one would expect good results from EGA. The
latter condition applies because the basis of the technique,
as implied by Eq. (5), is matching of band-pass integrated area
under the spectrum. The more featureless the combined spectrum

of all shells prior to the ith shell and the spectrum of the
ith shell, the more exact will be the match of broad band
emission curve shapes as well as areas. It was found that even
under less than these ideal conditions, remarkable agreement is
obtained. It should be noted that for monochromatic radiation,
i.e., $\phi(v)$ = delta function, the method is exact.

V. DERIVATION OF EMISSIVITY CURVES

In order to obtain a constituent profile solution, the
atmosphere is divided into altitude layers, as in Fig. 1, with
vertical thickness less than or equal to the desired vertical
resolution. The $d\tau$ increments, for the integration of Eq. (1)
correspond to the layer intersect segments of a ray. The EGA
method requires a data set containing a separate $\bar{E}_i(U)$ curve for
each layer traversed by the rays.

A data base has been developed from which these $\bar{E}_i(U)$ curves
can be found, by interpolation, for any realistic atmospheric
condition of temperature, pressure and mass path. This data
base consists of calculating many $\bar{E}_i(U)$ curves as defined by
Eqs. (3) and (4). Five curves corresponding to five temperatures
with 100 $(\bar{E}, U)$ points were calculated at 15 pressure levels.
The five temperatures were chosen as follows. Two consisted of
the upper and lower 1 percent probability temperatures for the
pressure level. One temperature was the average for the pressure
level and the last two were temperatures midway between the
1 percent and average. Each set of five $(\bar{E}, U)$ points in a given
layer (corresponding to the same pressure and mass path) was
fitted with a second-order polynomial.

The temperature and pressure values chosen for calculation
of the data base were obtained from the 1976 U.S. Standard
Atmosphere. The pressure levels are the average for 5-km
increments from 5- through 75-kilometers. The mass path values

were chosen to cover a range between the smallest cell mass path
expected to ten times the largest total ray mass path expected.
The mass path increments were chosen to be logarithmic because
of the approximate $\bar{E} \approx \text{Log}_{10}$ (U) nature of most broad bands.
Calculations comparing $\bar{E}$ values found exactly and those found
by interpolation on the data base have shown errors of
< 0.1 percent for pressures, temperatures, and mass paths within
the range of the data base.

It should be noted that the data-based range selected for P,
T, and U, the increments used for P, T, and U, and the interpolation techniques applied are not likely to be optimum.
However, the procedure described has given excellent results.
Further studies of data base accuracy, polynomial fit accuracy,
and possibilities of shrinking the data base will be done. It
is expected that a much smaller data base will be adequate, say
one tenth the size, if more sophisticated interpolating schemes
are used. This would further improve the efficiency of the
method.

A line-by-line method of calculating the data base is
preferred. With most available algorithms, this is very costly
considering the size of the data base. However, the efficiency
of line-by-line calculations is greatly improved by using
techniques like that recently developed by Mankin (9). Data
bases have been calculated with a Mankin type line-by-line
algorithm for all pertinent species and filters of the LIMS
instrument. Gases for which absorption line parameters are
unavailable are limited to band model data bases (e.g., HNO_3).

VI. RESULTS

The ultimate accuracy test of the EGA technique is the
comparison of multilayer radiances calculated with "exact"
line-by-line methods and those calculated by using EGA. This

has been done for limb radiances emitted by O_3 near 1100 cm^{-1} and H_2O near 1500 cm^{-1}. These results are listed in Table I along with results of a similar test on HNO_3 near 900 cm^{-1} using a band model developed by Goody and band model coefficients measured by Goldman *et al.* (10,11). In the O_3 and H_2O cases, a square filter was used over intervals listed in the table whereas for HNO_3 the actual LIMS filter was used. These intervals were chosen because of their relation to the LIMS experiment. Typical concentration profiles were used. Nearly all the EGA radiances are within 0.5 percent or less of the line-by-line values.

The main purpose of the EGA technique is to obtain rapid and accurate inversion of limb radiance data. The data bases were originally derived by using the random exponential band model for O_3, NO_2, and H_2O. A comparison of inversion solutions obtained by using EGA radiances with those obtained with radiances calculated by conventional techniques which employ the Curtis-Godson approximation (e.g., Ref. 12) is shown in Fig. 5. Note that the EGA inversion gives larger concentration values in the lower stratosphere. Table I indicates, however, that the EGA approach gives excellent agreement with line-by-line radiances in that altitude range when using line-by-line data bases. Similar comparisons of radiances determined with the random-exponential band model give larger radiance values than the line-by-line calculations. Thus, it is concluded that most of the differences in Fig. 5 are due primarily to use of the band model and, in particular, to the Curtis-Godson multilayer approximation which tends to overestimate absorption for large optical depth because of pressure weighting difficulties. If this is true, large differences are expected for gases having peaked altitude distributions or those which are optically thick. In these cases the mass-weighted pressure will be higher than required. This is shown in Fig. 5 by the fact that errors due to the band model are larger for ozone,

TABLE I. Calculated Limb Radiance Comparisons

Tangent altitude km	Temperature K	Pressure mb	O_3 100 cm^{-1} to 1050 cm^{-1}			H_2O 1370 cm^{-1} to 1570 cm^{-1}			HNO_3 847 cm^{-1} to 914 cm^{-1}		
			"Exact" radiance	EGA radiance	Per-cent error	"Exact" radiance	EGA radiance	Per-cent error	Band model radiance	EGA radiance	Per-cent error
40	253.1	2.51	1.1943	1.1941	-0.01	0.07146	0.07141	-0.08	0.00475	0.00475	-0.02
38	247.6	3.29	1.2793	1.2785	-0.07	0.07576	0.07559	-0.23	0.00808	0.00808	0.00
36	242.1	4.33	1.3118	1.3104	-0.11	0.07775	0.07745	-0.39	0.01439	0.01439	0.01
34	236.5	5.75	1.3207	1.3191	-0.12	0.07939	0.07903	-0.44	0.02637	0.02638	0.02
32	331.0	7.67	1.3204	1.3190	-0.10	0.08136	0.08094	-0.52	0.04666	0.04666	0.01
30	227.5	10.31	1.3230	1.3221	-0.07	0.08530	0.08495	-0.41	0.08435	0.08437	0.02
28	225.5	13.90	1.3267	1.3261	-0.05	0.09185	0.09157	-0.31	0.14061	0.14059	-0.01
26	223.5	18.80	1.3258	1.3254	-0.03	0.10064	0.10032	-0.32	0.20966	0.20957	-0.05
24	221.6	25.49	1.3211	1.3209	-0.02	0.11174	0.11140	-0.30	0.26381	0.26358	-0.09
22	219.6	34.67	1.3133	1.3132	-0.01	0.12528	0.12485	-0.34	0.31290	0.31232	-0.19
20	217.6	47.29	1.3051	1.3051	0.00	0.13567	0.13514	-0.39	0.37941	0.37834	-0.28
18	216.7	64.67	1.2943	1.2943	-0.01	0.14621	0.14550	-0.48	0.42786	0.42651	-0.32
16	216.7	88.50	1.2865	1.2863	-0.01	0.17748	0.17669	-0.45	0.44365	0.44189	-0.40
14	216.7	121.10	1.2784	1.2780	-0.03	0.22088	0.21991	-0.44	0.42751	0.42571	-0.42
12	216.7	165.80	1.2698	1.2689	-0.07	0.30380	0.30322	-0.22	0.38888	0.38737	-0.39

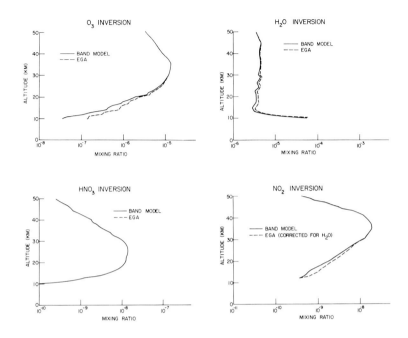

FIGURE 5.   Comparison of EGA to random exponential band model.

intermediate for water vapor, and least for HNO_3 which is
optically thin with little pressure dependence. This result
suggests that the EGA technique will provide a better inversion
than the random exponential band model approach even when the
band model is used to calculate the data base.

VII.   EGA INVERSION APPLICATIONS

    When employing EGA in an actual inversion, the procedure
in Figs. 3 and 4 is employed to calculate the total radiance
for each ray path. For an onion peeling method, which was
used, all cell mass paths are known except that of the tangent
layer.   The tangent layer mass path is approximated and the
total path radiance is calculated and compared with the measured

radiance. If the difference between measured and calculated radiance is not small enough (i.e., comparable with the noise level), the mass path of the tangent cell is adjusted and, starting at the tangent cell (since pretangent cell calculations are unaffected), the total ray radiance is recalculated and again compared with the measured radiance. When the convergence requirements are met, the inversion algorithm proceeds to the next lower level. This procedure continues until all altitudes are completed. This was a convenient algorithm to use in testing the method; however, the EGA is well suited for use in a variety of relaxation approaches.

Inversions which included 50 altitude increments were performed on a CDC Cyber 173 computer system and required under 1 second each once the data base was stored in the core. This time competes well with more inaccurate linear methods.

One restriction of the EGA technique is that multiple emitters cannot be handled directly. This problem was addressed by assuming that the radiance due to each constituent is independent of all other constituents. This is an excellent assumption for most of the cases studied. The procedure is simply to use the EGA technique to calculate the independent radiance due to each known emitter, total these, and subtract directly from the measured radiance. The remaining radiance is used for an inversion to obtain the unknown constituent profile. A correction procedure has been devised for those cases where direct subtraction is inaccurate and this will be described in a future paper.

The LIMS channel with the most interference is the NO_2 channel in which H_2O can contribute anywhere from 30 percent of the radiance where the NO_2 peaks, to nearly 100 percent at other altitudes. The subtraction procedure just described was applied, assuming the H_2O concentration was known, and by using band model derived data bases, the inversion results

obtained are shown in Fig. 5. The difference below the peak is
primarily due to the band model pressure weighting inaccuracies
discussed before. However, the linear correction error can be
observed at the lowest altitudes where the difference begins to
decrease. (The pressure weighting and linear correction errors
are of opposite direction.)

VIII. CONCLUDING REMARKS

   The EGA technique has been demonstrated to give forward
radiance calculations in the limb geometry which are more
accurate than those obtained with the random exponential band
model employing the Curtis-Godson approach. The errors are
much smaller than experimental errors due to instrument effects
and band parameter uncertainties. Comparisons with more
sophisticated many degree band models which employ special
weighting of temperature and pressure for multilayer calculations
have shown comparable accuracies. However, the band models must
integrate over the entire band at 1 cm^{-1} to 5 cm^{-1} intervals
for each layer path to obtain comparable results, whereas the
EGA technique needs only two simple interpolations per layer
path. This advantage makes EGA nearly two orders of magnitude
faster than band model techniques when working with spectral
widths of 100 cm^{-1} or more. Limb inversions for one constituent
require about 1 second per inversion at 1 km resolution for
EGA where band models commonly require minutes. The EGA method
allows a completely nonlinear relaxation type inversion to be
done at linear speeds.
   The EGA technique requires no temperature and pressure
weighting and makes no linear approximations. Although the
technique has only been tested for limb emission calculations,
it is expected to work well for nadir calculations (8). A

variation of the technique should also prove adequate for estimating occultation radiances, which would involve the use of transmission curves. This variation is currently being investigated.

REFERENCES

1. Gille, J. C., *J. Geophy. Res. 74(6)*, 1863-1868 (1968).

2. McKee, F. B., Whitman, R. I., and Lambiotte, J. J., Jr., A Technique to Infer Atmospheric Temperature from Horizon Radiance Profiles. NASA TN D-5068, National Technical Information Service, Springfield, Virginia (1969).

3. McKee, F. B., Whitman, R. I., and Lambiotte, J. J., Jr., A Technique to Infer Atmospheric Water-Vapor Mixing Ratio from Measured Horizon Radiance Profiles. NASA TN D-5252, National Technical Information Service, Springfield, Virginia (1969).

4. House, F. B., and Ohring, G. Inference of Stratospheric Temperature and Moisture Profiles from Observations of the Infrared Horizon. NASA CR-1419, National Technical Information Service, Springfield, Virginia (1969).

5. Gille, J. C., and House, F. B., *J. Atm. Sci. 28*, 1427-1442 (1971).

6. Russell, J. M. III, and Drayson, S. R., *J. Atm. Sci. 29*, 376-390 (1972).

7. Russell, J. M. III, and Gordley, L. L., *J. Atm. Sci. 36*, 2259-2266 (1979).

8. Weinreb, M. P., and Neuendorffer, A. C., *J. Atm. Sci. 30*, 662-666 (1973).

9. Mankin, W. G., Fourier Transform Method for Calculating Atmospheric Transmittances. Presented at the Topical Meeting on Atmospheric Spectroscopy, Keystone, Colorado (1978).

FAST AND ACCURATE RADIANCE ALGORITHM 607

10. Goody, R. M., "Atmospheric Radiation I: Theoretical
     Basis." Oxford University Press, New York (1964).
11. Goldman, A., Kyle, F. G., and Bonomo, F. S., *Appl. Opt. 10,*
     10(1), 65-73 (1971).
12. Armstrong, B. H., *J. Atm. Sci. 25,* 312-322 (1968).

DISCUSSION

Fleming: I think Weinreb and Neuendorffer[1] wrote a paper conceptually using a method just like this in connection with variable
mixing ratios for atmospheric transmission functions. Are you
aware of that paper?

Russell: No, I have not seen that.

Fleming: I was not sure exactly how you got your pseudo mass
path. Did you have to iterate? How did you finally decide to
determine it?

Russell: It is by interpolation actually. Using the mass path-
emissivity data base you interpolate to get the emissivity for
the first shell; you then go into the data base and interpolate
with that emissivity and find the mass path that goes along with
that emissivity in the second shell. So, you do two interpolations in each shell with this technique--once the pseudo mass
path, then a second time for the actual mass path of the layers
to get the emissivity.

Kaplan: In the HNO_3 and O_3 calculations, the model takes into
account the temperature dependence; so that is not a factor in
the matching. I notice on the three things that you showed, the
HNO_3 and O_3, the water vapor and the ozone that they all went
over a region that you would expect that the first order of the
temperature effect would cancel. Now suppose, in the CO_2, for
example, we were looking at 550 wavenumbers. In that wavenumber
interval, have you tried and made a match in there, because it
is a large temperature dependence? And, what I am wondering is,
how much of a difference does the temperature dependence make
with this strong correlation between opacity and temperature
contrast? And, in the other place, a weak line clearly will be
more temperature dependent than the others.

Russell: We have not done the calculation with weak lines for
CO_2. We did include, of course, in the ozone calculation 1,000
wavenumbers and you surely can get some weak lines in there. I
cannot comment on CO_2 until we do the calculation. One point I
should mention at this time is that the calculation for the
ozone inversion which covered the entire ozone band--all the way
out to the extremes of the band, 200 wavenumbers--agrees with
inversions done by our colleagues at NCAR, John Gille and Paul
Bailey, to less than 10 percent, so I do not think the temperature effect is very important.

[1]*Weinreb, M. P., and Neuendorffer, A. C., J. Atmos. Sci.
30, 662-666 (1973).*

Staelin: You have dN divided by dQ, where the radiance N is a function of Q. The derivative of the function with respect to the function cannot be done unless you assume, for example, that you know the shape of Q.

Russell: Well, what you do is take the derivative over a small increment around a mean profile, for example.

Staelin: That is correct. You have to assume that you know the profile, that it is a small division of the profile.

THERMAL STRUCTURE OF JUPITER'S ATMOSPHERE
OBTAINED BY INVERSION OF VOYAGER 1
INFRARED MEASUREMENTS

Barney J. Conrath

Goddard Space Flight Center
Greenbelt, Maryland

Daniel Gautier

Observatoire de Paris
Meudon, France

Data from the Voyager 1 infrared spectroscopy investigation
have been used to retrieve temperature profiles in the atmosphere
of Jupiter. An analysis of information content indicates good
vertical resolution with low measurement noise propagation in
the Jovian troposphere and reduced resolution in the stratosphere. Among the problems found in common with the sounding of
the terrestrial atmosphere are limited tropopause definition and
dependence on upper boundary constraints. Preliminary results
obtained by using a constrained linear algorithm and a filtered
Chahine approach are presented.

I. INTRODUCTION

The Voyager 1 and Voyager 2 spacecraft carry instrument
complements which include infrared spectrometers with a useful
spectral range of 180 cm^{-1} to 2500 cm^{-1} and a spectral resolution
of 4.3 cm^{-1} apodized. The spacecraft are both targeted for

flybys of Jupiter and Saturn with the possibility of Voyager 2 continuing on to Uranus. Encounter of Voyager 1 with Jupiter occurred on March 5, 1979, with approximately 30,000 spectra acquired during the time the spacecraft was in the vicinity of the Jovian system.

A primary objective of the Voyager Infrared Spectroscopy Investigation is to obtain information on the thermal structure of the upper layers of the Jovian atmosphere. Examples of spectra acquired by Voyager 1 are shown in Fig. 1 for the region between 180 cm^{-1} and 1400 cm^{-1}. The broad $S(0)$ and $S(1)$ collision-induced hydrogen lines are formed primarily below the Jovian tropopause, and data within these lines are employed to obtain tropospheric temperatures. Stratospheric temperatures are retrieved from the 1306 cm^{-1} CH_4 band, seen as a strong emission feature. As indicated by the brightness temperatures in Fig. 1, the portion of the Jovian atmosphere being sounded is over 100 K cooler than the lower terrestrial atmosphere and places severe demands on instrument sensitivity. Nevertheless, it has proven possible to achieve a noise equivalent radiance of about 7×10^{-9} $W\ cm^{-2}\ ster^{-1}\ cm^{-1}$ in individual spectra throughout most of the spectral region used for temperature sounding. This results in a signal to noise ratio which is several hundred near the low frequency end of the spectral range, but drops below 10 in portions of the CH_4 band. For purposes of temperature retrieval, however, the lower signal to noise ratio near 1300 cm^{-1} is partially offset by the strong temperature dependence of the radiance in this region at Jovian temperatures.

Thus far, only preliminary data analyses have been carried out. The present paper is essentially a progress report of the experiences to date in inverting these data. First

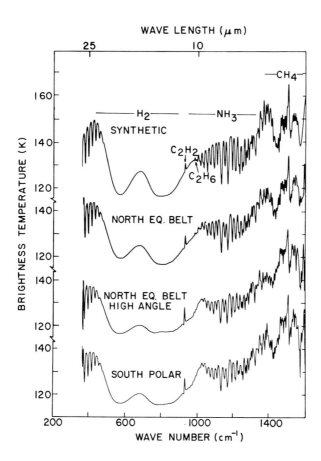

FIGURE 1.   Examples of long wavelength portions of infrared
spectra acquired by Voyager 1 (7).   A synthetic spectrum
calculated from a model atmosphere is shown for comparison in
the upper portion of the figure.

simple analyses of information content of the measurements are
considered and then the results obtained with two specific
inversion techniques are compared. Various problem areas which
must be dealt with will be considered.

II. INVERSION METHODS

In the preliminary analyses, all particulate and cloud
effects have been neglected, and it is assumed that the radiance
at the ith frequency can be written as

$$I_i = \int_{-\infty}^{\infty} B_i[T(Z)] \frac{\partial \tau_i}{\partial Z} \, dZ \qquad (1)$$

where $B_i(T)$ is the Planck radiance at temperature T, $Z = -\ln p$
is the vertical coordinate, and $\tau_i(Z)$ is the gaseous transmittance from level Z to the effective top of the atmosphere. It
will be found convenient to linearize Eq. (1) about some
reference profile $T^o(Z)$ resulting in

$$\Delta I_i \cong \int_{-\infty}^{\infty} K_i(Z) \, \Delta T(Z) \, dZ \qquad (2)$$

where

$$\Delta I_i = I_i - I_i^o \qquad (3)$$

$$K_i(Z) = \frac{dB_i}{dT}\bigg|_{T^o(Z)} \frac{d\tau_i}{dZ} \qquad (4)$$

$$\Delta T(Z) = T(Z) - T^o(Z) \qquad (5)$$

Defined in this way, the weighting function $K_i(Z)$ is a measure
of the sensitivity of the ith radiance to a small change in
temperature at level Z. Examples of $K_i(Z)$ are shown in Fig. 2

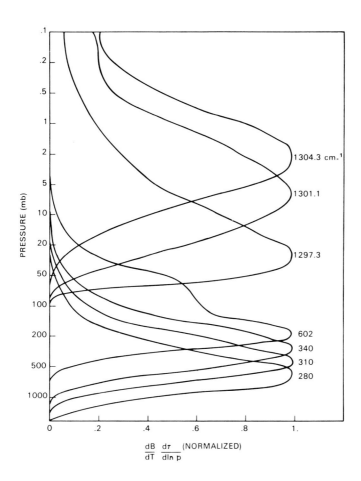

FIGURE 2. Weighting functions used in thermal sounding of
the Jovian atmosphere. The three upper weighting functions are
from the 1300 cm^{-1} band of CH_4 while the lower four are from the
S(0) and S(1) pressure-induced lines of hydrogen.

for four H_2 and three CH_4 frequencies. The H_2 weighting functions below the tropopause are relatively narrow and span the region nicely. (The Jovian tropopause lies between 100 mb and 200 mb or at about the same pressure level as the terrestrial tropopause.) In the stratosphere, however, the weighting functions tend to be broad, primarily because of the strong temperature dependence of the Planck function in the region of increasing temperature with height. This behavior is similar to that of weighting functions for the terrestrial stratosphere in the 4.3 μm CO_2 band. It should also be noted that the weighting functions tend to "avoid" the cold tropopause.

In comparison with the problem of temperature retrieval in the terrestrial atmosphere, there is little *a priori* information and few constraints available for Jovian profiles. Therefore, in the initial efforts with these data, it was decided to make use of relatively simple inversion techniques. Two inversion approaches have been developed and applied to the data independently; one is a constrained linear inversion or minimum information technique, and the second is based on Chahine's algorithm with explicit filtering.

The constrained linear inversion algorithm used provides an estimate $\widehat{\Delta T}(Z)$ of $\Delta T(Z)$ of the form

$$\widehat{\Delta T}(Z) = \underset{\sim}{K}(Z) \cdot [\underset{\approx}{S} + \underset{\approx}{\Gamma}]^{-1} \cdot \underset{\sim}{\Delta I} \qquad (6)$$

where $\underset{\sim}{K}(Z)$ is the vector of weighting functions $K_i(Z)$, $\underset{\sim}{\Delta I}$ is the vector of radiance residuals ΔI_i, the matrix $\underset{\approx}{S}$ is defined by

$$S_{ij} = \int_{-\infty}^{\infty} K_i(Z) \, K_j(Z) \, dZ$$

and $\underset{\sim}{\Gamma}$ is a matrix which depends on the constraint employed. Algorithms similar to Eq. (6) can be derived from several different points of view. With

$$\underset{\sim}{\Gamma} = \frac{\sigma_I^2}{\sigma_T^2} \underset{\sim}{1}$$

where σ_I^2 is the measurement noise variance and σ_T^2 is the temperature variance, Eq. (6) becomes the "minimum information" estimate of Foster (1). Other forms of constraint leading to Eq. (6) include minimization of the mean square departure of the Backus-Gilbert averaging kernel from a delta function (2).

The second inversion method, based on concepts developed by Gautier and Revah, makes use of a relaxation algorithm similar to that of Chahine

$$B_i[T^{n+1}(Z_i)] = \frac{I_i}{I_i^n} B_i[\bar{T}^n(Z_i)] \tag{7}$$

where the ith frequency is associated with an atmospheric level Z_i, usually taken near the peak of the weighting function, and superscripts denote iteration number (3,4). The filtered temperature profile is then given by

$$\bar{T}^n(Z) = \int_{-\infty}^{\infty} dZ' \, W(Z,Z') \sum_i T^n(Z_i) F_i(Z') \tag{8}$$

where the functions $F_i(Z)$ are chosen to give a linear inter-polation of the unfiltered temperature profile between the levels Z_i. The filtering applied to obtain the final solution is determined through the choice of $W(Z,Z')$. For this purpose, the Hamming function

$$W(Z,Z') = 0.54 + 0.46 \cos\left[\frac{(Z - Z')}{Z_o}\right] \tag{9}$$

has been used where the width Z_o can, in principle, be varied to take into account varying vertical resolution with height. However, only fixed values of Z_o between 0.5 and 1 scale height have been employed thus far.

III. INFORMATION CONTENT

    The information on the Jovian temperature profile extracted from the spectral measurements has been examined from the point of view of instrument noise propagation and vertical resolution. By considering the constrained linear inversion, if

$$\underset{\sim}{a}(Z) = \underset{\sim}{K}(Z) \cdot [\underset{\approx}{S} + \underset{\approx}{\Gamma}]^{-1} \cdot \underset{\sim\sim}{\Delta I} \tag{10}$$

then the estimate $\widehat{\Delta T}(Z)$ is related to $\Delta T(Z)$ by

$$\widehat{\Delta T}(Z) = \int_{-\infty}^{\infty} A(Z,Z') \, \Delta T(Z') \, dZ' \tag{11}$$

where

$$A(Z,Z') = \underset{\sim}{a}(Z) \cdot K(Z') \tag{12}$$

Thus, the degree to which $\widehat{\Delta T}$ can be expected to approximate ΔT depends on the behavior of $A(Z, Z')$, the so-called averaging kernel (5). The propagation of noise into the solution can be estimated to first order by using

$$\delta T^2(Z) = \underset{\sim}{a}(Z) \cdot \underset{\approx}{E} \cdot \underset{\sim}{a}(Z) \tag{13}$$

where $\underset{\approx}{E}$ is the measurement noise covariance matrix and $\delta T(Z)$ is the root mean square error in the solution profile. For the measurements considered here, it can be assumed that the noise associated with the various spectral intervals is uncorrelated and of constant variance; thus

$$\underset{\approx}{E} = \sigma_I^2 \, \underset{\approx}{1}$$

In preliminary inversions of the data $\Gamma = \underset{\approx}{\gamma} \; \underset{\approx}{1}$ in Eq. (6)
has been used with a value of γ which gives a noise propagation
as a function of height as shown in Fig. 3. It should be
emphasized that the calculated "error" is the formal precision
of the retrieved profile due to instrument noise, and the actual
departures from the true temperature profile may be substantially
larger in regions near the tropopause ($\approx$ 100 mb) where the
information content is low and the profile is poorly represented
in the retrieval. The noise propagation is relatively low in
the troposphere (below 100 mb) where most of the information is
coming from the hydrogen lines for which the signal to noise
ratio of the measurements is high. Above 100 mb the noise
propagation increases where the information is obtained primarily
from the low frequency wing of the 1306 cm^{-1} CH_4 band. With the
increasing height the stratospheric temperature increases and
results in a higher signal to noise ratio in the Q-branch of the
band, and the noise propagation decreases in the uppermost part
of the sounding region.

The corresponding averaging kernels for three selected
levels are shown in Figs. 4, 5, and 6. In the troposphere
(Fig. 4) the averaging kernel is well centered on the level for
which the temperature is being estimated and is relatively
narrow. In contrast, near the tropopause (Fig. 5) the averaging
kernel is broad, and the peak lies below the level for which a
temperature estimate is being made. Thus, the tropopause
height and shape cannot be precisely determined from the data.
In the stratosphere (Fig. 6), the averaging kernel is quite
broad and yields average temperatures over thick atmospheric
layers. Since the averaging kernels are in fact linear combinations of the weighting functions (Fig. 2), they simply reflect
the limitations of the weighting functions as a basis set for
representation in the sounding region.

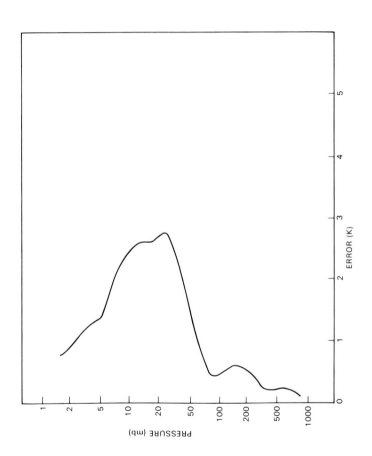

FIGURE 3. Root mean square error in retrieved profiles due to measurement noise propagation. The result is from a linear analysis; the actual error propagation may be larger because of strong nonlinearities.

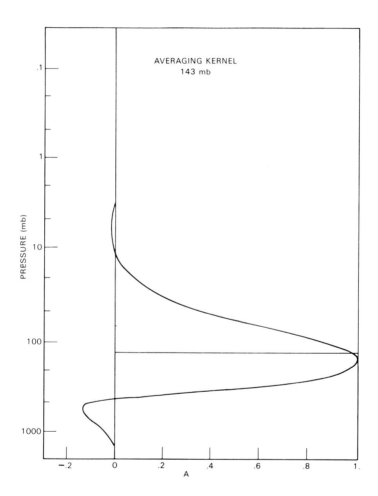

FIGURE 4. *Averaging kernel for the constrained linear inversion. The horizontal line indicates the level (427 mb) for which the temperature is being estimated.*

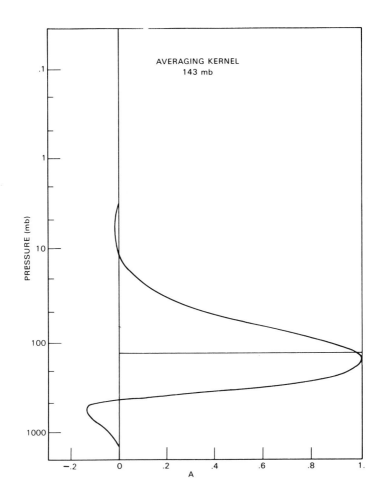

FIGURE 5.   Averaging kernal for the 143 mb level.

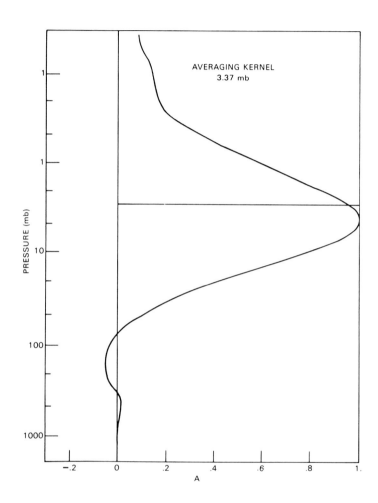

FIGURE 6.   Averaging kernel for the 3.37 mb level.

IV.   PRELIMINARY RESULTS

Because of the absence of *in situ* measurements, the problem
of directly assessing the quality of retrieved profiles from
Jovian data is much less straightforward than that for the
Earth's atmosphere.   Radio occultation measurements were obtained
from the Voyager spacecraft which provide temperature profile
measurements at two points, and comparisons with these results
should ultimately prove valuable (6).

A source of uncertainty in the retrieved profiles is the
current lack of precise knowledge of the mixing ratios of
optically active constituents.   The pressure-induced hydrogen
lines used for sounding the troposphere depend on both $H_2 - H_2$
and H_2 - He collisions; therefore, a knowledge of the
hydrogen-to-helium ratio is required.   In preliminary work,
hydrogen and helium volume mixing ratios of 0.88 and 0.12,
respectively, have been assumed.   These values are consistent
with currently accepted values of the solar abundances of these
elements.   The CH_4 mixing ratio thus far used has also been
chosen consistent with the solar C/H ratio.   Errors in the
mixing ratios result in the mapping of temperature onto incorrect
pressure levels in the retrieved profiles.   However, in both
spectral intervals considered here, there is an approximate
square root dependence on mixing ratio; thus, the resulting
profile errors are not expected to be large.

Examples of profiles retrieved from the same set of data
using the two different inversion methods are shown in Fig. 7.
The results are generally in good agreement except for a
systematic difference in the stratosphere which has been traced
to differences in methane band strengths and upper constraints
on the profile used in the two retrievals.

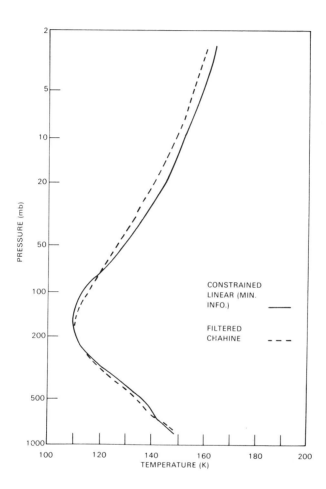

FIGURE 7. *Example of a temperature profile retrieved from Voyager 1 infrared data. The differences in temperatures obtained with the two inversion methods is primarily due to the use of different upper boundary constraints.*

The region of maximum information on the temperature profile lies essentially between the peaks of the uppermost and lower-most weighting functions (Fig. 2). However, there are signifi-cant contributions to the outgoing radiances from above and below these levels which, if incorrectly estimated, can affect the results within the region for which the solution is assumed to be valid. The effect is also encountered in sounding the terrestrial atmosphere and contributes to the general non-uniqueness of solutions. In the case of the constrained linear inversion, the solution tends to remain near the initial guess profile in those regions for which the amplitudes of the weight-ing functions become small. This effect was used to constrain the solution at the upper levels through choice of the initial profile. In the filtered Chahine inversion, a constraint was imposed through the choice of temperature at an upper anchor point at an arbitrarily chosen level above the peak of the uppermost weighting function. For the region below the lowest weighting function peak, the profile has been extrapolated along the adiabatic line in both inversion methods. The assumption of an adiabatic lapse rate is based on the presumed convective nature of the deeper layers of the Jovian atmosphere associated with the strong internal heat source known to exist.

A substantial number of preliminary profile retrievals have been obtained. These indicate considerable spatial variations in atmospheric temperature on both global and local scales. These results have been summarized by Hanel *et al.* (7).

V.   FUTURE WORK

The general approach adopted in retrieving temperature pro-files from Voyager data is to make use initially of a simplified approach with the introduction of greater sophistication and complexity as experience dictates. The work described here represents an initial step in this procedure.

Thus far the effects of clouds and particulates have been ignored. In the most transparent part of the far infrared spectrum near 225 cm^{-1}, brightness temperatures greater than 140 K are found essentially everywhere; therefore, any cloud decks which are opaque in this spectral region must lie deeper than about the 600-mb level. However, the data may still be significantly affected by clouds and possibly hazes which may exist above this level. Clouds and hazes are being investigated by using the infrared spectral data along with other Voyager data, and as understanding of this area improves, the effects of particulates can be incorporated in the profile retrievals.

The H_2/He ratio can, in principle, be determined from the infrared spectrum simultaneously with the temperature profile by taking advantage of the spectral dependence of the sensitivity of the gaseous absorption on the He abundance (8). A potentially more sensitive approach makes joint use of the radio occultation profiles and the infrared spectra. Applications of these techniques should substantially refine the estimates of H_2/He.

Infrared spectra have been acquired over a range of emission angles which should provide an additional source of information. At high emission angles, the weighting functions move upward, the peaks of those associated with the hydrogen lines moving into the tropopause and lower stratosphere. Thus, the use of high emission angle data along with normal viewing data may permit a better definition of the tropopause to be obtained. In addition, emission angle information combined with spectral information may aid in the study of atmospheric particulates.

REFERENCES

1. Foster, M., *J. Soc. Ind. Appl. Math. 9,* 387 (1961).
2. Conrath, B. J., *in* "Inversion Methods in Atmospheric
    Remote Sounding" (A. Deepak, ed.), p. 155. Academic
    Press, New York (1977).
3. Gautier, D., and Revah, I., *J. Atmos. Sci. 32,* 881 (1975).
4. Chahine, M. T., *J. Opt. Soc. Am. 58,* 1634 (1968).
5. Backus, G. E., and Gilbert, J. F., *Geophys. J. R.
    Astron. Soc. 16,* 169 (1968).
6. Eshleman, V. R., Tyler, G. L., Wood, G. E., Lindal, G. F.,
    Anderson, J. D., Levy, G. S., and Croft, T. A., *Science,
    204,* 976 (1979).
7. Hanel, R., Conrath, B., Flasar, M., Kunde, V., Lowman, P.,
    Maguire, W., Pearl, J., Pirraglia, J., Samuelson, R.,
    Gautier, D., Gierasch, P., Kumar, S., and Ponnamperuma, C.,
    Science, 204, 972 (1979).
8. Gautier, D., and Grossman, K., *J. Atmos. Sci. 29,* 788
    (1972).

DISCUSSION

Kaplan: In connection with the greater detail in the stratosphere relative to the troposphere, is there a possibility that there is a difference in mixing ratio of methane, or is it possible that there are some other components in the troposphere that affect the frequency you are measuring— what is your thinking?

Conrath: There could be opacity differences, but they would have to have a spatial variation comparable to the belt zone spacing.

Kaplan: Let us say there was a difference in methane concentration, where would you have more and where would you have less? Would you see deeper into the atmosphere? Were there breaks in the clouds? What is the correlation between the additional methane that you would need to correct it?

Conrath: We are seeing warmer temperatures over zones which are presumably cloudy areas, so in order to get artificially warmer temperatures in that region, the emission level would have to be moved upward. We have an increasing temperature with height, so more methane would be required. You would have to have a spatial variation comparable to the belt-zone structure which is a rather fine scale.

Kaplan: What is your feeling for that?

Conrath: I think they are real temperature variations, probably due to dynamic effects associated with the belt-zone structure. We may, in fact, be seeing the damping out of the counterflowing jets associated with the belt-zone system as we move up into the stratosphere, just as we are seeing the dying out of the red spot as we move up into the stratosphere. Those, in fact, would be the temperature gradients required to cancel out the thermal winds associated with the jets.

Kaplan: Why wouldn't this show up in the troposphere?

Conrath: In the troposphere, the vertical shear may be very small. Most of the action may be down below the cloud level with very little change in wind speed with height in the upper troposphere. The gradients maintained there may be related to phase changes of water, and the water cloud is down considerably deeper than we are seeing. But, again, this is speculation.

Staelin: In calibrating your pressure scale, you say you assume the composition to calibrate the pressure scale, can you use pressure broadening of lines or any other indications to help you?

Conrath: The main thing you would like to know is the hydrogen/ helium ratio. The Jupiter atmosphere is about 90 percent hydrogen so you peg everything to that.

PRELIMINARY RESULTS FROM NIMBUS 7
STRATOSPHERIC AND MESOSPHERIC
SOUNDER

C. D. Rodgers

Clarendon Laboratory
Oxford University
England

*The Nimbus 7 Stratospheric and Mesospheric Sounder is a limb
sounding pressure modulated radiometer, measuring temperature,
CO, NO, CH_4, N_2O, and H_2O. Some of the early results of
temperature and water vapor are presented.*

The Stratospheric and Mesospheric Sounder (SAMS) on Nimbus 7
is a limb scanning radiometer using pressure modulation
techniques to measure emission from carbon dioxide and a range
of trace constituents, so that the spatial distribution of
temperature and composition can be determined as a function of
time. Table I summarizes the emissions measured and the nature
of the information gathered by each channel.

The radiation in each spectral interval is modulated by both
a black chopper and a pressure modulator cell, so that two signals
can be obtained, described as "wide band" radiance, which
corresponds to a simple radiometer, and "PMR" radiance, which
originates in the centers of atmospheric spectral lines. The
pressure modulation allows signals to be obtained from much

TABLE I. *Summary of Emission Measurements and Nature of*
Information Gathered By Each Channel

Channel	Gas	Band	Data product
A1	CO_2	15.0 μm	Temperature; vibrational tempera-
A2	CO_2	4.0-5.0 μm	ture; attitude; CO_2 distribution; wind
A3	CO	4.0-5.0 μm	Distribution
A4	NO	4.0-5.0 μm	Distribution
B1	H_2O	2.7 μm	Distribution; wind
B2	H_2O	25.0-100.0 μm	Distribution
C1	CO_2	15.0 μm	Attitude; temperature
C2	N_2O	7.7 μm	Distribution
C3	CH_4	7.7 μm	Distribution

higher altitudes than is possible using conventional radiometric
techniques. A more detailed description can be found in the
Nimbus 7 Users' Guide (1).

Nimbus 7 was launched in October 1978 and SAMS has been
operating satisfactorily since it was turned on a few days
after launch. Data from the spacecraft is collected by the
National Aeronautics and Space Administration (NASA) and distributed among the experimenters, so that the SAMS data is
transmitted to Oxford on the day after collection. Preliminary
data analysis is carried out on line so that the performance of
the instrument can be monitored in detail and in near real time.

As with all limb sounders, the pointing direction must be
known to considerable accuracy, well beyond the capabilities of
the spacecraft attitude control system. Thus, the instrument
itself must be capable of making measurements of the tangent

height along the direction of view. This is achieved by means of
a comparison of the wide band and PMR radiances in the carbon
dioxide channels, in a similar way to the technique used by
LIMS and LRIR which use wide band and narrow band CO_2 channels
(2). The estimation of tangent height is carried out in two
stages, an approximate nonlinear estimator is used to provide a
starting value for a more accurate linear iterative method.

Retrieval of temperature profiles is complicated by two
factors in the case of SAMS. The first is that the attitude
control system allows the tangent height to vary considerably,
even if the instrument is not scanning, and the second is the
extremely flexible program control logic which allows arbitrary
limb scan patterns to be selected to meet any situation at any
time. This means that retrieval methods cannot rely on a predetermined scan pattern and must be able to cope with anything.
Consequently, a retrieval scheme has been designed which could
deal with a completely random scan pattern if desired. The
method retrieves a two-dimensional field of temperatures as a
function of height and latitude, rather than a vertical profile
for each scan. It is based on the sequential estimator
described by Rodgers (3). Every 2 seconds the instrument makes
four measurements of CO_2 15 μm radiance, wide band and PMR at
each of two different tangent heights. These radiances are
combined with *a priori* estimates of the temperature profile and
tangent height using a linear statistical estimator to produce
an optimal estimate of the temperature profile and tangent
height. The *a priori* profile used is a combination of
climatology and the temperature retrieval for 2 seconds earlier,
and has been found to be so close to the unknown profile that
the linearization is adequate and no iteration is required.
The estimate is further refined by carrying out the same process
in reverse time order and combining the two estimates.

This retrieval process also has the advantage that it automatically carries out the first stage of global gridding and analysis. It is not necessary to preserve the retrieved profile every 2 seconds of satellite time, so it is only retained at preselected grid points, every 2.5° of latitude. Transformation to a regular latitude-longitude grid is then very straightforward. Figure 1 shows an example of retrieved temperatures at the 30 mb level for January 24, 1979, compared with the Berlin analysis for that day, which is based on radio-sonde measurements.

Composition retrieval could, in principle, be carried out by the same process, but the linearization of the equation of transfer that is required in this case is prohibitively time consuming to carry out every 2 seconds. Therefore, a slightly

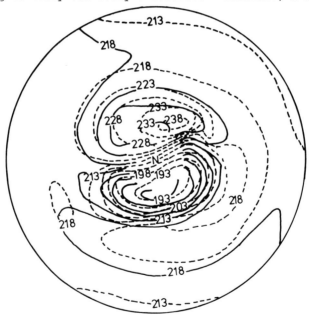

FIGURE 1. Comparison of SAMS 30 mb temperature for the northern hemisphere for January 24, 1978 (solid lines), with the Berlin analysis for that day (dashed lines). Polar stereographic projection.

different approach is used; the radiances themselves are
retrieved (i.e., gridded) at intervals of 2.5° of latitude by
using the same sequential estimator, then a more conventional
composition retrieval method is used on the radiance profiles
at the grid points. At present, various techniques are being
tested for speed and accuracy. A typical retrieved profile for
water vapor is shown in Fig. 2, using a simple onion peeling
method.

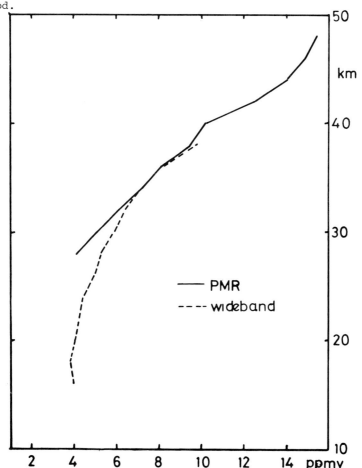

FIGURE 2.   Preliminary water vapor retrievals for April 23,
1979, using zonal mean radiances for the equator.

REFERENCES

1. Madrid, C. R., ed., "The Nimbus 7 Users' Guide." NASA
    Goddard Space Flight Center, Beltsville, Maryland (1978).
2. Gille, J. C., and Bailey, P., *in* "Inversion Methods in
    Atmospheric Remote Sounding" (A. Deepak, ed.), pp. 195-216.
    Academic Press, New York (1977).
3. Rodgers, C. D., *in* "Inversion Methods in Atmospheric Remote
    Sounding" (A. Deepak, ed.), pp. 117-138. Academic Press,
    New York (1977).

DISCUSSION

Rosenkranz: You showed a plot of your errors in the temperature retrieval. What was used as comparison?

Rodgers: The plot was the variance of the estimate, rather the root of the variance of the estimate, as produced by the Kalman filter. We have not done any ground truth comparisons with temperature yet. The plot is just there to give you an indication about the range of heights the information is coming from.

Smith: You said the "McClatchey" data, were you referring to the line atlas?

Rodgers: Yes.

INDEX